SEMI-MARKOV MODELS
Theory and Applications

SEMI-MARKOV MODELS
Theory and Applications

Edited by
Jacques Janssen

Free University of Brussels
Brussels, Belgium

PLENUM PRESS • NEW YORK AND LONDON

Library of Congress Cataloging in Publication Data

International Symposium on Semi-Markov Processes and Their Applications (1984: Brussels, Belgium)
   Semi-Markov models.

   "Proceedings of an International Symposium on Semi-Markov Processes and Their Applications, held June 4–7, 1984, in Brussels, Belgium"—T.p. verso.
   Bibliography: p.
   1. Markov, processes—Congresses. 2. Renewal theory—Congresses. I. Janssen, Jacques, 1939–   . II. Title.
QA274.7.I58   1984   519.2′33   86-12389
ISBN 0-306-42362-6

Proceedings of an International Symposium on Semi-Markov Processes and
Their Applications, held June 4–7, 1984, in Brussels, Belgium

© 1986 Plenum Press, New York
A Division of Plenum Publishing Corporation
233 Spring Street, New York, N.Y. 10013

Printed in the United States of America

PREFACE

This book is the result of the International Symposium on Semi-Markov Processes and their Applications held on June 4-7, 1984 at the Université Libre de Bruxelles with the help of the FNRS (Fonds National de la Recherche Scientifique, Belgium), the Ministère de l'Education Nationale (Belgium) and the Bernoulli Society for Mathematical Statistics and Probability.

This international meeting was planned to make a state of the art for the area of semi-Markov theory and its applications, to bring together researchers in this field and to create a platform for open and thorough discussion.

Main themes of the Symposium are the first ten sections of this book. The last section presented here gives an exhaustive bibliography on semi-Markov processes for the last ten years.

Papers selected for this book are all invited papers and in addition some contributed papers retained after strong refereeing.

Sections are :

I.   Markov additive processes and regenerative systems
II.   Semi-Markov decision processes
III. Algorithmic and computer-oriented approach
IV.   Semi-Markov models in economy and insurance
V.   Semi-Markov processes and reliability theory
VI.   Simulation and statistics for semi-Markov processes
VII. Semi-Markov processes and queueing theory
VIII. Branching
IX.   Applications in medicine
X.   Applications in other fields

XI. A second bibliography on semi-Markov processes

It is interesting to quote that sections IV to X represent
a good sample of the main applications of semi-Markov processes
i.e. Economy, Insurance, Reliability, Simulation, Queueing, Branching,
Medicine (including survival data), Social Sciences, Language Modelling, Seismic Risk, Analysis, Biology and Computer Science.

This strong interaction between recent theoretical results
presented in the first three sections and applications given above
was clearly pointed out by the participants.

I should like to thank all of them for their contribution and
their enthusiasm.

In particular, I should like to mention the strong support I
found by A.A. BOROVKOV (Institue of Mathematics, Novosibirsk),
E.ÇINLAR (Civil Engineering Department, Princeton University),
D.R.COX (Department of Mathematics, Imperial College, London),
V.S.KOROLYUK (Institute of Mathematics, Ukrainian Academy of Sciences,
Kiev), M.NEUTS (Department of Mathematical Sciences, University of
Delaware, Newark) and J.TEUGELS (Department of Mathematics, Katho-
lieke Universiteit te Leuven). This last one provides me with in-
valuable support.

Finally, I should mention the CADEPS (Centre d'Analyse des
Données et Processus Stochastiques, Ecole de Commerce Solvay,
Brussels) and the CEME (Centre d'Economie Mathématique et d'Econo-
métrie, Brussels) which give me effective support by taking in hands
the local organization.

Here too, I have to retain particularly the strong help of
Mr ABIKHALIL (CADEPS) for the final preparation of this volume.

J. Janssen

CONTENTS

SECTION XI

SOME LIMIT THEOREMS FOR MARKOV ADDITIVE PROCESSES

Peter Ney [1] and Esa Nummelin [2]

[1] Mathematics Department, University of Wisconsin
Wisconsin 53706, USA

[2] Department of Mathematics, University of Helsinki

At the Semi-Markov Symposium we presented some new results on Markov-additive processes which will be published in Ney and Nummelin (1984), Ney and Nummelin (1985). In their proofs we used regeneration constructions similar to those in several previous papers (Athreya and Ney, 1978; Nummelin, 1978; Iscoe, Ney and Nummelin, 1984). The proof of the particular regeneration used in Ney and Nummelin (1984) was omitted, and we will now provide it here. We also prove a slight extension of the results in Iscoe, Ney and Nummelin (1984); Ney and Nummelin (1984), and summarize the results we announced at the symposium.

Let $\{X_n; n = 0,1,\ldots\}$ be a Markov chain on a state space $\mathbb{E}$ with σ-field $\mathscr{E}$, which is irreducible with respect to a maximal irreducibility measure ϕ on $(\mathbb{E},\mathscr{E})$, and is aperiodic. Define an associated sequence of $\mathbb{R}^d$-valued random variables $\{\xi_n; n = 0, 1,\ldots\}$ such that $\{(X_n,\xi_n); n = 0,1,\ldots\}$ is a M.C. on $E \times \mathbb{R}^d$ having transition function

$$P\{(X_{n+1},\xi_{n+1}) \in A \times \Gamma \mid X_n = x,\mathscr{F}_n\}$$

$$= P\{(X_{n+1},\xi_{n+1}) \in A \times \Gamma \mid X_n = x\} \equiv P(x,A \times \Gamma),$$

for all $x \in \mathbb{E}$, $A \in \mathcal{E}$, $\Gamma \in \mathcal{R}^d$, $n \geq 0$, where $\mathcal{F}_n$ is the σ-field generated by $(X_0, \ldots, X_n, \xi_1, \ldots, \xi_n)$. Set $S_n = S_0 + \xi_1 + \ldots + \xi_n$. The sequence $\{(X_n, S_n); n = 0, 1, \ldots\}$ is a Markov-Additive (MA) process.

Consider the condition

$(M^{(1)})$ Minorization. There exists a family of measures $\{h(x, \cdot) \leq 1; x \in \mathbb{E}\}$ on $\mathbb{R}^d$, and a probability measure $\nu(\cdot \times \cdot)$ on $\mathbb{E} \times \mathbb{R}^d$, such that

$$(1) \quad (h(x, \cdot) * \nu(A \times \cdot))(\Gamma) \leq P(x, A \times \Gamma) \tag{1}$$

for all $x \in \mathbb{E}$, $A \in \mathcal{E}$, $\Gamma \in \mathcal{R}^d$ = Borel sets of $\mathbb{R}^d$. (* = convolution).

Abbreviate the kernel on the left side of (1) by $h \circledast \nu$. Thus

$$(M^{(1)}) \qquad h \circledast \nu \leq P.$$

Later we use the less restrictive condition

$$(M^{(n_0)}) \qquad h \circledast \nu \leq P^{n_0}$$

for some h, ν and integer $0 < n_0 < \infty$. Here,

$$P^2(x, A \times \Gamma) \equiv \int_{\mathbb{E}} \int_{\mathbb{R}^d} P(x, dy \times ds) P(y, A \times \Gamma - s),$$

and P^k, $k \geq 1$, is defined inductively.

<u>Regeneration Lemma</u>

Assume $M^{(1)}$. Then there exist r.v.'s $0 \leq T_0 \leq T_1 \leq \ldots$, and a decomposition $\xi_{T_i} = \xi'_{T_i} + \xi''_{T_i}$, $i = 0, 1, \ldots$, with the following properties :

(i) $\{T_{i+1} - T_i; i = 0, 1, \ldots\}$ are i.i.d. random variables;

(ii) The random blocks

$$\{X_{T_i}, \ldots, X_{T_{i+1}-1}, \xi''_{T_i}, \xi_{T_i+1}, \ldots, \xi_{T_{i+1}-1}, \xi'_{T_{i+1}}\}, \quad i = 0, 1, \ldots,$$

are independent;

(iii) $\mathbb{P}_x\{(X_{T_i}, \xi''_{T_i}) \in A \times \Gamma'' \mid \mathcal{F}_{T_i-1}, \xi_{T_i}\} = \nu(A \times \Gamma'')$

for $A \in \mathcal{E}$, $\Gamma'' \in \mathcal{R}^d$.

<u>Proof</u>

Let

$$\bar{P}(x, A \times \Gamma) = \frac{P(x, A \times \Gamma) - (h \otimes \nu)(A \times \Gamma)}{1 - h(x, \mathbb{R}^d)}$$

and note that due to $(M^{(1)})$, $\bar{P}$ is an MA-transition kernel. Adjoin to $\{(X_n, \xi_n)\}$ the sequence of 0 or 1-valued random variables $\{Y_n\}$, such that

(2) $P\{X_{n+1} \in A, \xi_{n+1} \in \Gamma, Y_n = 1 \mid \mathfrak{C}_n\} = h(X_n, \cdot) * \nu(A \times \cdot)(\Gamma),$

$\qquad P\{X_{n+1} \in A, \xi_{n+1} \in \Gamma, Y_n = 0 \mid \mathfrak{C}_n\} = (1 - h(X_n, \mathbb{R}^d))\bar{P}(X_n, A \times \Gamma),$

where $\mathfrak{C}_n = \sigma\{X_0, \ldots, X_n, \xi_1, \ldots, \xi_n, Y_0, \ldots, Y_{n-1}\}$. Now define

$\qquad T_0 = \inf\{n \geq 1 : Y_{n-1} = 1\},$

$\qquad T_i = \inf\{n > T_{i-1} : Y_{n-1} = 1\}, \quad i = 1, 2, \ldots .$

If $P\{X_0 \in A, S_0 \in \Gamma\} = \nu(A \times \Gamma)$, define $T_0 = 0$. $\xi_0' = 0$, $\xi_0'' = S_0$. Then $\{T_i; i = 0, 1, \ldots\}$ satisfy (i) of the lemma. The random variables $(\xi_{T_i}', \xi_{T_i}'')$ are determined by the construction

$\qquad P\{X_{T_i} \in A, \xi_{T_i}' \in \Gamma', \xi_{T_i}'' \in \Gamma'' \mid \mathcal{F}_{T_{i-1}}\}$

(3)

$$= \frac{h(X_{n-1}, \Gamma')}{h(X_{n-1}, \mathbb{R}^d)} \cdot \nu(A \times \Gamma''), \quad A \in \mathcal{E}, \ \Gamma', \ \Gamma'' \in \mathcal{R}^d.$$

Now (ii) and (iii) follow from (2) and (3). $\square$

<u>Corollary</u>

If $h(x, \mathbb{R}^d) \geq c > 0$ then

$$P\{T_{i+1} - T_i > k\} \leq c^k .$$

(This is equivalent to the Doeblin condition).

Two special cases of $(M^{(1)})$ should be noted; one where $\xi'_n \equiv 0$, the other where $\xi''_n \equiv 0$. When $\xi''_n \equiv 0$, $n = 1,2,\ldots$, $(M^{(1)})$ becomes

$$(M_1^{(1)}) \qquad h(x,\Gamma)\nu(A) \leqslant P(x, A \times \Gamma),$$

where ν is now a measure on $(\mathbb{E}, \mathcal{E})$. When $\xi'_n \equiv 0$, $(M^{(1)})$ becomes

$$(M_2^{(1)}) \qquad h(x)\nu(A \times \Gamma) \leqslant P(x, A \times \Gamma),$$

where now $h(\cdot) : \mathbb{E} \to \mathbb{R}'$. (We use the same symbols h and ν in all these cases, though they denote different objects). Under $(M_1^{(1)})$ the independent blocks are

$$\{X_{T_i}, \ldots, X_{T_{i+1}-1}, \xi_{T_i+1}, \ldots, \xi_{T_{i+1}}\}, \quad i = 0,1,2,\ldots,$$

with

$$P_x\{X_{T_i} \in A \mid \mathcal{F}_{T_i-1}, \xi_{T_i}\} = \nu(A).$$

Similarly under $(M_2^{(1)})$, the independent blocks are

$$\{X_{T_i}, \ldots, X_{T_{i+1}-1}, \xi_{T_i}, \ldots, \xi_{T_{i+1}-1}\},$$

and

$$P_x\{(X_{T_i}, \xi_{T_i}) \in A \times \Gamma \mid \mathcal{F}_{T_i-1}\} = \nu(A \times \Gamma).$$

Thus under either $(M_1^{(1)})$ or $(M_2^{(1)})$, the "uncoupling" of the process into independent blocks does not require the further decomposition $\xi_{T_i} = \xi'_{T_i} + \xi''_{T_i}$.

Heuristics

As in other such regenerations, the present construction can most easily be thought of in terms of a coin tossing (randomization) scheme as follows. At time n, if $X_n = x$, toss a coin with probability of a head $= h(x, \mathbb{R}^d)$. Then

(a) If the coin comes up head let ξ'_{n+1}, ξ''_{n+1}, X_{n+1} have the distribution

$$P\{\xi'_{n+1} \in \Gamma', \xi''_{n+1} \in \Gamma'', X_{n+1} \in A \mid X_n = x \text{ and coin} = H\}$$

$$= h(x,\Gamma')\nu(A \times \Gamma'')/h(x,\mathbb{R}^d).$$

Namely the conditional distribution of ξ'_{n+1} is $h(x,\cdot)/h(x,\mathbb{R}^d)$, and it is independent of (ξ''_{n+1}, X_{n+1}), which has conditional distribution ν. Let $\xi_{n+1} = \xi'_{n+1} + \xi''_{n+1}$.

(b) If the coin comes up tails, take

$$P\{X_{n+1} \in A, \xi_{n+1} \in \Gamma \mid X_n = x, \text{Coin} = T\} = \bar{P}(x, A \times \Gamma).$$

Then clearly

$$P\{X_{n+1} \in A, \xi_{n+1} \in \Gamma \mid X_n = x\}$$

$$= P\{X_{n+1} \in A, \xi_{n+1} \in \Gamma \mid X_n = x \text{ and Coin} = H\} \cdot P\{H \mid X_n = x\}$$

$$+ P\{X_{n+1} \in A, \xi_{n+1} \in \Gamma \mid X_n = x \text{ and Coin} = T\} \cdot P\{T \mid X_n = x\}$$

$$= h(x,\mathbb{R}^d) \frac{h(x,\cdot)}{h(x,\mathbb{R}^d)} * \nu(A \times \cdot)(\Gamma)$$

$$+ [1 - h(x,\mathbb{R}^d)] \bar{P}(x, A \times \Gamma) = P(x, A \times \Gamma).$$

Thus the underlying transition probabilities of the chain $\{(x_n, \xi_n); n = 0, \ldots\}$ remain unchanged by the randomization scheme. The T_i's are now defined as the times following the occurrence of heads. The event $[\text{regeneration at time } n] \equiv [T_i = n \text{ for some } i] \equiv [Y_{n-1} = 1] \equiv [\text{head at time } n - 1]$.

Now by a result of Niemi and Nummelin (1984) stated as the Proposition below, it follows that in fact a mild non-singularity condition on P is sufficient to assure either $(M_1^{(n_0)})$ or $(M_2^{(n_0)})$ (hence of course also $(M^{(n_0)})$).

<u>Proposition</u>

Assume that there exists a ϕ-positive set $B \in \mathcal{E}$ such that for every $x \in B$, $P^n(x, \cdot \times \cdot)$ is non-singular with respect to $\phi \times$

Lebesgue measure for some $n = n(x) \geq 1$. Then there exists an integer n_0, a function $f(x,s) : \mathbb{E} \times \mathbb{R}^d \to \mathbb{R}^+$ and a probability measure ν on $(\mathbb{E}, \mathcal{E})$ such that $\iint f(x,s)\phi(dx)ds > 0$ and

$$f(x,s)ds\nu(dy) \leq P^{n_0}(x, dy \times ds).$$

There also exists a integer n_1, a function $h(\ ) : \mathbb{E} \to \mathbb{R}^+$ and a probability measure $\nu(dy \times ds) = \nu(dy \times s)ds$, such that

$$h(x)\nu(dy \times ds) \leq P^{n_1}(x, dy \times ds).$$

In (Ney and Nummelin, 1984) we proved the following large deviation theorems under the condition $M_1^{(1)}$. Let $\tau \overset{D}{=} T_{i+1} - T_i$, $S_\tau \overset{D}{=} \xi_{T_i} + \ldots + \xi_{T_{i+1}-1}$, for any $0 \leq i < \infty$, and define the generating function

$$\psi(\alpha,\zeta) = E_\nu e^{<\alpha, S_\tau> - \zeta\tau}, \quad \alpha \in \mathbb{R}^d, \quad \zeta \in \mathbb{R}',$$

with domain $\mathbb{W} = \{(\alpha,\zeta) \in \mathbb{R}^{d+1} : \psi(\alpha,\zeta) < \infty\}$. In addition to $M_1^{(1)}$, our main hypothesis is that $\mathbb{W}$ is an open set. We also assume that the interior of the convex hull of the support of $\dfrac{S_\tau}{\tau} \equiv \overset{\circ}{\mathcal{S}} \neq \phi$. Then

$$\psi(\alpha,\zeta) = 1$$

(as an equation in ζ) has a unique solution $\zeta = \Lambda(\alpha)$. Let $\mathbb{D} = \{\alpha : \Lambda(\alpha) < \infty\}$. Then we proved in Ney and Nummelin (1984) that

(i) $e^{\Lambda(\alpha)}$ is an eigenvalue of $\hat{P}(\alpha) = \int P(x, A \times ds)e^{<\alpha,s>}$, with (right) eigenfunction $r(\cdot;\alpha)$ and (left) eigenmeasure $\ell(\cdot;\alpha)$, for $\alpha \in \mathbb{D}$.

(ii) $\Lambda(\cdot)$ is analytic on $\mathbb{D}$, and there is a set $\mathbb{F}$ with $\phi(\mathbb{F}^c) = 0$ such that $r(x;\cdot)$ is analytic on $\mathbb{D}$ for all $x \in \mathbb{F}$.

(iii) $\hat{P}^n(x, A; \alpha) = \ell(A;\alpha)e^{\Lambda(\alpha)n}r(x;\alpha)[1 + O(\delta^n(\alpha))]$ $0 < \delta < 1$.

From these facts there followed :

<u>Theorem 1</u>

(i) Assume that $M_1^{(1)}$ holds, $\mathbb{W}$ is open, $\overset{\circ}{\mathcal{S}} \neq \phi$, $x \in \mathbb{F}$, $A \in \mathcal{E}$ satisfies $\phi(A) > 0$ and is small enough in a sense made precise in Ney and Nummelin (1984). (We call this an s-set). Then for closed F and open G

$$\overline{\lim} \frac{1}{n} \log P^n(x, A \times nF) \leqslant - \overline{\Lambda}(F),$$

and

$$\underline{\lim} \frac{1}{n} \log P^n(x, A \times nG) \geqslant - \overline{\Lambda}(G),$$

where $\overline{\Lambda}(B) = \inf\{\Lambda^*(v) : v \in B\}$, and $\Lambda^* =$ the convex conjugate of Λ.

(ii) If furthermore B is convex with $B \cap \overset{\circ}{\mathcal{S}} \neq \phi$ and if $\{S_n\}$ is lattice valued, then there exist $0 < c' < c'' < \infty$ such that

$$(4) \quad c' n^{-d/2} e^{-n\overline{\Lambda}(B)} \leqslant P^n(x, A \times nB) \leqslant c'' e^{-n\overline{\Lambda}(B)}.$$

We observed in Ney and Nummelin (1984) that the entire argument could be carried out similarly under $(M^{(1)})$ or $(M_2^{(1)})$ in place of $(M_1^{(1)})$. We now indicate that $(M_1^{(k_0)})$, $(M_2^{(k_0)})$ or $(M^{(k_0)})$ will suffice. We give the argument when $M_1^{(k_0)}$ holds. Consider the embedded MA-process $\{(X_{nk_0}, S_{nk_0}); n = 0, 1, \ldots\}$. By the above results there will exist a regeneration for this sequence, with an associated generating function $\psi_{k_0}(\alpha, \zeta)$ with domain W_k, which we assume open. There will then also exist an eigenvalue $\lambda_{k_0}(\alpha) = e^{\Lambda_{k_0}(\alpha)}$ for $\hat{P}^{k_0}(\alpha)$, with eigenfunction $r_{k_0}(x; \alpha)$ and eigenmeasure $\ell_{k_0}(A; \alpha)$, $A \in \mathcal{E}$, having all the regularity properties of the theorem above. Hence

$$(5) \quad \overline{\lim} \frac{1}{n} \log (P^{k_0})^n(x, A \times nF) \leqslant - \overline{\Lambda}_{k_0}(F),$$

and

$$(6) \quad \underline{\lim} \frac{1}{n} \log (P^{k_0})^n(x, A \times nG) \geqslant - \overline{\Lambda}_{k_0}(G).$$

Now let $\lambda(\alpha) = (\lambda_{k_0}(\alpha))^{1/k_0}$. Then one easily checks that $\Lambda_{k_0}^*(v) = k_0 \Lambda^*(v/k_0)$ and $\overline{\Lambda}_{k_0}(B) = k_0 \overline{\Lambda}(B/k_0)$ for any $B \subset \mathcal{R}^d$. Hence replacing F by $k_0 F$ in (5)

$$(7) \quad \overline{\lim} \frac{1}{nk_0} \log P^{nk_0}(x, A \times nk_0 F) \leqslant - \bar{\Lambda}(F)$$

and similarly for (6).

Now apply P to both sides of the inequality $h \otimes \nu \leqslant P^{k_0}$. We get

$$(8) \quad \tilde{h} \otimes \nu \leqslant P^{k_0+1}$$

where $\tilde{h}(x,\Gamma) = (P*h)(x,\Gamma)$. Thus we have a minorization $M_1^{(k_0+1)}$ which yields a regeneration and all the associated functions above, provided W_{k_0+1} is open. Now if we consider $\hat{P}^{k_0(k_0+1)}(\alpha)$, then by the $M_1^{(k_0)}$ minorization $\hat{P}^{k_0(k_0+1)}(\alpha)$ will have eigenvalue $\lambda_{k_0}^{k_0+1}(\alpha)$. By $M_1^{(k_0+1)}$, it will have eigenvalue $\lambda_{k_0+1}^{k_0}(\alpha)$. Hence

$$\lambda_{k_0}^{k_0+1} = \lambda_{k_0+1}^{k_0}, \quad \text{i.e.} \quad \Lambda_{k_0+1} = \frac{k_0+1}{k_0} \Lambda_{k_0},$$

and with $\Lambda_{k_0}(\alpha) = k_0 \Lambda(\alpha)$, we have $\Lambda_{k_0+1}(\alpha) = (k_0 + 1)\Lambda(\alpha)$. Furthermore the associated eigenfunctions and measures must be the same, since the recurrence of $\hat{P}(\alpha)$ assures uniqueness. Thus we obtain under $M^{(k_0+1)}$:

$$(9) \quad \overline{\lim} \frac{1}{n(k_0 + 1)} \log^{n(k_0+1)}(x, A \times n(k_0 + 1)F) \leqslant - \bar{\Lambda}(F).$$

Together with (7) this implies

$$(10) \quad \overline{\lim} \frac{1}{n} \log P^n(x, A \times nF) \leqslant - \bar{\Lambda}(F).$$

A similar argument works for the lower bound.

If the initial minorization is $(M_2^{k_0})$. Then in place of (8) use

$$(7) \quad h \otimes \tilde{\nu} \leqslant P^{k_0+1},$$

where $\nu(A \times \Gamma) = (\nu * P)(A \times \Gamma)$. Also $M^{(k_0)}$ is treated similarly.

In the same way we prove under the stronger hypothesis of part (ii) of Theorem 1 that (4) holds with n replaced by nk_0 or by $n(k_0 + 1)$, hence holds in its original form. Thus we have

Theorem 2

If M_1, M_2, or M are replaced by $M_1^{(k_0)}$, $M_2^{(k_0)}$ or $M^{(k_0)}$ respectively, if W_{k_0} and W_{k_0+1} are open, and if the other hypothesis of Theorem 1 hold, then its conclusions remain the same.

Remark

It seems W_{k_0} open might imply W_{k_0+1} open, but we have not proved this. In most applications we discussed in Iscoe, Ney and Nummelin (1984) and Ney and Nummelin (1984) W_{k_0} and W_{k_0+1} are clearly open together. Most important, in the proof of the theorem stated below, $W_{k_0} = W_{k_0+1} = \mathbf{R}^{d+1}$ (trivially).

We conclude by stating another theorem wich we announced at the symposium. Proofs and details will appear elsewhere.

Let $\{(X_n, S_n)\}$ be as defined before. Denote the convergence parameter of $\widehat{P}(\alpha)$ by $\exp\{-\Lambda(\alpha)\}$. This always exists (it may $= 0$), but $\exp\{-\Lambda(\alpha)\}$ may not be an eigenvalue as under W open.

Theorem 3

Assume $M^{(k_0)}$. Then

(i) $\varliminf \dfrac{1}{n} \log P_x\{X_n \in A, \dfrac{S_n}{n} \in G\} \geqslant - \bar{\Lambda}(G)\}$ for G open.

(ii) If $d = 1$ then for all s-sets A and closed sets F, and for almost all $x[\phi]$

$$\varlimsup \frac{1}{n} \log P_x\{X_n \in A, \frac{S_n}{n} \in F\} \leqslant - \bar{\Lambda}(F).$$

Example

All atoms are s-sets. Thus if E is a countable space the upper bound (ii) holds for all closed F.

(iii) If $d > 1$ and $0 \in \mathring{D}$, where $D = \{\alpha : \Lambda(\alpha) < \infty\}$ then (ii) holds.

References

Athreya, F. and P.Ney (1978). Limit theorems for semi-Markov processes. <u>Bull.Austral.Math.Soc.</u> 19, 283-294.

Iscoe, I., P. Ney and E. Nummelin (1984). Large deviations of
uniformly recurrent Markov additive processes. Adv. in Appl.
Math. To appear.

Ney, P. and E. Nummelin (1984). Markov additive processes I : Eigenvalue properties and limit theorems. University of Wisconsin
technical report.

Ney, P. and E.Nummelin. (1985). Markov additive processes II: Large
deviations. Univ.of Wisconsin. Report.

Niemi, S. and E.Nummelin (1984). On nonsingular renewal kernels
with an application to a semigroup of transition kernels.
University of Helsinki technical report.

Nummelin, E. (1978).Uniform and ratio limit theorems for Markov
renewal and semi-regenerative processes on a general state
space. Ann.de l'Inst.H.Poincaré, B 14, 119-143.

STATIONARY REGENERATIVE SYSTEMS

H. Kaspi [1] and B. Maisonneuve [2]

[1] Technion, Haifa, Israel

[2] IMSS, 47 X-38040 Grenoble, France

1. <u>Introduction</u>

The theory of regenerative systems on the real line unifies the notions of strong Markov processes indexed by $\mathbb{R}$, recurrent events, regenerative processes, semi-Markov processes, etc. A regenerative system, as it will be defined in this paper consists of a random set $M \subset \mathbb{R}$ and a right continuous process $X = (X_t)_{t \in \mathbb{R}}$ such that

$$((M - T) \cap (0,\infty), (X_{T+s})_{s \geqslant 0}) \quad \text{and} \quad (M \cap (-\infty,T], (X_{s \wedge T})_{s \in \mathbb{R}})$$

are conditionally independent given X_T for each stopping time T falling in M. Under some regularity assumptions, we establish the existence of a local time for M and of an exit system from M. We also establish a correspondence between stationary regenerative systems and invariant measures for the process $(X_t)_{t \geqslant 0}$ time changed by using the local time of M. In particular we extend the notion of stationarity and the formulae for the stationary distributions of the backward and forward recurrence times which appear in the context of classical renewal theory and in the context of regenerative sets (Taksar, 1980; Maisonneuve, 1983). Finally we give a general result for constructing a stationary regenerative system via

a Markov additive process. These results rely on representations
of invariant measures obtained by Kaspi (1983, 1984) for Markov
processes.

In this paper we have restricted our attention to the case
where M is a.s. perfect and has a continuous local time, and we
have mainly dealt with the process X restricted to M. In a
forthcoming work (Kaspi and Maisonneuve) we shall investigate more
thoroughly the structure of the set M, which will allow us to
treat the more general case and the incursion process. For this
reason most proofs in this presentation will be omitted.

2. Preliminaries

We first define a canonical setting for regenerative systems.
Consider $\Omega = \Omega^0 \times \Omega^1$, where Ω^0 is the set of all closed subsets
of $\mathbb{R}$ and Ω^1 is the set of right continuous functions from $\mathbb{R}$
into a Lusinian space E. The space E is equipped with its Borel
field E. The projections $(\omega^0,\omega^1) \to \omega^0$ and $(\omega^0,\omega^1) \to \omega^1$ are
denoted by M and X respectively, so that (M,X) is just the
canonical map on Ω. For $t \in \mathbb{R}$ we set

$$D_t = \inf\{s > t : s \in M\}, \quad X_t = X(t),$$

$$H^0_t = \sigma\{D_{s \wedge t}, \; X_{s \wedge D_t}, \; s \in \mathbb{R}\},$$

$$H^0 = \sigma\{D_t, \; X_t, \; t \in \mathbb{R}\}.$$

The future of t is represented by

$$\phi_t = \overline{((M - t) \cap (0,+\infty))}, \; (X_{t+s^+}, \; s \in \mathbb{R})),$$

where $s^+ = s \vee 0$. Note that $(\phi_t)_{t \in \mathbb{R}}$ is a measurable process on
(Ω, H^0) with values in (Ω, H^0) and that $\phi_{t+s} = \phi_s \circ \phi_t$ for
$t \in \mathbb{R}, \; s \in \mathbb{R}_+.$

We now introduce a measurable family $(P^x)_{x \in E_o}$ of probability measures on (Ω, H^0). Here E_o is a Borel subset of E equipped with its Borel field E_o (E_o is the trace of E on E_o). We assume that

$$P^x\{M \subset \mathbb{R}_+, X_0 = x\} = 1, \; x \in E_o.$$

__Definition.__ (1) Given a stochastic basis $(\bar{\Omega}, \bar{H}, (\bar{H}_t)_{t \in \mathbb{R}}, \bar{P})$ and a random variable $(\bar{M}, \bar{X})$ from $(\bar{\Omega}, \bar{H})$ into (Ω, H^0), the system $(\bar{\Omega}, \bar{H}, \bar{H}_t, \bar{M}, \bar{X}, \bar{P})$ is called regenerative with regeneration law $(P^x)_{x \in E_o}$ provided (with the notation $\bar{Z} = Z \circ (\bar{M}, \bar{X})$ for a function Z^0 on Ω)

(i) (M,X) is $\bar{H}_t / H_t^0$ measurable (or equivalently $\bar{D}_{s \wedge t}$, $\bar{X}_{s \wedge D_t}$, $s \in \mathbb{R}$ are $\bar{H}_t$ measurable) for each $t \in \mathbb{R}$,

(ii) the process $(\bar{X}_{D_t}, t < \sup \bar{M})$ is $\bar{P}$ a.s. E_o valued,

(iii) for every $t \in \mathbb{R}$

$$\bar{P}_{\overline{\phi D_t} \,|\, \bar{H}_t} = P^{\bar{X}_{D_t}} \quad \text{on} \quad \{\bar{D}_t < \infty\}$$

(2) The system $(\Omega, H^0, H_t^0, M, X, (P^x)_{x \in E_o})$ (or simply the family $(P^x)_{x \in E_o}$) is called regenerative provided $(\Omega, H^0, H_t^0, M, X, P^x)$ is regenerative for each $x \in E_o$ with regeneration law $(P^x)_{x \in E_o}$.

3. Examples

(1) If $(\bar{\Omega}, \bar{H}, \bar{H}_t, \bar{M}, \bar{X}, \bar{P})$ is regenerative with respect to $(P^x)_{x \in E_o}$ and if E_o is reduced to one point x_o, then $\bar{M}$ is a regenerative set relative to $\bar{H}_t$ in the sense of Maisonneuve (1983) (provided the definition of τ_t in this paper is replaced, as it should be, by $\tau_t = (\bar{M} - t) \cap (0, \infty))$ and on $\{\bar{D}_t < \infty\}$ $\overline{\phi_{D_t}}$ is independent of $\bar{H}_t$ with a fixed distribution P^{x_o}. If further $\bar{M}$ is $\bar{P}$ a.s. discrete then $\bar{X}$ is a regenerative process in the sense of Çinlar (1975).

(2) If $(\bar{\Omega},\bar{F},\bar{F}_s,\bar{X}_s,\theta_s,\bar{P}^x;\ s \in \mathbb{R}_+,\ x \in E)$ is a right continuous E valued strong Markov process and if $\bar{M}$ is a homogeneous optional closed random set of $\mathbb{R}_+$, the system $(\bar{\Omega},\bar{F},\bar{F}_{D_t},\bar{M},\bar{X},\bar{P}^x)$ is regenerative for each $x \in E$ (we set $\bar{X}_t = \bar{X}_o$ if $t < 0$ and $\bar{X} = \bar{X}.$), with regeneration law $(P^x)_{x\in E}$, where P^x denotes the distribution of $(\bar{M},\bar{X})$ under $\bar{P}^x$. The family (P^x) can be restricted to $x \in E_o$ provided $\overline{X_{D_t}} \in E_o$ for $t < \sup \bar{M}$.

(3) Let $(\bar{\Omega},M,M_s,E_s,S_s,\theta_s,\bar{P}^x;\ s \in \mathbb{R}_+,\ x \in E)$ be a Markov additive process in the sense of E.Çinlar (1972), with S additive increasing and such that the closure $\bar{M}$ of the range of S is a.s. perfect (e.g. when S is continuous or strictly increasing). Let $C_t = \inf\{s \in \mathbb{R}_+ : S_s > t\}$, $t \in \mathbb{R}$ and let $X_t = E_{C_t}$, $\bar{X} = \bar{X}..$ Then $(\bar{\Omega},M,M_{C_t},\bar{M},\bar{X},\bar{P}^x)$ is regenerative for each $x \in E$ with regeneration laws $(P^x)_{x\in E}$, where $P^x = (\bar{M},\bar{X})(\bar{P}^x)$.

4. The Markov process $Y_t = (R_t,X_{D_t})$

We consider a (canonical) regenerative system $(\Omega,H^o,H^o_t,M,X,$ $(P^x)_{x\in E_o})$ as defined in Section 2 and let

$$Y_t = (R_t,X_{D_t}),\ t \in \mathbb{R}$$

with $R_t = D_t - t$ and $X_\infty = \Delta \notin E$. Note that $Y_{t+s} = Y_s \circ \phi_t$ for $t \in \mathbb{R}$, $s \in \mathbb{R}_+$ and that the process (Y_t) is right continuous provided its state space

$$U = \mathbb{R}_+ \times E_o \cup \{(+\infty,\Delta)\}$$

is equipped with the obvious topology.

Let us now introduce probability measures $(P^y)_{y\in U}$ on (Ω,H^o) in the following way. First we set $P^{o,x} = P^x$ for $x \in E_o$ and

$$P^{+\infty,\Delta} = \varepsilon_{(\emptyset,[\Delta])} \; ; \text{ then for } \quad (r,x) \in (0,\infty) \times E_o \quad \text{we set}$$

$$P^{r,x} = ((\{0\} \cup M) + r, \; [x]/r/X)(P^x).$$

As usual $[x]$ denotes the constant map $t \to x$ and for $w, w' \in \Omega^1$ $w/r/w'$ denotes the function which agrees with w on $(-\infty,r)$ and with $w'(\cdot - r)$ on $[r,\infty)$. The argument that led to theorems (2.3) and (2.5) of Maisonneuve (1983) proves the following theorem. Here and in the sequel we "<u>restrict</u>" Ω to the points of the form (ω^o,ω^1), where $\omega^o \in \Omega^o$ and ω^1 is constant on each interval $[\alpha,\beta)$ contiguous to ω^o, with $\omega^1(t) = \Delta$ for $t \geqslant \sup \omega^o$. Due to this last condition the new Ω is not a subset of the previous one, but all previous notations carry over in the obvious manner. The reason for this restriction is that from now on we shall deal mainly with the process (Y_t).

<u>Theorem.</u> (1) $(\Omega,H^o,H^o_t,Y_t,\phi_t,P^y : t \in \mathbb{R}, y \in U)$ is an a.s. U valued Markov process with shifts (ϕ_t). That is, for each $y \in U$ and $t \in \mathbb{R}$ one has

$$P^y_{\phi_t} \big| H^o_t = P^{Y_t}.$$

(2) Suppose that P is a probability on (Ω,H^o) that satisfies $P_{\phi_t} \big| H^o_t = P^{Y_t}$, $t \in \mathbb{R}$. Then (Ω,H^o,H^o_t,M,X,P) is a regenerative system with respect to $(P^x)_{x \in E_o}$.

Note that the semi group $(P_s)_{s \in \mathbb{R}_+}$ of the process (Y_t), given by $P_s(y,\cdot) = P^y\{Y_s \in \cdot\}$, has some branch points, namely the points $(0,x)$ such that $x \in E_o \setminus F$. Here $F = \{x \in E_o : P^x\{R = 0\} = 1\}$ with $R = R_0$. They will be eliminated under the following assumption.

<u>Assumption (R)</u>. For each $s > 0$ and each bounded continuous function f on U the mapping $t \to P^{X_{D_t}}(f \circ Y_s)$, $t < \sup M$ is a.s. right continuous.

Here and in the sequel a.s. means P - a.s. for every probability P on (Ω, H^o) under which (M,X) is regenerative with respect to (P^x). The filtration obtained by completing (H^o_{t+}) with respect to the family of all such P's will be denoted by (H_t).

Assuming (R) we can use arguments of (1.9) and (2.3) of Maisonneuve (1983) to prove that for each stopping time T of (H_t) and for $y \in U$

$$P^y_{\phi_{D_T}} \Big|_{H_T} = P^{X_{D_T}} \quad \text{on} \quad \{D_T < \infty\},$$

$$P^y_{\phi_T} \Big|_{H_T} = P^{Y_T} \quad \text{on} \quad \{T < \infty\}.$$

Therefore $P^{X_{D_T}}\{R = 0\} = 1$ a.s. on $\{D_T < \infty, R_T = 0\}$ and by the section theorem (Dellacherie, 1972) the set of branch points $\{(0,x) : x \in E_o \setminus F\}$ is polar. If we get rid of it (without changing the notation U), $(\Omega, H, H_{t+}, Y_s, \phi_s, P^y : s \geq 0, y \in U)$ becomes a strong Markov process which is a.s. right continuous and U valued, without branch points. Since its semi-group $(P_s)_{s \in \mathbb{R}_+}$ is Borel, the process $(Y_s)_{s \in \mathbb{R}_+}$ is a right process.

5. <u>Exit systems</u>

Let $(P^x)_{x \in E_o}$ be a regenerative family on (Ω, H^o), satisfying (R). Consider the random set

$$G = \{ t \in \mathbb{R} : R_{t-} = 0, R_t > 0 \} .$$

Note that G is homogeneous in the following sense :

$$(G - t) \cap (0,\infty) = G \circ \phi_t \qquad t \in \mathbb{R}$$

It follows from the definition of G that there exists a sequence (T_n) of (H_t) stopping times such that $G = \{T_n, n \geqslant 0\} \cap \mathbb{R}$. We shall assume in this work that these stopping times are <u>totally inaccessible</u>. This implies that M is a.s. perfect and that the process (X_t) has a.s. left limits in F for t in G (recall that Ω has been restricted in § 4). For this and the more general case we refer the reader to Kaspi and Maisonneuve, (forthcoming paper). The above assumptions lead to the following.

<u>Theorem</u>. There exists a continuous (H_t) adapted additive functional $(L_t)_{t \in \mathbb{R}}$ (a.s. one has $L_0 = 0$, $t \to L_t$ is increasing and $L_{t+s} = L_t + L_s \circ \phi_t$, $t \in \mathbb{R}$, $s \in \mathbb{R}_+$), a positive E_0 measurable function ℓ and a kernel $*P$ from (E_0, E_0) into (Ω, H^0) such that

(i)   the set of increase points of L is M a.s. and
$$1_M(t)dt = \ell(X_t)dL_t \quad \text{a.s.,}$$

(ii) $*P^{X_t}dL_t$ is the (H_t) dual predictable projection of
$$\sum_{t \in G} \varepsilon_{\phi_t},$$

(iii) $\ell + *P \cdot (1 - e^{-R}) = 1_F$.

Let $S_s = \inf\{t : L_t > s\}$ for $s \in \mathbb{R}_+$. The process $(\Omega, H, H_{S_s}, S_s, \phi_{S_s}, P^X)$ is then a Markov additive process in the sense of Çinlar (1972). The semi-group of (X_{S_s}) will be denoted by $(\widetilde{P}_s)$ in the sequel.

6. <u>Stationarity</u>

With the previous assumptions and notations one can prove the following results (see Kaspi and Maisonneuve, forthcoming paper).

<u>Theorem</u>. Each of the following formulae establishes a $1 - 1$ correspondence between (P_s) invariant probabilities η on $\mathbb{R}_+ \times E_o$ and $(\widetilde{P}_s)$ invariant measures μ on E_o such that $\mu(\ell + {}^*P \cdot (R)) = 1$ (in particular μ is finite) :

$$(1) \quad \mu(f) = P^\eta \int_{(0,1]} f(X_s)dL_s, \quad f \in E_o^+,$$

$$(2) \quad \eta(h) = \mu(\ell h(0,\cdot) + {}^*P \cdot \int_0^R h(s,X_R)ds), \quad h \in \mathbb{R}_+ \times E_o^+$$

Assuming that η and μ are associated by this correspondence, let P be the probability on (Ω, H^o) under which (Y_t) is a stationary Markov process with semi group (P_s) and one dimensional distribution η. Then for positive appropriately measurable functions f,g one has

$$(3) \quad P \sum_{t \in G} f(X_{t-},t,\phi_t) = \int_{E_o \times \mathbb{R}} \mu(dx){}^*P^x f(x,t,\cdot),$$

$$(4) \quad P g(X_{G_t^-},A_t,R_t,X_{D_t}) = \int_{E_o} \mu(dx)\ell(x)f(x,0,0,x)$$

$$+ \int_{E_o \times \mathbb{R}} \mu(dx)da \int_{(a,\infty) \times E_o} {}^*P^x(R \in dr, X_R \in dy)f(x,a,r-a,y)$$

here $G_t = \sup\{s \leqslant t : s \in M\}$ and $A_t = t - G_t$.

<u>Remark</u>. If E_o has only one point x_o, (S_s) is a subordinator with drift $\alpha = \ell(x_o)$ and Levy measures $\lambda = R({}^*P^{x_o})$. Then we have, with $m = \alpha + \int_{(0,\infty)} x\lambda(dx)$ like in Maisonneuve (1983),

$$P g(A_t,R_t) = \frac{\alpha}{m} g(0,0) + \int_{\mathbb{R}_+} da \int_{(a,\infty)} \frac{1}{m} \lambda(dr)g(a,r-a).$$

If $\alpha = 0$ and λ is finite we get

$$P\{A_t > u, R_r > v\} = \frac{1}{\bar{v}} \int_{u+v}^{\infty} v(a,\infty)da$$

with $v = \frac{\lambda}{\lambda(0,\infty)},\ \bar{v} = \int_{(0,\infty)} xv(dx)$. This is the known formula for the joint distribution of the backward and forward recurrence

times in renewal theory, ν being the distribution of the inter-
renewal epochs (see Çinlar, 1975).

7. Construction of stationary regenerative systems through MAP's

Let $(\bar{\Omega},M,M_s,E_s,S_s,\theta_s,\bar{P}^x)$ be a MAP like in example (3) of
§ 3. Consider the associated regenerative system $(\bar{\Omega},M,M_{C_t},\bar{M},\bar{X},\bar{P}^x)$.
The corresponding process $(\bar{Y}_t)$ is equal to $(S_{C_t} - t, E_{C_t})$ and
its semi group (P_s), given by

$$
P_s f(r,x) = \begin{cases} f(r-s,x) & \text{if } s < r \\[2ex] \bar{P}^x f(\bar{Y}_{s-r}) & \text{if } s \geq r \end{cases} ,
$$

is a Borel right semi group due to the strong Markov property of
the MAP (E,S).

We shall assume that (E,S) is (M_s) quasi left continuous.
It was shown in Çinlar (1972) that there exist continuous additive
functionals A, B of (E_s) and a kernel K from (E,E) into
$(E \times \mathbb{R}_+, E \times \mathbb{R}_+)$ such that for $t \in \mathbb{R}_+$

(i) $S_t = A_t + \sum_{s \leq t} (S_s - S_{s-})$,

(ii) $\bar{P}^x \sum_{s \leq t} 1_{S_{s-} \neq S_s} f(E_{s-},E_s,S_s-S_{s-})$

$$
= \bar{P}^x \int_0^t dB_s \int_{E \times \mathbb{R}_+} K(E_s,dy,du) f(E_s,u,u)
$$

The following is proved in Kaspi and Maisonneuve (forthcoming
paper).

<u>Theorem.</u> Let μ be a finite invariant measure for (E_s) and let

$$
\mu_A = \bar{P}^\mu \int_0^1 1_{E_s \in .} dA_s , \quad \mu_B = \bar{P}^\mu \int_0^1 1_{E_s \in .} dB_s .
$$

Then the measure η defined on $\mathbb{R}_+ \times E$ by

$$\eta(f) = \int_E \mu_A(dx)f(0,x) + \int_E \mu_B(dx) \int_{E \times \mathbb{R}_+} K(x,dy,(a,\infty))f(a,y)da$$

is invariant for (P_s). If it is finite one may use it to construct a measure P on (Ω, H^O) under which (M,X) is a stationary
regenerative system, that is Y is a stationary Markov process.

Remark. Formula (2) of section 6 can be understood as a particular
case of the above formula, written for the MAP (X_{S_s}, S_s) with
$A_t = \int_0^t \ell(X_{S_s})ds$, $B_t = t$, $K(x,du,dy) = {}^*P^x(R \in du, X_R \in dy)$.

References

E. Çinlar, Markov additive processes. II. Z.Wahrscheinlichkeits-
theorie verw. Geb. <u>24</u> 95-121 (1972).

E. Çinlar, Introduction to Stochastic Processes. Prentice-Hall Inc.
(1975).

C. Dellacherie, Capacités et Processus Stochastiques. Springer-
Verlag. Berlin, Heidelberg, New York (1972).

H. Kaspi, Excursions of Markov processes : an approach via Markov
additive processes. Z.Wahrscheinlichkeitstheorie verw. Geb.
<u>64</u> 261-268 (1983).

H. Kaspi, On invariant measures and dual excursions of Markov
processes. Z.Wahrscheinlichkeitstheorie verw. Geb. <u>66</u> 185-204
(1984).

H. Kaspi and B. Maisonneuve, Regenerative systems on the real line.
Forthcoming paper.

B. Maisonneuve, Ensembles régénératifs de la droite.
Z.Wahrscheinlichkeitstheorie verw. Geb. <u>63</u> 501-510 (1983).

M.I. Taksar, Regenerative sets on real line. Séminaire de
Probabilités XIV. Lecture Notes in Mathematics N° <u>784</u>. (1980)
Springer Verlag, Berlin, Heidelberg, New York.

ASYMPTOTIC ANALYSIS OF SOME NON-HOMOGENEOUS SEMI-MARKOV PROCESSES

R.V. Benevento

Dipartimento di Matematica dell'Università della Calabria

87036 - Arcavacata di Rende (CS), Italy

1. Introduction

Semi-Markov models turn out to be inadequate in some applications, see for instance (Janssen, Volpe, 1986), because the
independence of the transition lengths given all the states visi-
ted, is too restrictive an assumption to be practical.
In 1972, A.Iosifescu-Manu pointed out the convenience of
allowing the sojourn in a state between two transitions to depend
on the time when that sojourn began. She adapted the Markov renewal
equation to these circumstances. A more general situation arises
when the dependence is also on the number of the transitions preceding that sojourn. The foundations for the resulting non-homogeneous semi-Markov theory are in (De Dominicis, Janssen, 1984),
but asymptotic theorems are still missing.

Here we begin to fill this gap. To this purpose we consider
some generalized semi-Markov processes with finite state space
fulfilling a <u>strong-law condition</u>. For such processes we find an
explicit formula for the limit of the expected ratio of the time
spent in a given state up to time t to the total time t, under

sufficient conditions which ensure the existence of that limit as
t goes to infinity.

When the state space is finite, every irreducible aperiodic
semi-Markov process is included in our class of generalized semi-
Markov processes and fulfills the required conditions. Included in
the same class are also many <u>non-homogeneous semi-Markov</u> processes,
see definition 2 below, and two examples are given here.

2. Definitions

In a probability space $(\Omega, \mathcal{F}, P)$ suppose we have defined, for
each $n \in N$, a random variable X_n, taking values in a finite set
E, and an r.v. T_n, taking values in $\mathbb{R}_+ = [0, \infty[$, such that
$0 = T_0 \leqslant T_1 \leqslant T_2 \leqslant \ldots < \infty.$

We will call the resulting two-dimensional process $\{X_n, T_n;$
$n \in N\}$ an (X,T) process. Therefore a Markov renewal process
(MRP), see (Çinlar, 1975) for definitions, is a particular (X,T)
process, provided it has a finite state space for the chain X.

For any state $i \in E$, the conditional probability measure
$P\{\cdot \mid X_0 = i\}$ will be denoted by $P_i(\cdot)$, and similar notation will
be used for the conditional expectation.

Definition 1

An (X,T) process is said to be a strong-law process (SLP) if, for
any $i \in E$,

$$E_i T_n / n \to \alpha_i > 0$$

and the sequence $\{T_n/n; n \in N\}$ converges P_i - a.s. to the same
constant α_i. ∎

In relation to an (X,T) process we introduce the following
notations :

$$^{(n)}Q_{ij}(T_n,t) = P\{X_{n+1} = j, \; T_{n+1} \leqslant t \mid X_n = i, T_n\}$$

$$^{(n)}F_i(T_n,t) = P\{T_{n+1} - T_n > t \mid X_n = i, T_n\}.$$

Thus the functions $^{(n)}Q_{ij}(s,t)$, $^{(n)}F_i(s,t)$ are defined when s is in the range of T_n and $T \in \mathbb{R}$. In particular, when (X,T) is an MRP with transition kernel $Q_{ij}(x)$, we have

$$(1) \quad \begin{cases} ^{(n)}Q_{ij}(s,t) = Q_{ij}(t - s); \\[2ex] ^{(n)}F_i(s,t) = 1 - \sum_{j \in E} Q_{ij}(t). \end{cases}$$

With an (X,T) process we always associate a continuous time parameter process $Z = \{Z_t; \; t \in \mathbb{R}_+\}$ defined on the event $(\sup_n T_n = \infty)$ by $Z_t(\omega) = X_n(\omega)$ if $T_n(\omega) \leqslant t < T_{n+1}(\omega)$; $\omega \in \Omega$, $n \in N$. Clearly Z is a.s. defined in Ω when (X,T) is an SLP.

We now extend the notion of semi-Markov process by the following

<u>Definition 2</u>

The process Z associated with an (X,T) process is called a non-homogeneous semi-Markov process (NHSMP) if (X,T) is a two-dimensional Markov chain with possibly non-stationary transitions.

Putting

$$(2) \quad \begin{cases} ^{(n)}P_{X_n j}(T_n) = \lim_{t \to \infty} {}^{(n)}Q_{X_n j}(T_n,t) \\[2ex] ^{(n)}\mathbf{P}(x) = \{^{(n)}P_{ij}(x); \; i,j \in E\} \end{cases}$$

we now derive from the Markov property

$$P\{X_{n+1} = j \mid X_0, \ldots, X_n; \; T_0, \ldots, T_n\} = {}^{(n)}P_{X_n j}(T_n)$$

and hence

$$P(X_{n+1} = j) = E_i \, {}^{(n)}P_{X_n j}(T_n).$$

Thus the process X can be viewed as a conditional multiple Markov chain; this means that given the sequence T each transition from X_n to X_{n+1} is ruled by the stochastic matrix ${}^{(n)}\mathbb{P}(T_n)$. In (De Dominicis, Janssen, 1984) X is called the imbedded multiple Markov chain and when ${}^{(n)}\mathbb{P}(x)$ is independent of x [resp. of n, x] the multiple chains X behaves as an one-dimensional non-homogeneous [resp. homogeneous] Markov chain.

If for all $i, j \in E$ there is $\pi_i(j)$ such that $\sum_j \pi_i(j) = 1$ and $\lim_n P_i(X_n = j) = \pi_i(j)$, we say that X has an equilibrium distribution $\pi_i(\cdot)$ given the initial state i.

3. Results

Let 1_j be the indicator function of $j \in E$, we will prove

Theorem 1

If (X,T) is an SLP and if, for $i, j \in E$ and $n \to \infty$, the Cesaro limit of the sequence $\{E_i(T_{n+1} - T_n)1_j(X_n); n \in N\}$ exists and is a finite number $\mu_i(j)$, then the continuous time parameter process Z associated with (X,T) is such that for any $i, j \in E$

$$\lim_{t \to \infty} E_i\left\{\frac{1}{t} \int_0^t 1_j(Z_x)\,dx\right\} = \mu_i(j) \bigg/ \sum_{k \in E} \mu_i(k).$$

Theorem 2

If (X,T) is an SLP and if, for $i, j \in E$

(a) X has an equilibrium distribution $\pi_i(\cdot)$;

(b) the sequence $\{{}^{(n)}F_j(a_n,t); n \in N\}$ of functions of $t \in \mathbb{R}$ converges to a function $F_j(t)$ as $n \to \infty$, $\{a_n; n \in N\}$ being any divergent sequence of reals with a_n in the domain of ${}^{(n)}F_j(\cdot,t)$;

(c)   for any $\{a_n;\ n \in N\}$ as in (b)

$$\lim_n \int_0^\infty {}^{(n)}F_j(a_n,t)\,dt = \int_0^\infty F_j(t)\,dt < \infty,$$

then the Cesaro limit in theorem 1 exists and is given by

$$\mu_i(j) = \pi_i(j) \int_0^\infty F_j(t)\,dt.$$

Corollary 1

If (X,T) is an irreducible aperiodic MRP with finite state space E and stationary distribution π, then the semi-Markov process Z, associated with (X,T) is such that

$$\lim_{t\to\infty} P_i(Z_t = j) = \frac{\pi(j)E_j T_1}{\sum\limits_{k\in E} \pi(k)E_k T_1} = \lim_{t\to\infty} E_i\left\{\frac{1}{t}\int_0^t 1_j(Z_x)\,dx\right\}.$$

To prove theorem 1 we need two lemmas

Lemma 1

If (X,T) is an SLP, then the r.v.'s T_n/n converge also in $L_1(\Omega,\mathcal{F},P_i)$ for any $i \in E$.

Proof of Lemma 1

For brevity we drop the subscript i appearing in definition 1 and put $Y_n = T_n/n$. Since $Y_n \to \alpha$ and $0 \leqslant (\alpha - Y_n)^+ \leqslant \alpha$, then by dominated convergence $E(\alpha - Y_n)^+ \to 0$. Hence $E|\alpha - Y_n| \to 0$ because of the equality $E|\alpha - Y_n| = 2E(\alpha - Y_n)^+ - E(\alpha - Y_n)$. $\square$

Lemma 2

Under the hypothesis of theorem 1, for any $i \in E$, the number α_i in definition 1 is given by $\sum\limits_{j\in E} \mu_i(j)$.

Proof of Lemma 2

Dropping i again, we have

$$\alpha = \lim_n ET_{n+1}/n + 1$$

$$= \lim_n \frac{1}{n+1} E \sum_{m=0}^{n} (T_{m+1} - T_m)$$

$$= \lim_n \frac{1}{n+1} \sum_{m=0}^{n} E \sum_{j \in E} 1_j(X_m)(T_{m+1} - T_m)$$

$$= \sum_{j \in E} \lim_n \frac{1}{n+1} \sum_{m=0}^{n} E(T_{m+1} - T_m)1_j(X_m)$$

$$= \sum_{j \in E} \mu(j). \quad \square$$

<u>Proof of theorem 1</u>

Again $i \in E$ is fixed and dropped wherever it should appear as a subscript. We put for $n \in N$

$$\beta_n = E\left\{\frac{1}{n} \int_{n\alpha}^{T_n} 1_j(Z_x)\,dx\right\},$$

where α is given by lemma 2. Because of lemma 1 we have $|\beta_n| \leqslant E|T_n/n - \alpha| \to 0$ as $n \to \infty$. Hence

$$\frac{\mu(j)}{\sum_{k \in E} \mu(k)} = \alpha^{-1} \lim_n \frac{1}{n+1} \sum_{m=0}^{n} E(T_{m+1} - T_m)1_j(X_m)$$

$$= \alpha^{-1} \lim_n E\left\{\frac{1}{n+1} \int_0^{T_{n+1}} 1_j(Z_x)\,dx\right\}$$

$$= \alpha^{-1} \lim_n E\left\{\frac{1}{n+1} \int_0^{(n+1)\alpha} 1_j(Z_x)\,dx\right\} + \alpha^{-1} \lim_n \beta_{n+1}$$

$$= \lim_n E\left\{\frac{1}{(n+1)\alpha} \int_0^{(n+1)\alpha} 1_j(Z_x)\,dx\right\}.$$

Through this chain of equalities we have proved the assertion when t increases to infinity along the sequence $\{(n+1)\alpha\}$. Let $\{t_n\}$ be any other sequence such that $n\alpha \leqslant t_n < (n+1)\alpha$ and so that $t_n/n\alpha \to 1$. The inequalities

$$E\left\{\frac{1}{n\alpha}\int_0^{n\alpha} 1_j(Z_x)\,dx\right\} \leq \frac{t_n}{n\alpha}\, E\left\{\frac{1}{t_n}\int_0^{t_n} 1_j(Z_x)\,dx\right\}$$

$$\leq \frac{(n+1)}{n}\, E\left\{\frac{1}{(n+1)\alpha}\int_0^{(n+1)\alpha} 1_j(Z_x)\,dx\right\}$$

prove the assertion for t spanning the sequence $\{t_n\}$, also. The extension to general sequences increasing to infinity is almost immediate.

<u>Proof of theorem 2</u>

Now $i \in E$ is fixed and kept as a subscript. Let
$$\Omega^1 = \bigcup_{n \in N} (T_n = T_{n+1} = \ldots) \quad \text{and for} \quad n \in N, \quad s_n = \inf_{\omega \notin \Omega^1} T_n(\omega).$$

Since $P_i(\sup_n T_n = \infty) = 1$, then $P_i(\Omega^1) = 0$ and $s_n \uparrow \infty$ as n increases. From (b), as $n \to \infty$, both functions of $t \in R$

$$f_n(t) = \inf_{s > s_n} {}^{(n)}F_j(s,t) \quad \text{and} \quad g_n(t) = \sup_{s > s_n} {}^{(n)}F_j(s,t)$$

converge everywhere to $F_j(t)$. Putting

$$h_n(t) = E_i\{{}^{(n)}F_j(T_n,t)1_j(X_n)\}$$

since $P_i(T_n \geq s_n) = 1$, we have

$$\int_0^\infty P_i(X_n = j)f_n(t)\,dt \leq \int_0^\infty h_n(t)\,dt \leq \int_0^\infty P_i(X_n = j)g_n(t)\,dt.$$

From (a) and (c) both first and last term of these inequalities converge to

$$\pi_i(j)\int_0^\infty F_j(t)\,dt.$$

The proof is completed by

$$\pi_i(j)\int_0^\infty F_j(t)\,dt = \lim_n \int_0^\infty h_n(t)\,dt$$

$$= \lim_n \int_0^\infty E_i\{P(T_{n+1} - T_n > t\,|\,X_n = j, T_n)1_j(X_n)\}\,dt$$

$$= \lim_n \int_0^\infty P_i(T_{n+1} - T_n > t, X_n = j)dt$$

$$= \lim_n E_i(T_{n+1} - T_n)1_j(X_n)$$

$$= \mu_i(j).$$

Proof of corollary 1

The first equality is a classical result of W.Feller (1964). The
second comes from a general strong-law for MRP's (Janssen, 1982),
which applies to our MRP and shows that

$$\lim_n T_n/n = E_\pi T_1 = \sum_{j \in E} \pi(j)E_j T_1, \quad P_i - a.s. \quad \text{for any} \quad i \in E.$$

Homogeneity yields $E_i(T_{n+1} - T_n)1_j(X_n) = P_i(X_n = j)E_j T_1$ for any
$n \in N$, and thus $\mu_i(j) = \pi(j)E_j T_1$; $\alpha_i = E_\pi T_1$ according to lemma 2.
Now theorem 1 is in force, since $\mu_i(j)$ exists and $\lim_n T_n/n = \alpha_i$
$= \lim_n E_i T_n/n.$

4. Examples

By only modifying the transition times of an irreducible
aperiodic MRP $(\widetilde{X},\widetilde{T})$ with finite state space E and stationary
distribution π for the regular chain $\widetilde{X}$, we obtain two examples
of (X,T) processes which originate NHSMP. Hereafter $Q_{ij}(x)$
will denote the transition kernel of $(\widetilde{X},\widetilde{T})$ and P the transition
matrix of $\widetilde{X}.$

Example 1

Let $U_1, U_2, \ldots$ be a sequence of independent r.v.'s, non-negative,
with finite mean, finite variance and independent of $(\widetilde{X},\widetilde{T})$, just
introduced above. If we put $U_0 = 0$ and

$$X = \widetilde{X} \; ; \; T_n = \widetilde{T}_n + \sum_0^n U_m, \quad n \in N,$$

it is not difficult to prove that

$$P\{X_{n+1} = j, T_{n+1} \leq t \mid X_0, \ldots, X_n; T_0, \ldots, T_n\}$$

$$= E\{Q_{X_n j}(t - T_n - U_{n+1}) \mid X_n, T_n\}$$

Thus the two-dimensional process (X,T) is a non-homogeneous Markov chain, and since now, recalling (2), $^{(n)}P(t) = P$ the multiple chain X behaves as the regular chain $\tilde{X}$.

Theorem 3

Any two-dimensional Markov chain (X,T) constructed as in example 1 and such that

$$\frac{1}{n} \sum_1^n EU_m \to \mu < \infty$$

$$\frac{1}{n} \sum_1^n \text{Var } U_m \to \sigma < \infty$$

meets the hypothesis of theorem 1 with

$$\mu_i(j) = \pi(j)[E_j\tilde{T}_1 + \mu].$$

Corollary 2

If (X,T) fulfills theorem's 3 conditions and Z is the NHSMP associated with, then

$$\lim_{t \to \infty} E_i \left\{ \frac{1}{t} \int_0^t 1_j(Z_x)dx \right\} = \frac{\pi(j)[E_j\tilde{T}_1 + \mu]}{\sum_{k \in E} \pi(k)[E_k\tilde{T}_1 + \mu]} , \quad i,j \in E.$$

Proof of theorem 3

The sequence $\{\sum_0^n U_m; n \in N\}$ is a submartingale integrable, which we split into the sum $\{M_n + A_n; n \in N\}$ where

$$M_n = \sum_0^n (U_m - EU_m); \quad A_n = \sum_0^n EU_m.$$

Now $\sum_1^\infty \text{Var } U_n$ can be finite or infinite.

When it is finite the martingale $\{M_n\}$ converges to a finite limit, by proposition VII-2.3 (Neveu, 1972), and thus $M_n/n \to 0$ a.s..

When $\sum_1^\infty \text{Var } U_n$ is infinite, proposition VII-2.4 (Neveu, 1972) tells us that

$$M_n \Big/ \left(1 + \sum_1^n \text{Var } U_m\right) \to 0, \text{ a.s..}$$

Therefore $M_n/n \to 0.\sigma = 0$, a.s..

In any case $\lim_n \dfrac{1}{n} \sum_1^n U_m = \lim_n \dfrac{1}{n} M_n + \lim_n \dfrac{1}{n} A_n = \mu$. Since $\lim_n E_i T_n/n = \mu + \lim_n E_i \widetilde{T}_n/n$, we have also proved that

$$\lim_n T_n/n = \lim_n E_i T_n/n \quad \text{for} \quad i \in E;$$

thus (X,T) is an SLP.

To complete the proof we notice that for $i, j \in E$ and $n \in N$

$$E_i(T_{n+1} - T_n)1_j(X_n) = E_i(\widetilde{T}_{n+1} - \widetilde{T}_n)1_j(\widetilde{X}_n) + EU_{n+1}P_i(X_n = j).$$

We recall now that the product term by term of one sequence of non-negative numbers Cesaro convergent and another sequence ordinarily convergent is Cesaro convergent to the product of the limits. Thus the Cesaro limit of the second summand, on the right of the equality above, is the number $\mu\pi(j)$. Because of the homogeneity of the chain $(\widetilde{X},\widetilde{T})$ the first summand converges to $E_j\widetilde{T}_1 \cdot \pi(j)$. This completes the proof. $\square$

Example 2
===

Let $(\widetilde{X},\widetilde{T})$, X, $Q_{ij}(x)$ and π be as in example 1, but now let us take

$$T_n = \log(\widetilde{T}_n + 1) + n, \; n \in N.$$

Putting $\alpha = \sum_{j \in E} \pi(j)E_j\widetilde{T}_1$, since we know that $\widetilde{T}_n/n \xrightarrow{a.s.} \alpha$, we know also that $T_n/n \xrightarrow{a.s.} 1$. Moreover, by lemma 1 we have $\tilde{T}_n/n \xrightarrow{L_1} \alpha$ and thus $\sqrt{\tilde{T}_n}/n \xrightarrow{L_1} 0$. Since $\log(x + 1) < 2\sqrt{x}$, $x \in \mathbf{R}_+$, we have now $E \log(\tilde{T}_n + 1)/n \to 0$ and $ET_n/n \to 1$. Therefore (X,T) is an SLP. Through calculations we obtain

$$P\{X_{n+1} = j, T_{n+1} \leq t \mid X_0,\ldots,X_n; T_0,\ldots,T_n\}$$
$$= Q_{X_n j}(e^{-n}[e^{t-1} - e^{T_n}]).$$

Thus the new process (X,T) is a two-dimensional non-homogeneous Markov chain and again the multiple chain X behaves as the regular chain $\tilde{X}$.

Theorem 4

The (X,T) process of example 2 meets all the conditions of theorem 2 with $F_j(t) = 1_{]\infty, 1]}(t)$.

Corollary 3

When Z is the NHSMP associated with (X,T) of example 2, then for any $i, j \in E$

$$\lim_{t \to \infty} E_i \left\{ \frac{1}{t} \int_0^\infty 1_j(Z_x)\,dx \right\} = \lim_n P_i(X_n = j) = \pi(j).$$

Proof of theorem 4

We only need to prove conditions (b) and (c) since now the equilibrium distribution of X is π. By calculations we have

$$^{(n)}F_j(a_n,t) = 1 - \sum_{k \in E} Q_{jk}([e^{t-1} - 1]e^{a_n - n}).$$

We introduce a sequence $\{\tilde{s}_n\}$ with respect to the process $(\tilde{X},\tilde{T})$, by a definition analogous to that of the sequence $\{s_n\}$ introduced in the proof of theorem 2. Thus we have $\tilde{T}_n \geq \tilde{s}_n$ a.s. and $T_n \geq \log(\tilde{s}_n + 1) + n$, a.s. . Therefore in order to belong to the range of T_n the number a_n mus be such that $a_n - n \geq \log(\tilde{s}_n+1) \uparrow \infty$. This yields

$$\lim_{n} {}^{(n)}F_j(a_n,t) = \begin{cases} 1 - \sum_{k \in E} Q_{jk}(0) & \text{if } t = 1 \\[2ex] 1 - \sum_{k \in E} Q_{jk}(+\infty) & \text{if } t > 1 \\[2ex] 1 - \sum_{k \in E} Q_{jk}(-\infty) & \text{if } t < 1 \end{cases}$$

$$= 1_{]\infty, 1]}(t) \quad \text{everywhere}$$

independently of the sequence $\{a_n\}$ and of $j \in E$. Thus (b) is proved, and $F_j(t) = 1_{]\infty, 1]}(t)$.

To prove (c) we do more calculations and we obtain

$$\int_0^\infty {}^{(n)}F_j(a_n,t)\,dt = 1 + E_j \log(1 + e^{\frac{n-a_n}{n}\tilde{T}_1}).$$

The expectation above goes to zero, since $e^{\frac{n-a_n}{n}} \downarrow 0$ and the r.v. inside is dominated by $\sup_n e^{\frac{(n-a_n)}{n}\tilde{T}_1}$. The proof is now completed because

$$\lim_{n} \int_0^\infty {}^{(n)}F_j(a_n,t)\,dt = 1 = \int_0^\infty F_j(t)\,dt. \quad \square$$

Acknowledgement

I would like to thank Professor J.Janssen for his suggestions and also the Institute for Advanced Study in Princeton for providing me with the best working conditions while writing this paper.

References

Çinlar, E. (1975). Introduction to stochastic processes - Prentice Hall.

De Dominicis, R., J.Janssen (1984). Finite non-homogeneous semi-Markov processes, Insurance : Math. & Econ., 3.

Feller, W. (1964). On semi-Markov processes, Proc.Nat.Acad.Sci. 51, 653-659.

Iosifescu-Manu, A. (1972). Non-homogeneous semi-Markov processes, Stud.Lere.Mat. 24, 529-533.

Janssen, J. (1982). Strong laws for semi-Markov chains with a general state space, Ricerche di Matematica 31, 405-414.

Janssen, J., E.Volpe (1986). Stochastic modelling for private pension funds, technical report, CADEPS-ULB, to appear.

Neveu, J. (1972). Martingales à temps discret. Masson et Cie.

SECTION II. SEMI-MARKOV DECISION PROCESSES

MARKOV AND SEMI-MARKOV DECISION MODELS AND OPTIMAL STOPPING

Manfred Schäl

Institut für Angewandte Mathematik
Universität Bonn

1. The general Markov decision model

We consider a system with a finite number of states $i \in S$.
Periodically we observe the current state of the system, and then
choose an action a from a set A of possible actions. As a result
of the current state i and the chosen action a, the system moves to
a new state j with the probability $p_{ij}(a)$. As a further consequence, an immediate reward $r(i,a)$ is earned. If the control process
is stopped in a state i, then we obtain a terminal reward $u(i)$.
Thus, the underlying model is given by a tupel $M = (S,A,p,r,u)$.

(i)   S stands for the <u>state space</u> and is assumed to be finite.

(ii)   A is the <u>action space</u>.

(iii) $p_{ij}(a)$ are <u>the transition probabilities</u>, where
       $\Sigma_{j \in S} \, p_{ij}(a) = 1$ for all $i \in S$, $a \in A$.

(iv) $r(i,a)$ is the real valued <u>one-step reward</u>.

(v) $u(i)$ is the real valued <u>terminal reward</u> or the utility func-
       tion.

Some of the results can be extended to the case where S may be denu-
rable by the use of recent results by Deppe (1984), Hordijk &
Dekker (1983), Mann (1983), Zijm (1982).

Moreover, the results will easily extend to the case where the set A of available actions depends on the current state. We do not explicitly introduce a _discount factor_. A possible discount factor can be looked upon as the probability of staying in the system at each stage (that is not going to an extra absorbing state) and may be incorporated in the transition law.

The one-step reward can be negative, e.g. $r(i,a) = -c(i,a)$ where $c(i,a)$ are the _observation cost_. We write $H_m = S \times \ldots \times S$ (m + 1 factors) for the set of all _histories_ $(i_o, \ldots, i_m)$ at stage m. A _policy_ $\delta = (\delta_m)$ is defined as a sequence of decision functions $\delta_m : H_m \to A$. We write Δ for the set of all policies. In the present paper, we will mainly be concerned with the set $\mathbf{F}$ of _stationary policies_. They can be identified with decision functions $f : S \to A$ where $\delta \overset{\wedge}{=} f$ iff $\delta_m(i_o, \ldots, i_m) = f(i_m)$ for all m.

Given the initial state i, any policy $\delta \in \Delta$ defines a stochastic process (X_n), where X_n describes the state of the system at stage n. The process (X_n) is a Markov chain if the system is controlled by a stationary policy $f \in \mathbf{F}$. We write $\mathbb{P}_f = (p_{ij}(f(i)))$ for the corresponding transition matrix, $\mathbb{P}_f^{\star}$ for the matrix of the _Cesaro limits_ $\lim_{n\to\infty} \frac{1}{n}(\mathbb{P}_f + \ldots + \mathbb{P}_f^n)$, and H_f for the _fundamental matrix_ $(I - \mathbb{P}_f + \mathbb{P}_f^{\star})^{-1} - \mathbb{P}_f^{\star}$. We agree to identify real-valued functions on S with the corresponding column vectors. As an example, we mention the one-step reward r_f under policy $f \in \mathbf{F}$ defined by $r_f(i) = r(i,f(i))$.

For any stopping time τ with $E_{i\delta}\tau < \infty$, the expected total reward up to time τ is well-defined as

$$v_{\tau,\delta}(i) = E_{i\delta}\left[\Sigma_{m=0}^{\tau-1} r(X_m, \delta_m(X_o, \ldots, X_m)) + u(X_\tau)\right] \tag{1.1}$$

If the stopping time τ is not finite, the definition of the expected total reward is more problematic.

In the framework of gambling models (cf. Dubins & Savage, 1965; Sudderth, 1971) and optimal stopping models (cf. Chow, Robbins & Siegmund, 1971) where $r = 0$, $v_\delta(i) = v_{\infty,\delta}(i)$ is defined as

$$E_{i\delta} [\overline{\lim}_n u(X_n)] = \overline{\lim}_\tau \ v_{\tau,\delta}(i)$$

This definition was also used by Bodewig (1985) for the case $r \neq 0$.

In the present paper, the expected total reward v_δ for the infinite stage case will be defined as the (upper) limit of $v_{n,\delta}$ in the sense of Abel or Cesaro. This definition will enable us to build on methods and results from dynamic programming with the average reward criterion.

If v_δ is defined by an Abel limit, then v_δ may be interpreted as the limit of the expected total discounted reward v_δ^β as the discount factor β tends to one.

For $\beta < 1$, the expected total discounted reward is well-defined as

$$v_\delta^\beta(i) = E_{i\delta} [\sum_{m=0}^\infty \beta^m \{\beta r(X_m,\delta_m(X_0,\ldots,X_m))+(1-\beta)u(X_m)\}]$$

$$= (1-\beta)\sum_{n=0}^\infty \beta^n v_{n,\delta}(i). \tag{1.2}$$

This formula looks a bit strange because of the presence of the terminal reward function u, but agrees with the usual formula in the more familiar case $u \equiv 0$. A similar formula was used by Schäl (1983). The reason for not assuming the finite-stage expected total rewards $v_{n,\delta}$ to be convergent (as $n \to \infty$) is that we want to cover also models for optimal stopping problems (where r may vanish). Instead, we make the following general assumption.

<u>Assumption B</u>

1) $\sup_n v_{n,\delta}(i) < \infty$ for all $i \in S$, $\delta \in \Delta$;

2) for every $i \in S$, there is some $f \in \mathbf{F}$ such that
$\inf_n v_{n,f}(i) > -\infty$.

Defining as usual the average one-step reward for some $f \in \mathbf{F}$ by
$$g_f = \mathbb{P}_f^{\star} \, r_f$$
one may include from Assumption B1 that $g_f \leq 0$. Further, from B2,
it follows that for any i there is some f that $g_f(i) = 0$. Now, it
is clear that

$$g = \sup_f g_f = 0. \tag{1.3}$$

Thus we are concerned with a model where the optimal average return
g is constant and vanishes.

Now we can define our <u>objective function</u>

$$v_\delta = \overline{\lim}_{\beta \uparrow 1} \, v_\delta^\beta \, .$$

Another possible definition would be

$$\bar{v}_\delta = \overline{\lim}_{n \to \infty} \frac{1}{n}(v_{1,\delta} + \ldots + v_{n,\delta}) \, .$$

From (1.2) and B1 it follows that

$$v_\delta \leq \bar{v}_\delta < \infty \quad , \quad \delta \in \Delta \, . \tag{1.4}$$

From the theory of Markov chains one knows that $v_f = \bar{v}_f$ for stationary policies $f \in \mathbf{F}$ and $\overline{\lim}$ can be replaced by $\lim$. Moreover one
has the following representation (cf. Blackwell, 1962):

$$v_f = \infty \cdot g_f + H_f r_f + \mathbb{P}_f^\star u = \bar{v}_f \,, \qquad f \in \mathbb{F} \,. \tag{1.5}$$

We are only interested in policies for which g_f vanishes. Then the term $\infty \cdot g_f$ vanishes, too. We write

$$\mathbb{F}^\star = \{f \in \mathbb{F} \,, \quad g_f = 0\}.$$

Since matrix inversion is a continuous operation, we know that H_f depends continuously on $\mathbb{P}_f$ and $\mathbb{P}_f^\star$. The same is true for the discounted fundamental matrix

$$H_f^\beta = (I - \beta \, (\mathbb{P}_f - \mathbb{P}_f^\star) \,)^{-1} - \mathbb{P}_f^\star$$

which is even jointly continuous in $(\beta, \mathbb{P}_f, \mathbb{P}_f^\star)$, and

$$v_f^\beta = (H_f^\beta + \frac{1}{1-\beta} \, \mathbb{P}_f^\star) \, (\beta r_f + (1 - \beta)u). \tag{1.6}$$

Henceforth, we make the following compactness and continuity assumption which imply that $\mathbb{F}$ is a compact metric space and $f \to \mathbb{P}_f$, $f \to r_f$ are continuous mappings on $\mathbb{F}$.

Assumption C

1) A is a compact metric space;

2) $p_{ij}(a)$ and $r(i,a)$ depend continuously on a for all i,j.

It is well-known that Assumption C does not imply the continuity of $f \to \mathbb{P}_f^\star$. More exactly we have the following lemma.

Lemma 1.1   The following conditions are equivalent:

(i) $f \to n_f$ is continuous where n_f is the number of recurrent classes for $\mathbb{P}_f$;

(ii) $f \to \mathbb{P}_f^\star$ is continuous ;

(iii) $\sup_{f \in \mathbf{F}} \|H_f\| < \infty$.

<u>Proof</u>: The equivalence of the first two conditions was shown by Schweitzer (1968). The third condition was used by Chitashvili (1975). It is clear that (ii) implies (iii) because of Assumption C. In order to prove the converse one can use the following relation (cf. Blackwell, 1962; Miller & Veinott, 1969)

$$(1 - \beta)H_f^\beta = (1 - \beta)\sum_{m=0}^{\infty} \beta^m \mathbb{P}_f^m - \mathbb{P}_f^\star = \sum_{m=1}^{\infty} \rho^m H_f^m \qquad (1.7)$$

where $\beta = (1 + \rho)^{-1}$.

Here $\sum_{m=1}^{\infty} \rho^m H_f^m$ can be made arbitrarily small uniformly in $f \in \mathbf{F}$ by the choice of β and $(1 - \beta)\sum_{m=0}^{\infty} \beta^m \mathbb{P}_f^m$ is continuous in f for all β by Assumption C.$\square$

Now, in addition to Assumption C, we make the following assumption:

<u>Assumption C</u>* : One of the conditions in Lemma 1.1. is satisfied.

This assumption is fulfilled if A is finite or more generally if $\{p_{ij}(a)$, $a \in A$, $p_{ij}(a) > 0\}$ are bounded away from zero for each tupel (i,j). Another well-known example is a discounted model with discount factor $\alpha < 1$. Then after incorporating α in the transition law by the introduction of an extra state, the extra state forms the only ergodic class and hence $n_f = 1$ for all f.

<u>Theorem 1.2</u>

(a) $\sup_{\delta \in \Delta} v_\delta = \sup_{f \in \mathbf{F}} v_f = \sup_{f \in \mathbf{F}^\star} v_f;$

b) there exists some $f_o \in \mathbb{F}^\star$ such that $v_{f_o} = \sup_{\delta \in \Delta} v_\delta$.

Proof

For any $i \in S$, $\delta_o \in \Delta$, choose a sequence $\beta_n \uparrow 1$ such that $v_{\delta_o}(i) = \lim_{n \to \infty} v_{\delta_o}^{\beta_n}(i)$. It is well-known that under Assumption C there exist optimal stationary policies for every discounted problem. Thus, we may choose for any β_n some $f_n \in \mathbb{F}$ such that

$$v_{f_n}^{\beta_n} \geq v_\delta^{\beta_n} \quad \text{for all} \quad \delta \in \Delta . \tag{1.8}$$

Since $\mathbb{F}$ is a compact metric space, we may assume that (f_n) converges to some $f_o \in \mathbb{F}$; otherwise we replace (f_n) by a subsequence. Then by (1.6)

$$(1 - \beta_n)v_f^{\beta_n} \to g_f \quad \text{for all } f \in \mathbb{F}$$

and

$$(1 - \beta_n)v_{f_n}^{\beta_n} \to g_{f_o} \quad .$$

Now, by (1.8)

$$g_{f_o} \geq g_f \quad \text{for all} \quad f \in \mathbb{F}$$

and by (1.3) $g_{f_o} = 0$, i.e., $f_o \in \mathbb{F}^\star$. Further, again by (1.6) and (1.8),

$$\underline{\lim}_{n \to \infty} v_{f_n}^{\beta_n} = H_{f_o} r_{f_o} + \underline{\lim}_{n \to \infty} g_{f_n} / (1 - \beta_n) + \mathbb{P}_{f_o}^\star u$$

$$\geq \sup_{f \in \mathbb{F}^\star} \underline{\lim}_n v_f^{\beta_n} = \sup_{f \in \mathbb{F}^\star} (H_f r_f + \mathbb{P}_f^\star u)$$

Hence $\underline{\lim}_n g_{f_n} / (1 - \beta_n) \geq 0$ and by (1.3) $\lim_n g_{f_n} / (1 - \beta_n) = 0$.

Thus we can conclude that

$$v_{f_o}(i) = \lim_{n\to\infty} v_{f_n}^{\beta_n}(i) \geq \lim_{n\to\infty} v_{\delta_o}^{\beta_n}(i) = v_{\delta_o}(i) \quad ,$$

which implies part a), and further

$$v_{f_o} = \lim_{n\to\infty} v_{f_n}^{\beta_n} \geq \lim_{n\to\infty} v_f^{\beta_n} = v_f \quad \text{for all} \quad f \in \mathbb{F}^\star \quad ,$$

which implies $v_{f_o} = \sup_{f \in \mathbb{F}^\star} v_f$ Now part b) follows from a). $\square$

We do not know whether $\sup_{\delta \in \Delta} v_\delta$ is equal to $\sup_{\delta \in \Delta} \bar{v}_\delta$ in general. In Theorem 1.9 below we will however give a sufficient condition for the equality. Henceforth, we use the following notation for the value function

$$v = \sup_{\delta \in \Delta} v_\delta \quad .$$

<u>Lemma 1.3</u>   (<u>optimality equation</u>)

$$v(i) = \max_{a \in A} \{ r(i,a) + \Sigma_i \, p_{ij}(a) v(j) \} \quad .$$

<u>Proof</u>

According to theorem 1.2, we may choose $f_o \in \mathbb{F}$ such that $v = v_{f_o}$ and define $\delta = (f, f_o, f_o, \ldots)$ for any $f \in \mathbb{F}$.   Then

$$v_{n,\delta} = r_f + \mathbb{P}_f \, v_{n-1,f_o}$$

and hence

$$v_\delta = r_f + \mathbb{P}_f \, v_{f_o} \quad \text{by (1.2).}$$

Since $v_\delta \leq v$, we have $v \geq r_f + \mathbb{P}_f v$.   Moreover $v = r_{f_o} + \mathbb{P}_{f_o} v$ and the proof is complete. $\square$

For convenience, we write for any real function w on S

$$L_f w = r_f + \mathbb{P}_f w \ , \quad f \in \mathbb{F} \ .$$

Then we know by Lemma 1.3 that $v \geqslant L_f v$ for all $f \in \mathbb{F}$ and

$$\frac{1}{n}(L_f w + \ldots + L_f^n w) \to g_f \cdot \infty + H_f \, r_f + \mathbb{P}_f^{\star} w \ , \quad f \in \mathbb{F}. \qquad (1.9)$$

The latter formula follows from (1.5) replacing u by w.

<u>Lemma 1.4</u>

$$\mathbb{P}_f^{\star} v(i) \geqslant \mathbb{P}_f^{\star} u(i) \quad \text{if} \quad g_f(i) = 0$$

<u>Proof</u>

By (1.5), we have
$$v \geqslant v_f = \mathbb{P}_f^{\star} \, r_f \cdot \infty + H_f \, r_f + \mathbb{P}_f^{\star} u. \quad \text{Since}$$
$\mathbb{P}_f^{\star} \, H_f = 0$ and $\mathbb{P}_f^{\star} \, \mathbb{P}_f^{\star} = \mathbb{P}_f^{\star}$ (cp. Blackwell, 1962), the result
follows. $\square$

<u>Proposition 1.5</u>: for $f \in \mathbb{F}$

$$v \geqslant L_f v \geqslant \frac{1}{n} \{L_f v + \ldots + L_f^n v\} \downarrow g_f \cdot \infty + H_f \, r_f + \mathbb{P}_f^{\star} v \geqslant v_f.$$

<u>Proof</u>

The first inequality is Lemma 1.3, the second one follows by induction, the limit can be evaluated by (1.9), and the last inequality
follows from Lemma 1.4 if $g_f(i) = 0$ and is clear if $g_f(i) < 0.\square$

<u>Lemma 1.6</u>

If $w = L_f w$ for some $f \in \mathbb{F}$ and some real-valued function w on S, then $g_f = 0$ and $w = H_f \, r_f + \mathbb{P}_f^{\star} w$.

Proof
______

From (1.9) we conclude

$$g_f \cdot \infty + H_f \, r_f + \mathbb{P}_f^{\star} w = w. \,\square$$

Theorem 1.7
____

The following conditions are equivalent for $f \in \mathbf{F}$.

(i) $v = v_f$

(ii)   (a) $g_f = 0$ (i.e. $f \in \mathbf{F}^{\star}$)

       (b) $v = H_f \, r_f + \mathbb{P}_f^{\star} v$

       (c) $\mathbb{P}_f^{\star} v = \mathbb{P}_f^{\star} u$

(iii) (a) $v = L_f v$

       (b) $\mathbb{P}_f^{\star} v = \mathbb{P}_f^{\star} u$

Proof
_____

The relations (i) $\Longleftrightarrow$ (ii) $\Rightarrow$ (iii) follow from Proposition 1.5.
The relation (iii) $\Rightarrow$ (ii) follows from Lemma 1.6. $\square$

If condition (i) is satisfied, f is called _optimal_. If condition (iiia) is satisfied, f is called conserving. If condition
(iiib) is satisfied, f is called equalizing. Then the theorem states
that f is optimal iff f is conserving and equalizing which is strongly related to the characterization of optimal policies by Dubins &
Savage (1965). A real-valued function w on S is called excessive iff

$$w \geqslant L_f w \quad \text{and} \quad \mathbb{P}_f^{\star} w \geqslant \mathbb{P}_f^{\star} u \qquad \text{for all} \quad f \in \mathbf{F}^{\star}$$

<u>Theorem 1.8</u>

The value function v is the smallest excessive function on S.

<u>Proof</u>

By Lemma 1.3 and Lemma 1.4 v is excessive. Now let w be another excessive function, the by (1.9)

$$w \geq H_f\, r_f + \mathbb{P}_f^{\star} w \geq H_f\, r_f + \mathbb{P}_f^{\star} u = v_f \quad , \quad f \in \mathbb{F}^{\star}.$$

Now the result follows from theorem 1.2.$\Box$

<u>Theorem 1.9</u>

If $\mathbb{P}_f^{\star} v \geq \mathbb{P}_f^{\star} u$ for all $f \in \mathbb{F}$ (e.g. if $g_f = 0$ for all $f \in \mathbb{F}$ or if $v \geq u$) then

$$v = \sup_{\delta \in \Delta} \bar{v}_\delta \; .$$

The property $\mathbb{P}_f^{\star} v \geq \mathbb{P}_f^{\star} u$ or $\mathbb{P}_f (v - u) \geq 0$ for all f, is strongly related to the property that v − u has the property anne (asymptotic non-negative expectation) introduced by Hordijk (1974).

<u>Proof</u>

By assumption and Theorems 1.2 and 1.7,

$$\inf_{f \in \mathbb{F}} \mathbb{P}_f^{\star}(v - u) = 0 \tag{1.10}$$

Now we consider the average cost problem with one-step reward function v − u. This problem inherits the assumption C and $C^{\star}$. Thus, the two optimality equations for average cost problems are valid. By (1.10),

the value function for the average cost problem is constant, hence the
first optimality equation is trivial and the second optimality equation reads

$$w(i) = \inf_{a \in A} \{v(i) - u(i) + \Sigma_p p_{ij}(a)w(j)\} \qquad (1.11)$$

for some real-valued function w on S. The equation (1.11) is essentially the equation of Lemma 1.3 with $r(i,a) = u(i) - v(i)$. However,
the proof of Lemma 1.3 makes use of instationary policies. Another
proof can be given by using Howard's policy improvement routine starting with an optimal stationary policy f_o. This proof can be carried
through within the framework of stationary policies and only needs the
assumption $\sup_{f \in F} g_f = 0$ (cp. 1.10) instead of our general assumption B (cf. Mann, 1983).

From (1.11) one may derive that for $n = 1, 2, \ldots$

$$w(i) \leq E_{i\delta} [\Sigma_{m=0}^{n-1} (v(X_m) - u(X_m)) + w(X_n)] , \quad i \in S, \delta \in \Delta$$

and hence

$$\underline{\lim} \frac{1}{n} E_{i\delta} [\Sigma_{m=0}^{n-1} (v(X_m) - u(X_m))] \geq 0 , \quad i \in S, \delta \in \Delta \qquad (1.12)$$

Similarly, from Lemma 1.3 one may derive that

$$v(i) \geq E_{i\delta} [\Sigma_{m=0}^{n-1} r(X_m, \delta_m(X_o, \ldots, X_m)) + v(X_n)] ,$$

hence

$$v(i) \geq v_{n,\delta}(i) + E_{i\delta} [v(X_n) - u(X_n)] , \quad i \in S, \delta \in \Delta \qquad (1.13)$$

Combining (1.12) and (1.13), one obtains

$$v(i) \geq \bar{v}_\delta(i) , \quad i \in S, \delta \in \Delta . \square$$

It is possible to characterize an optimal stationary policy for the
present model also by two optimality equations (c . Schäl 1984a)
as it is known for average reward models.

2. The leavable Markov decision model

In this section, we will assume that the underlying model is
leavable, i.e., we make the following general assumption.

Assumption A'

$A = A_o + \{e\}$ where $p_{ij}(e) = \delta_{ij}$ and $r(i,e) = 0$ for all $i, j \in S$.
Choosing action e means that one pause for one period, i.e. the
system remains in the same state and one obtains no reward. As Dubins
& Savage (1965), we interpret stopping at time n by continuation such
that one always chooses action e from stage n onwards. Then, with
probability one, the system remains in the state of stage n time
after time. In the case where A_o is a singleton set, one has an
ordinary stopping model. Now we can replace assumption B by

Assumption B'

$$\sup_n v_{n,\delta}(i) < \infty \text{ for all } i \in S, \delta \in \Delta .$$

The assumption B' is satisfied if $r = -c$ where c are nonnegative
observation cost. Now consider the stationary policy f which stops
immediately, i.e., $f(i) = e$ for all $i \in S$. We will write e for this
policy. Then obviously $\mathbb{P}_e = I$, $\mathbb{P}_e^* = I$, $H_e = 0$, $r_e = 0$, $v_{n,e} = 0$ for
all $n, g_e = 0$; hence

$$v_e = u \text{ and } v \geqslant u. \qquad (2.1)$$

Thus condition B2 is automatically satisfied and Assumption B' implies Assumption B.   Assumption C will be replaced with

<u>Assumption C'</u>

1) A_o is a compact metric space and e is made an isolated point.

2) $p_{ij}(a)$ and $r(i,a)$ are continuous functions in a on A_o for all i, j.

It is obvious that Assumption C' implies condition C. The condition C^* is more delicate in a leavable model. We premise the following Lemma which is easy to prove under Assumption C'.

<u>Lemma 2.1</u>

The following conditions are equivalent for all i, j :
(i) $\{a \in A_o \; ; \; p_{ij}(a) > 0\}$ is closed ;
(ii)   sign $p_{ij}(a)$ is continuous on A_o ;
(iii) $\inf\{p_{ij}(a) \; ; \; a \in A_o, \; p_{ij}(a) > 0\} > 0$.

<u>Assumption $C^{*}{}'$</u>

For all i, j $\in$ S one of the conditions in Lemma 2.1 is satisfied. Two important examples where Assumption $C^{*}{}'$ is satisfied are the cases where A is finite or where (for all i, j $\in$ S) $p_{ij}(a) = 0$ either for all a or for none.

<u>Lemma 2.2</u>

Condition C^* is satisfied.

<u>Proof</u>

The class structure, in particular the number n_f of recurrent classes of $\mathbb{P}_f$ depends only on (sign $p_{ij}(f(i))$). Since the latter matrix depends continuously on f, so does n_f. $\square$

<u>Theorem 2.3</u>

a) There exists an optimal stationary policy f in the sense that
$v_f = \sup_{\delta \in \Delta} v_\delta = \sup_{\delta \in \Delta} \tilde{v}_\delta = \bar{v}_f$.

b) A stationary policy f is optimal iff
(i) $v = r_f + \mathbb{P}_f v$
(ii) $v(i) = u(i)$ for all states i which are recurrent under f.

<u>Proof</u>

Part a) follows Theorem 1.2 and Theorem 1.9 combined with (2.1). To prove part b) it is sufficient, by Theorem 1.2, to prove that 2.3b (ii) is equivalent with $\mathbb{P}_f^*(v - u) = 0$. But this is clear since, again by (2.1), $v - u \geqslant 0$.

<u>Theorem 2.4</u>

There exists an optimal stationary policy f^* such that

$f^*(i) = e$ if $v(i) = u(i)$
$f^*(i) \in A_o$ if $v(i) \neq u(i)$

<u>Proof</u>

Let f be any optimal stationary policy f and define f^* through

$f^*(i) = f(i)$ if $v(i) \neq u(i)$,
$f^*(i) = e$ if $v(i) = u(i)$.

Passing from f to f^* means that one makes the states in $[v = u]$

absorbing, but then the states, which are transient under f and are
not made absorbing, obviously remain transient under $f^\star$. Let us write
R for the set of recurrent states. Then $R_{f\star} = R_f \cup [v = u]$. Now, by
Theorem 2.3b $R_f \subset [v = u]$, hence $R_{f}\star \subset [v = u]$. By construction of
$f^\star$ we know that $[v = u] \subset R_{f}\star$, hence

$$R_{f\star} = [v = u] \tag{2.2}$$

and $f^\star$ satisfies condition (ii) of Theorem 2.3b. Condition (i) is
also satisfied for $f^\star$ since it is satisfied for f and e. $\square$

Corollary 2.5

For the optimal policy $f^\star$ of Theorem 2.4 one has

$$v_{f\star} = v_{\tau\star}, \, f\star$$

where $\tau^\star$ is the entrance time in the set $[v = u]$ and $E_{if\star}\tau^\star < \infty$
for all $i \in S$.

The Corollary obviously follows from Theorem 2.4 and (2.2) since
$v_{n,f\star} = v_{n\wedge\tau\star}, \, f\star$ for all n. It is well-known for the classical optimal stopping problem where A_o is singleton set.

In Schäl (1985) it is shown for the case r = 0 that the policy
$f^\star$ of Theorem 2.4 is also optimal for the corresponding leavable
gambling model (in the sense of Dubins & Savage (1965) with the
objective function

$$E_{i\delta}\overline{\lim} \, u(X_n) \, .$$

Finally, we remark that for the leavable model of this section the
value function can be obtained by the value iteration (cf. Schäl,
1984b).

3. The leavable semi-Markov decision model

In this section we consider a semi-Markov model given by a Tupel $SM = (S,A,p,r,u,t,1)$ where S,A,p,r,u are defined as in section 2 and t and 1 as follows.

Assumption A''

(vi) $t(i,a)$ is the expected <u>holding time</u> in state i if action a has been chosen, where $0 \leqslant t < \infty$ and $t(i,e) = 0$, $i \in S$.

(vii) $1(i)$ is the expected <u>inoperative time</u> after the system has been stopped in state i, where $0 < 1 < \infty$.

Let us write T for the set of all (not necessarily finite) stopping times τ and T_δ, $\delta \in \Delta$, for the set of all $\tau \in T$ such that $E_{i\delta}\tau < \infty$ for all i.   Then

$$s_{\tau,\delta}(i) = E_{i\delta}[\, \Sigma_{m=0}^{\tau-1} t(X_m,\sigma_m(X_o,\ldots,X_m))+1(X_\tau)]\ ,\quad \tau \in T_\delta \quad (3.1)$$

is the expected total time of the process stopped at time τ.   We define s_δ in analogy with v_δ.   There v_δ is well-defined if r is non-positive.   As long as no optimization is involved, the case $r \leqslant 0$ is symmetric to the case $r \geqslant 0$.   Since $t \geqslant 0$, s_δ is hence well-defined. Now we can define the case $\tau \notin T_\delta$

$$s_{\tau,\delta} = s_{\delta\wedge\tau} \quad \text{where } \delta \wedge \tau = (\delta_{n,\tau}) \in \Delta \text{ such that} \qquad (3.2)$$

$$\delta_{n,\tau} = \begin{cases} \delta_n & \text{on } [\tau > n] \\[2mm] e & \text{otherwise .} \end{cases}$$

More exactly we should replace $[\tau > n]$ by the projection of $[\tau > n]$ on H_n.   The policy $\delta \wedge \tau$ let the process stagnate from time τ onwards. Obviously, the two formulas (3.1) and (3.2) are consistent.

The definition of $v_{\tau,\delta}$ is extended in the same way where v_δ is defined because of the following general assumption.

Assumption B''

Condition B' holds.

Of course, since $\infty \in T$ and $\delta \wedge \infty = \delta$, we have

$$\sup\nolimits_{\delta \in \Delta} v_\delta = \sup\nolimits_{\delta \in \Delta, \tau \in T} v_{\tau,\delta}$$
$$\sup\nolimits_{\delta \in \Delta} s_\delta = \sup\nolimits_{\delta \in \Delta, \tau \in T} s_{\tau,\delta} \qquad (3.3)$$

In this section we prefer the notation $v_{\tau,\delta}$ instead of v_δ. By Corollary 2.5, we may replace T with T_δ in (3.3) under the following general assumptions of this section.

Assumption C''

Condition C' holds and, in addition, $t(i,a)$ is a continuous function in a on A_o for all $i \in S$.

Assumption $C^{\star}$''

Condition $C^{\star}$', holds.

Since $1 > 0$, $t \geqslant 0$, we know that $0 < s_{\tau,\delta} \leqslant \infty$ and in the case $\tau \in T_\delta$ even $0 < s_{\tau,\delta} < \infty$. As usual, we define c/∞ as 0 for $-\infty \leqslant c < +\infty$. Then the objective function of this section

$$\frac{v_{\tau,\delta}}{s_{\tau,\delta}} \to \sup\nolimits_{\delta \in \Delta, \tau \in T} \qquad (3.4)$$

is well-defined. Our aim is to maximize the expected total reward devided by the expected total time. Breiman (1964) and Miller (1981) show that our objective function can be looked upon as an average reward.

Moreover, they explain how the problem (3.4) can be reduced to the problem treated in section 2. Then it will follow that the supremum in (3.4) is attained by some $f \in \mathbf{F}$ and some $\tau \in T_f$. The reduction to leavable Markov decision problems in the sense of section 2 runs as follows.

Focus on some fixed state $i \in S$, say $i = 0$, and put

$$\lambda^\star = \sup_{\delta \in \Delta, \tau \in T} v_{\tau,\delta}(0) \cdot s_{\tau,\delta}^{-1}(0) \; .$$

Since $v_{\tau,\delta}(0)$ is bounded from above and $s_{\tau,\delta}(0)$ is bounded from below by $\inf_{i \in S} 1(i) > 0$, $\lambda^\star$ is a real number. Now consider the one parameter family of models $M^{(\lambda)}$, $\lambda \in \Lambda$, where Λ is some subset of the real line. A model $M^{(\lambda)}$ has the same state space S, the same action space $A = A_o + \{e\}$ and the same transition probabilities $p_{ij}(a)$. However, the one-step reward and the terminal reward are now

$$r^{(\lambda)} = r - \lambda t \quad , \quad u^{(\lambda)} = u - \lambda t.$$

Let us write (if $\lambda > 0$ or if $s_{\tau,\delta} < \infty$ for all τ,δ)

$$v_{\tau,\delta}^{(\lambda)} = v_{\tau,\delta} - \lambda s_{\tau,\delta} \text{ and } v^{(\lambda)} = \sup_{\delta \in \Delta, \tau \in T} v_{\tau,\delta}^{(\lambda)}$$

for the corresponding objective function and value function. We will consider the cases $\lambda^\star > 0$ and $\lambda^\star \leq 0$ separately. In the case $\lambda^\star > 0$, one can make a profit and one wants to get the profit as soon as possible. In the case $\lambda^\star \leq 0$, one can only lose money and one wants to distribute the losses on a time period as large as possible. We make the following general assumption:

Assumption D

Either (1) $\lambda^\star > 0$ and $\Lambda = (0,\infty)$ or (2) $\lambda^\star \leq 0$, $\Lambda (-\infty,+\infty)$, and $s_{\tau,\delta}(i) < \infty$ for all $i \in S$, $\delta \in \Delta$, $\tau \in T$.

Obviously, we have the case $\lambda^\star > 0$ if $u > 0$ since $\lambda^\star \geqslant u(0)/1(0)$.
The condition $s_{\tau,\delta}(i) < \infty$ is satisfied in replacement models where
the breakdown state b, say, is absorbing (that implies
$p_{bb}(a) = 1$, $t(b,a) = 0$ for $a \in A$) and is reached in a time with finite
expectation.

The case $\lambda^\star \leqslant 0$, $s_{\tau,\delta}(i) = \infty$ for some i,τ,δ seems not be very
interesting.

Lemma 3.1

Each Model $M^{(\lambda)}$, $\lambda \in \Lambda$, satisfies the conditions B', C' and
$C^{\star}{}'$ of section 2.

Proof

The conditions C' and $C^{\star}{}'$ are obviously implied by the Assumption
C'' and $C^{\star}{}''$. In the case $\Lambda = (0, \infty)$ we have $r^{(\lambda)} \leqslant r$, $u^{(\lambda)} \leqslant u$ and B'
follows B''. For the second case $\Lambda = (-\infty, +\infty)$, we write $s'_{n,\delta}$ instead
of $s_{n,\delta}$ if 1 is replaced by zero and L for the upper bound of 1.
Then $s'_{n,\delta}$ is increasing in n and

$$\sup_n s_{n,\delta} \leqslant \sup_n s'_{n,\delta} + L = s'_{\infty,\delta} + L \leqslant s_{\infty,\delta} + L < \infty \qquad (3.5)$$

by assumption. Hence we know from B'' that

$$\sup_n v^{(\lambda)}_{n,\delta} \leqslant \sup_n v_{n,\delta} + |\lambda| \sup_n s_{n,\delta} < \infty. \ \square$$

Proposition 3.2

Let be $\lambda \in \Lambda$. Then $v^{(\lambda)}(0) = 0$ iff $\lambda = \lambda^\star$.

Proof

We first prove the "if" direction. From $v_{\tau,\delta}(0) s^{-1}_{\tau,\delta}(0) \leqslant \lambda^\star$

we conclude $v_{\tau,\delta}(0) - \lambda^{\star} s_{\tau,\delta}(0) \leqslant 0$ for all δ,τ in both cases, hence $v^{(\lambda\star)}(0) \leqslant 0$. Now choose for $\varepsilon > 0$ some δ,τ such that $v_{\tau,\delta}(0) s_{\tau,\delta}^{-1}(0) \geqslant \lambda^{\star} - \varepsilon$ and $0 < s_{\tau,\delta}(0) < \infty$. Then

$$v_{\tau,\delta}(0) - \lambda^{\star} s_{\tau,\delta}(0) \geqslant \varepsilon_{\tau,\delta}$$

In the case $\lambda^{\star} \leqslant 0$ we know from the assumption and (3.5) that the model $(S,A,p,t,1)$ satisfies all conditions of section 2. Hence Corollary 2.5 applies and there exists some $f' \in \mathbb{F}$, $\tau' \in T_f$, such that $s_{\tau,\delta} \leqslant s_{\tau',f'}$ for all τ, δ.

In the case $\lambda^{\star} > 0$, we may assume that $\lambda^{\star} - \varepsilon > 0$. Then $s_{\tau,\delta}(0) \leqslant v_{\tau,\delta}(0)/(\lambda^{\star} - \varepsilon) \leqslant v(0)/(\lambda^{\star} - \varepsilon)$.

Therefore, in both cases, there exists some $m > 0$ independent of τ,δ such that $v_{\tau,\delta}^{(\lambda\star)}(0) \geqslant -\varepsilon m$. Hence $v^{(\lambda\star)}(0) = 0$. The "only if" direction can be proved by similar but easier arguments. $\square$

<u>Theorem 3.3</u>

(a) For the model $M^{(\lambda\star)}$ there exists a stationary policy which can be described by a pair $f^{\star} \in \mathbb{F}$, $\tau^{\star} \in T_{f\star}$ and which is optimal in the sense that

$$v_{\tau\star,f\star}^{(\lambda\star)}(i) = v^{(\lambda\star)}(i) \quad \text{for all} \quad i \in S.$$

(b) For any $f \in \mathbb{F}$, $\tau \in T$ the following statements are equivalent

(i) $v_{\tau,f}^{(\lambda\star)}(0) = v^{(\lambda\star)}(0)$, i.e., (τ,f) is optimal for $M^{(\lambda\star)}$ and initial state 0 ;

(ii) $v_{\lambda,f}(0) s_{\tau,f}^{-1}(0) = \lambda^{\star}$, i.e. (τ,f) is optimal for the semi-Markov decision model and initial state 0.

(c) There exists a pair $f^{\star} \in \mathbb{F}$, $\tau^{\star} \in T_{f\star}$ which is optimal in the sense of (bii).

<u>Proof</u> :

Part (a) follows from Lemma 3.1 and Corollary 2.5 Part (b) is
clear in view of Proposition 3.2 and its proof. Part (c) follows from
(a) and (b). $\square$

Since λ^{*} depends on the initial state, so does the optimal pair
(τ^{*},f^{*}) of Theorem 3.3c in opposition to the optimal pair (τ^{*},f^{*}) of
Corollary 2.5.

Miller (1981) treats a model which can be reduced to an essentially negative total reward problem. Then condition C^{*}" is unnecessary and the state space can be denumerable. $\square$

<u>References</u>

D. Blackwell (1962), Discrete dynamic programming. Ann. Math.
     Statist. <u>33</u>, 719-726.

L. Breiman (1964), Stopping-rule problems. Chapter 10 of Applied
     Combinatorial Mathematics, ed. E. Beckenbach, Wiley, New York.

H.-H. Bodewig (1985), Markow-Entscheidungsmodelle mit nichtkonvergenten
     Gesamtkosten. Doctoral thesis, Inst. Angew.Math., Univ. Bonn.

R.Ya. Chitashvili (1975), A controlled finite Markov chain with an
     arbitrary set of decisions. Siam Theory Prob. Appl. <u>20</u>, 839-
     847.

Y.S. Chow, H. Robbins, D. Siegmund (1971), Great expectations:the
     theory of optimal stopping. Boston:Houghton Mifflin Company.

H. Deppe (1984), On the existence of average optimal policies in
     semi-regenerative decision models. Math.Op.Res.<u>9</u>, 558-575.

L.E. Dubins, L.J. Savage (1965), How to gamble if you must. New
     York, McGraw-Hill.

A. Hordijk (1974), Dynamic Programming and Markov Potential Theory.
     Amsterdam, Mathematical Centre Tracts 51.

A. Hordijk, R. Dekker (1983), Denumerable Markov Chains, sensitive optimality criteria. Operations Research Proc. 1982, Springer-Verlag Berlin Heidelberg.

E. Mann (1983), Optimality equations and bias-optimality in bounded Markov decision processes. Univ. Bonn, Inst. Angew. Math. SFB 72, preprint n° 574, to appear in : optimization 16 (1985).

B.L. Miller (1981), Countable-state average-cost regenerative stopping problems. J. Appl. Prob. 18, 361-377.

B.L. Miller, A.F. Jr. Veinott (1969), Discrete dynamic programming with a small interest rate.

M. Schäl (1983), Stationary policies in dynamic programming models under compactness assumptions. Math. Op. Res. 8, 336-372.

– (1984a), Markovian decision models with bounded finite-state rewards. Operations Research Proc. 1983, Springer-Verlag Berlin Heidelberg.

– (1984b), On the value iteration in Markov decision models. Zamm 65.

– (1985), Optimal stopping and leavable gambling models with the average return criterion. To be published.

W.D. Sudderth (1971), On measurable gambling problems. Ann. Math. Statist. 42, 260-269.

P.J. Schweitzer (1968), Perturbation theory and finite Markov chains. J. Appl.Prob. 5, 401-413.

W.H.M. Zijm (1982), The optimality equations in multichain denumerable state Markov decision processes with the average cost criterion: the bounded cost case. A & E report 2182, Univ. of Amsterdam.

MARKOV DECISION DRIFT PROCESSES

Frank A. Van der Duyn Schouten

Department of Actuarial Sciences and Econometrics
Postbox 7161, 1007 MC Amsterdam, The Netherlands

1. Introduction

In Markov decision theory we distinguish

(a) discrete-time Markov decision processes
(b) semi-Markov decision processes
(c) continuous time Markov decision processes.

In class (a) changing the current action is only allowed at
discrete points in time, in class (b) only at random epochs at
which a jump in the state of the system occurs, while in class (c)
there are only measurability conditions on the evolution over time
of the actions that are chosen. The available theory and techniques
for discrete time and semi-Markov decision processes are rather
well unified and highly developed. This also holds for the class
of controlled diffusion processes, which can be seen as a subclass
of (c). In recent years there have been made several attempts to
achieve some unification in the heterogeneous collection of models
and methodologies within continuous time non-diffusion Markov decision theory (Davis, 1984; Hordijk and Van der Duyn Schouten,

1984; Yushkevich, 1983). In this paper we give a global overview
of the results of these attempts.

One might ask for the reason to deal with continuous time
Markov decision processes when well developed theory for the discrete time and semi-Markov case is available. The answer to this
question is threefold. In the first place there are many applications for which the continuous time description is the most
natural one. Discretizing the time parameter in these models is in
fact an approximation for the model under consideration. Although
many of these applications fit into the semi-Markov description,
there are important exceptions. We mention the problems with
finite planning horizon and those in which the state of the system
changes not only by jumps but also continuously between jumps. In
the next section we shall give several examples of this type. A
second argument for the analysis of continuous time Markov decision
processes is of computational kind. Computation or approximation
of the optimal policy, within a class of well structured policies
(bang-bang, switch-over, control limit) may be more straightforward
in continuous time than in discrete time (cf. Hordijk and Van der
Duyn Schouten, 1983; De Kok, Tijms and Van der Duyn Schouten,
1984). As a final reason we mention that in discrete time processes the decision epochs are assumed to be predetermined, although
in practical situations the determination of the decision epochs is
part of the decision process itself. The continuous time model
seems to be an appropriate description to deal with this aspect.

The examples given in (Van der Duyn Schouten, 1983) and the
next section clarify that in modelling continuous time Markov decision processes we have to deal with two kinds of control, viz.
control of the infinitesimal generator of a stochastic process and
impulsive control operating directly on the state of the system.
The first type of control has an indirect impact on the evolution
of the process (controlling the arrival rate of shocks in a

maintenance model, or the service rate in a queueing model), whereas impulsive controls cause instantaneous jumps in the state of the system (ordering in an inventory model or replacement in a maintenance model). It turns out that in model building we have to distinguish between these two types of control because of their different impact on the evolution of the process. It is an advantage of the discrete time processes that for these a distinction of this kind is not necessary. For this reason it can be profitable in analysing continuous time models to consider a sequence of approximating discrete time models. (cf. Hordijk and Van der Duyn Schouten, 1984 and 1985; Van der Duyn Schouten, 1983).

The first publications on continuous time Markov decision theory dealt with models in which only generator control is allowed. For an overview of the literature we refer to (Hordijk Duyn Schouten, 1984; Yushkevich, 1983). On the other hand there are various results on the impulsive control problem of stochastic differential equations (see (Yushkevich, 1983), for references).

In section 2 we shall give some examples which may clarify the concept of a Markov decision drift process (MDDP). The formal definition of a MDDP is given in section 3, while in section 4 we summarize some results obtained by several authors on the existence and the structure of the optimal policy of a MDDP.

2. Examples

Before presenting the formal definition of a MDDP we consider some examples of decision processes which can be represented as a MDDP.

2.1. A cash management model

We consider the cash management of a company with well pre-

dictable income rate and stochastic outflow of cash. A typical
example is an insurance company with income from the premium paid
by its customers and outflow due to claims by these customers.
Suppose that the cash inflow is deterministic and constant over
time at a rate $\sigma > 0$. The outflow of cash is described by a
compound Poisson process, i.e. claims arrive according to a Poisson
process with rate λ, while the sizes of the claims are i.i.d.
random variables, independent of the arrival process and with probability distribution function $F(\cdot)$. There is a holding- or
penalty cost at a rate $h(s)$ when the cash level equals s. The
cost of transferring funds and changing the cash level from s to
t is $K(s,t)$.

A modification of this model in which the total cash flow is
modelled by a Wiener process with drift has been studied by
Constantinides (1976) and Constantinides and Richard (1978). In
(Constantinides and Richard, 1978) it is shown that under certain
assumptions on the cost functions the optimal policy is characte-
rized by four parameters $d \leqslant D \leqslant U \leqslant u$, such that no transaction
is made when the cash level is within the open set (d,u), while
the cash level is adjusted to D (resp. U) whenever it falls down
to d (rises to u resp.). In figure 1 we have drawn a sample
path of the stochastic process describing the evolution of the cash
level over time under the (d,D,U,u)-policy. In this picture t_1,
t_3, t_4 and t_5 are arrival epochs of claims, at t_2 the cash
level is decreased by intervention and at t_5 the cash level
is increased by intervention immediately upon a claim arrival.

2.2. A production-inventory control model

Consider a single product production-inventory control problem
in which the production rate can be continuously adapted in order
to cope with random fluctuations in the demand. Moreover there is
control on the demand by price adjustment. The demand process

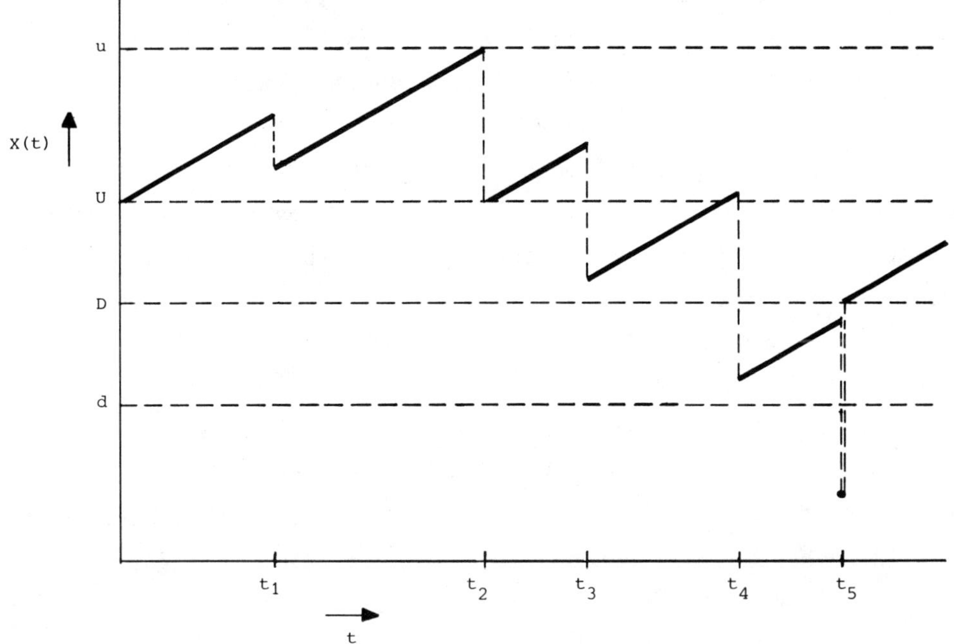

Fig. 1. Sample path of the cash management model under a (d,D,U,u)-
         policy.

for the product is a compound Poisson process with arrival rate λ
and demand distribution $F(\cdot)$. We assume that by adjusting the
price of the product we have control on λ within the boundaries
$0 \leqslant \lambda_1 \leqslant \lambda \leqslant \lambda_2 < \infty$. Maintaining arrival λ yields an expected
profit of $c(\lambda)$ per unit time. At any point in time the produc-
tion is governed by one of two possible production rates σ_1 and
σ_2, with $0 \leqslant \sigma_1 < \sigma_2$. The production costs are $b(\sigma)$ per unit
time when production rate σ is used. Thus the inventory level
decreases by jumps at arrival epochs of customers and between arrival epochs it increases linearly with slope depending on the production rate used. A holding- or penalty cost rate $h(s)$ is incurred when the inventory level equals s. Moreover a switch-
over cost K is incurred whenever the production rate is switched
from σ_1 to σ_2 or v.v.

A variant of this model with uncontrollable λ and focussing on service level constraints instead of cost functions has been treated by De Kok, et.al (1984)(cf. also references in De Kok et. al). In (De Kok, Tijms and Van der Duyn Schouten, 1984) this production-inventory control model is studied under a so-called switch-over rule. A switch-over rule is determined by two critical numbers $0 \leqslant y_1 \leqslant y_2$, such that the production rate is switched from σ_1 to σ_2 only when the inventory level has fallen below y_1 and from σ_2 to σ_1 only when the inventory level reaches y_2. In (De Kok, Tijms and Van der Duyn Schouten, 1984) approximations are given for several operating characteristics of this model under a switch-over policy. Assuming that excess demand is backlogged and switching the production rate is instantaneously a sample path of the inventory process under a switch-over policy is drawn in figure 2, where t_1 and t_2 denote switch-over epochs.

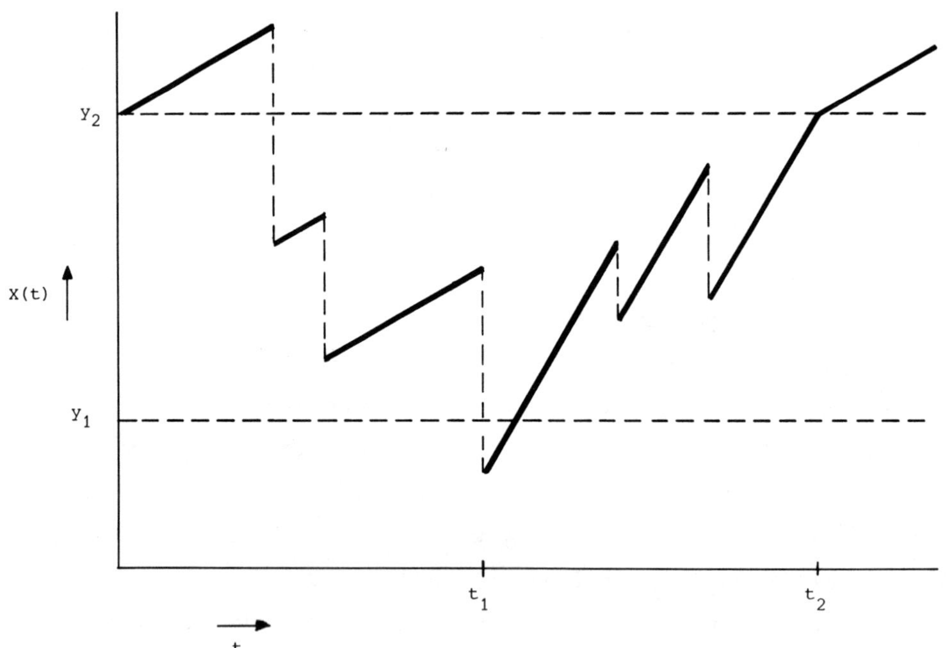

Fig. 2. A sample path of the production-inventory process under
          switch-over policy.

2.3. A reliability model with partial information

Consider a gradually deteriorating equipment whose actual degree of deterioration can be revealed by inspections only. At inspection the system can be observed in one of the working conditions $1, 2, \ldots, N$, which describe increasing degrees of deterioration. Working condition 1 represents a new system and N denotes a malfunction. The equipment undergoes deterioration according to a continuous time Markov process with transition probability matrices $P(t)$, $t \geq 0$. At any point in time the process is observed to be in a state (i, t), $1 \leq i < N$, $t \geq 0$, which means that t time units ago the operator knew that the system had working condition i and he has not obtained any information about the working condition since then. At any epoch the operator has the option to do an inspection or a revision. An inspection, which is assumed to be instantaneously, reveals the exact working condition of the system at a cost $J > 0$. After revision, which is also instantaneously, the system has working condition 1 again. Upon revision in state (i, t) an expected revision cost $\sum_{j=1}^{N} p_{ij}(t) b_j$ is incurred, where b_j denotes the revision cost when the exact working condition turns out to be j. By a_j we denote the operating cost rate of the equipment when it has working condition j. The problem is to find a maintenance schedule (prescribing inspections and revisions) that minimizes the sum of operating, inspection and revision costs.

A discrete time version of this model has been studied by Rosenfield (1976). Under straightforward assumptions on the one-step Markov transition matrix and the costs he shows that a monotonic four region policy is optimal for the discounted and average cost criteria. A monotonic four region policy is determined by a set of critical numbers $t_1(i) \leq t_2(i) \leq t(i)$, $1 \leq i \leq N$, with $t(i)$ non-increasing in i, such that the policy prescribes revision in state (i,t) iff $t \geqslant t(i)$, inspection iff $t_1(i) \leqslant t \leqslant t_2(i)$ and no action otherwise. Luss (1976) and Tijms et al. (1985) gave computational methods for finding the optimal critical values.

A sample path of the stochastic process describing the evolution of the observed state of the system under a four region policy is given in figure 3.

To conclude this section we mention several other applications which are modelled as a MDDP : a maintenance-replacement model wich perfect information (Hordijk and Van der Duyn Schouten, 1984; Hordijk and Van der Duyn Schouten, 1983), an inventory model with

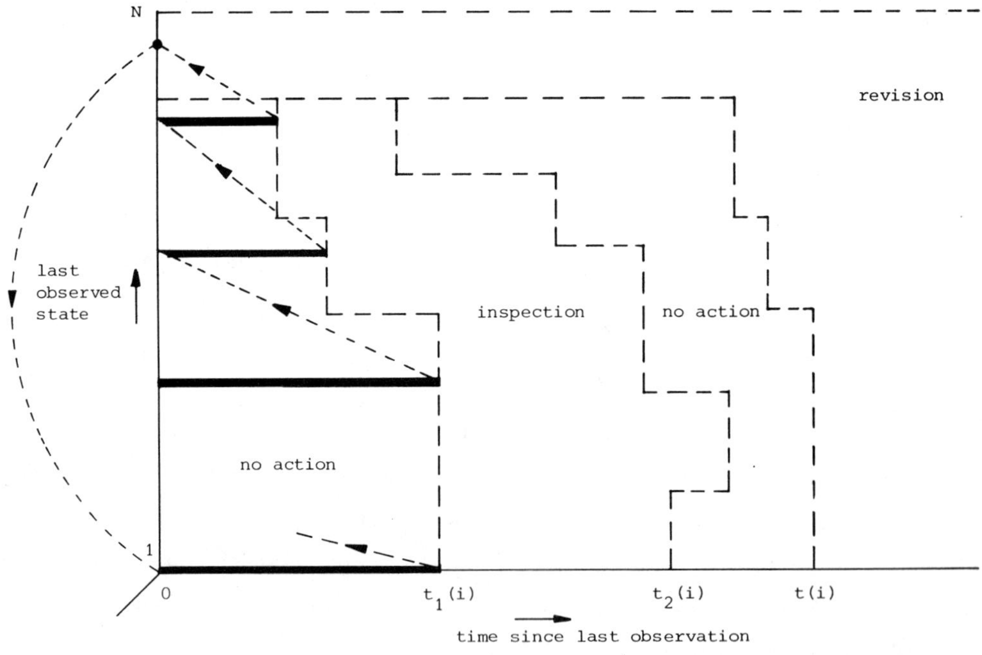

Fig. 3. A sample path of the reliability model under a four region policy.

reorder decisions (Van der Duyn Schouten, 1983), various queueing control models (Davis, 1984; Hordijk and Van der Duyn Schouten, 1983 and 1985) and a model for the adaptive control of queues (Lam and Thomas, 1983).

3. Definition of a Markov (decision) drift process

A Markov drift process, to be defined below, can be seen as a generalization of a pure jump Markov process in continuous time. A Markov process $\{X(t), t \geqslant 0\}$ is a pure jump process on the state space S if, starting from any point $s \in S$, all sample paths of the process are right continuous and constant except for isolated jumps. $\{X(t), t \geqslant 0\}$ is a _Markov drift process_ on S if, starting from any point $s \in S$, all sample paths are right continuous and deterministically determined except for isolated jumps. Hence a Markov drift process is representable by four elements (S,q,Π,f), where

- S is a complete metric space (_the state space_)

- q is a real-valued, bounded, non-negative mapping on S (_the jump rate_)

- Π is a transition probability from S to S (_the jump distribution_)

- f is a continuous mapping from $S \times [0,\infty)$ into S, such that $f(s,0) = s$ for $s \in S$ and $f(s,t) = f(f(s,u),t - u)$ for $s \in S, 0 \leqslant u \leqslant t$ and $t \geqslant 0$ (_the drift function_).

In addition to the examples given in the previous section we note that Markov drift processes arise when semi-Markov processes are "markovianized" (see also Davis, 1984). Consider a semi-Markov process on the state space $\tilde{S}$. Let $\tilde{F}(s)$ denote the probability distribution function of the sojourn time in state $s \in \tilde{S}$

and $\widetilde{\Pi}(s)$ the distribution of the system immediately after a jump from state $s \in \widetilde{S}$. To markovianize this semi-Markov process we enlarge the state space to $S := \widetilde{S} \times [0,\infty)$, where (s,t) represents the situation that the system is in state s and the elapsed sojourn time in that state equals t. If we define for $(s,t) \in S$

$$q(s,t) := \lim_{\Delta \to 0} \frac{1}{\Delta} \cdot \frac{\widetilde{F}(s)(t + \Delta) - \widetilde{F}(s)(t)}{1 - \widetilde{F}(s)(t)}$$

(failure rate of $\widetilde{F}(s)$ in t);

$$\Pi(s,t) := \widetilde{\Pi}(s)$$

$$f(s,t),u) := (s, t + u)$$

then the Markov drift process (S,q,Π,f) is a "Markov-equivalent" of the semi-Markov process.

A <u>Markov decision drift process</u> (MDDP) arises by introducing controls in a Markov drift process (S,q,Π, f). In our definition (see also Hordijk and Van der Duyn Schouten, 1984) we have control on the parameters q and Π. Moreover, there are impulsive controls operating directly on the state of the induced stochastic processes. A MDDP is represented by nine elements $(S;A_1,q,\Pi,f,c_1; A_2,p,c_2)$, where

- S is a complete metric space with Borel field S (<u>the state space</u>)

- A_1 is a complete metric space (<u>the set of generator controls</u>)

- q is a real-valued bounded, non-negative mapping on $S \times A_1$ (<u>the jump rate</u>)

- Π is a transition probability from $S \times A_1$ to S (<u>the jump distribution</u>)

- f is a continuous function from $S \times [0,\infty)$ to S, with

$f(s,0) = s$ for $s \in S$ and $f(s,t) = f(f(s,u), t - u)$ for $s \in S$, $0 \leqslant u \leqslant t$, $t \geqslant 0$ (the drift function)

- c_1 is a real valued function on $S \times A_1$ (the cost rate)

- A_2 is a complete metric space (the set of impulsive controls)

- p is a transition probability from $S \times A_2$ to S (the impulsive jump distribution)

- c_2 is a real-valued function on $S \times A_2$ (the lump cost)

If at an arbitrary point in time the system is in state s and impulsive control a is chosen then an immediate lump cost $c_2(s,a)$ is incurred and an instantaneous jump occurs to $B \in S$ with probability $p(s,a,B)$.

An appropriate sample space for a Markov drift process (S,q,Π,f) is the space of all S-valued functions on $[0,\infty)$, which are right continuous and have left hand limits at every $t > 0$. From figure 1 follows that the sample paths of a MDDP do not necessarily belong to this space, since the right continuity might be violated. Therefore Hordijk and Van der Duyn Schouten (1984 and 1983) introduced and analysed a new sample space $J[0,\infty)$, in which the sample paths of all MDDP's are contained. For details we refer to (Hordijk and Van der Duyn Schouten, 1984).

Let $(S;A_1,q,\Pi,f,c_1;A_2,p,c_2)$ be a MDDP. A policy is a pair $R = (R_1,R_2)$ with R_1 a transition probability from $J[0,\infty) \times [0,\infty)$ to A_1 (the generator control rule) and R_2 a transition probability from $J[0,\infty) \times [0,\infty)$ to $A_2 \cup \{\theta\}$ (the impulsive control rule), such that R_i is non-anticipative, $i = 1,2$ and $R_2(x,t)$ is either a probability measure on A_2 or equal to θ. $R_1(x,t)$ denotes the randomized control at time t given that the history of the process is represented by the restriction of the sample path $x \in J[0,\infty)$ to $[0,t]$. If $R_2(x,t) = \theta$ this implies that no impulsive control is chosen at time t, otherwise $R_2(x,t)$ denotes the randomized impulsive control chosen at t under history x. As illustration we define the appropriate MDDP's for the examples of section 2.

3.1. A cash management model

The cash management problem described in section 2.1. can be modelled as a MDDP as follows :

$$S := \mathbb{R}$$

$$A_1 := \{\lambda\}$$

$$q(s,\lambda) := \lambda$$

$$\Pi(s,\lambda,[s,s-t]) := F(t), \; t \geqslant 0$$

$$f(s,t) := s + \sigma t$$

$$c_1(s,\lambda) := h(s)$$

$$A_2 := \mathbb{R}$$

$$p(s,a,\{s+a\}) := 1$$

$$c_2(s,a) := K(s,s+a).$$

In this definition the impulsive control a represents an instantaneous change of the cash level by an amount a (positive or negative).

3.2. A production)-inventory control model

For the production-inventory control model from section 2.2. we define :

$$S := \mathbb{R} \times \{\sigma_1,\sigma_2\}$$

$$A_1 := [\lambda_1,\lambda_2]$$

$$q(s,\sigma),\lambda) := \lambda$$

$$\Pi((s,\sigma),\lambda,([s,s-t],\sigma)) := F(t), \ t \geqslant 0$$

$$f((s,\sigma),t) := s + \sigma t$$

$$c_1((s,\sigma),\lambda) := h(s) + b(\sigma) - c(\lambda)$$

$$A_2 := \{1\}$$

$$p((s,\sigma_i),1,(s,\sigma_j)) := 1, \ i \neq j, \ i,j \in \{1,2\}$$

$$c_2((s,\sigma),1) := K.$$

The single impulsive control, denoted by 1, represents a switch of the production rate.

3.3. A reliability model with partial information

The reliability model of section 2.3. is representable as a MDDP in the following way.

$$S := \{1,\dots,N\} \times [0,\infty)$$

$$A_1 := \{0\}$$

$$q((i,t),0) := 0$$

$$\Pi \quad \text{undefined}$$

$$f((i,t),u) := (i,t+u)$$

$$c_1((i,t),0) := \sum_{j=1}^{N} p_{ij}(t)a_j$$

$$A_2 := \{1,2\}$$

$$p((i,t),1,(j,0)) := p_{ij}(t)$$

$$p((i,t),2,(1,0)) := 1$$

$$c_2((i,t),1) := J$$

$$c_2((i,t),2) := \sum_{j=1}^{N} p_{ij}(t)b_j.$$

Note that in this case only jumps caused by impulsive controls occur. (The jump rate is zero and hence the jump distribution is

undefined). There are two possible impulsive controls : inspection
(denoted by 1) and revision (denoted by 2).

4. Existence and structure of optimal policies

In this section we summarize verbally some results obtained
in the literature on the existence and structure of optimal policies
for a MDDP. The concept of MDDP was introduced by Hordijk and
Van der Duyn Schouten in (1983 and 1984). The analysis in (Hordijk
and Van der Duyn Schoute, 1984 and 1985; Van der Duyn Schouten,
1983) is concerned with discrete time approximations for
MDDP's.

For any MDDP a sequence of discrete time Markov decision processes is defined with decreasing length of the intervals between
subsequent decision epochs. The analysis in (Hordijk and Van
der Duyn Schoute, 1984 and 1985; Van der Duyn Schouten,
1983) results in a theorem which provides sufficient conditions
under which a limit point of a sequence of discounted optimal policies for the approximating discrete time processes is a discounted eptimal policy for the MDDP. (See Hordijk and Van der Duyn
Schouten, 1985, theorem 414). Using this theorem the authors
were able to prove for several MDDP's that the optimal policy belongs to a special class of well-structured policies. In
obtaining these structural results time discretization turned out
to be a powerful instrument. It is clear that this method is only
practicable if we are able to prove (e.g. by the usual method of
induction) that the optimal policy for the approximating discrete
time processes has the desired structure. It should be admitted
that this in general is not an easy job even for fairly simple
models.

In Hordijk and Van der Duyn Schouten (1983) MDDP's are considered under the average cost criterion. Suffcent conditions are
given under which a limit point of a sequence of discounted optimal

policies is average optimal. A more direct analysis of MDDP's is
given by Yushkevich (1983). He obtains sufficient conditions in
the form of quasi-variational inequalities for a function and a
Markov policy to be the value and an optimal policy for a MDDP. In
the case of positive or negative rewards these conditions are shown
to be also necessary.

A class of processes very similar to MDDP's has been introduced
by Davis (1984) under the name "piecewise deterministic (PD) processes". Davis provides a complete characterization of the extended
generator and its domain for a PD process under a fixed policy.
Besides control of jump rate and jump distribution Davis assumes
that the drift function can be controlled in a deterministic way.
However, the absence of impulsive controls is assumed. Finally we
mention a paper by Vermes (1985) in which he provides necessary and
sufficient conditions in form of a Hamilton-Jacobi-Bellman equation
for the value function and an optimal policy controlling a PD
process.

<u>References</u>

Constantinides, G.M., (1976). Stochastic cash management with fixed
    and proportional transaction costs. Management Sci. $\underline{22}$,
    1320-1331.
Constantinides, G.M. and S.F. Richard (1978). Existence of optimal
    simple policies for discounted-cost inventory and cash manage-
    ment in continuous time. Operations Res. $\underline{26}$, 620-636.
Davis, M.H.A., (1984). Piecewise-deterministic Markov processes :
    A general class of non-diffusion stochastic models. J.R.
    Statist.Soc. B. $\underline{46}$, 353-388.
Hordijk, A. and F.A. Van der Duyn Schouten (1984). Discretization
    and weak convergence in Markov decision drift processes. Math.
    Oper.Res. $\underline{9}$, 112-141.

Hordijk, A. and F.A.Van der Duyn Schouten (1985). Markov
     decision drift processes; conditions for optimality obtained
     by discretization. Math.Oper.Res. 10, 160-173.

Hordijk, A. and F.A.Van der Duyn Schouten (1983). Average
     optimal policies in Markov decision drift processes with
     applications to a queueing and a replacement model. Advances
     in Appl.Probability 15, 274-303.

De Kok, A.G., H.C.Tijms and F.A.Van der Duyn Schouten (1984).
     Approximations for the single product production-inventory
     problem with compound Poisson demand and service-level
     constraints. Advances in Appl.Probability 16, 378-401.

Lam, Y. and L.C.Thomas (1983). Adaptive control of M/M/1
     queues-continuous time Markov decision process approach.
     J.Appl.Probability 20, 368-379.

Luss, H., (1976). Maintenance policies when deterioration can be
     observed by inspections. Operations Res. 24, 359-366.

Rosenfield, D., (1976). Markovian deterioration with uncertain
     information. Operations Res. 24, 141-155.

Tijms, H.C. and F.A.Van der Duyn Schouten (1985). A Markov
     decision algorithm for optimal inspections and revisions in a
     maintenance system with partial information. European J. of
     Oper.Res. 21, 245-253.

Van der Duyn Schouten, F.A., (1983). Markov decision processes with
     continuous time parameter. Math.Centre Tract 164. Mathematical
     Centre, Amsterdam.

Vermes, D., (1985). Optimal control of piecewise deterministic
     processes. Stochastics 14, 165-208.

Yushkevich, A.A., (1983). Continuous time Markov decision processes
     with interventions. Stochastics 9, 235-274.

THE FUNCTIONAL EQUATIONS OF UNDISCOUNTED DENUMERABLE STATE MARKOV
RENEWAL PROGRAMMING

Elke Mann

Institut für Angewandte Mathematik der Universität Bonn

Wegerlerstr.6, 5300 Bonn 1, Federal Republic of Germany

1. Introduction

In this paper we want to establish conditions for the existence of a finite solution of the functional equations

$$(1) \quad g(i) = \max_{a \in A(i)} \sum_{j \in I} p(i,a,j)g(j), \quad i \in I,$$

$$(2) \quad v(i) = \max_{a \in A_o(i)} [r(i,a) - t(i,a)g(i) + \sum_{j \in I} p(i,a,j)v(j)], \quad i \in I$$

$$\text{with} \quad A_o(i) := \{a \in A(i) \mid g(i) = \sum_{j \in I} p(i,a,j)g(j)\},$$

where I is a denumerable set, $A(i)$ is a compact metric space for all $i \in I$, $0 < t(i,a) < \infty$, $-\infty < r(i,a) < \infty$, $p(i,a,j) \geq 0$ for all $i, j \in I$, $a \in A(i)$ and $\sum_{j \in I} p(i,a,j) = 1$ for all $i \in I$, $a \in A(i)$.

These equations arise in undiscounted multichain Markov Renewal Programmes with denumerable state space I. Here $A(i)$ denotes the prescribed set of admissible actions in state i, while

$r(i,a)$, $t(i,a)$ and $p(i,a,j)$ denote the prescribed expected one-step reward, expected holding time and transition probability to state j, respectively, if action a is chosen in state i.

As on the information concerning the rewards and holding times, we shall only require the expected one-step rewards and expected holding times, we consider the semi-Markov decision model $(I, A(i)_{i \in I}, p, t, r)$ with transition function t and reward function r.

For our purposes it is further possible to stay within the framework of stationary policies and we will identify each stationary policy with the corresponding decision rule. Let F denote the set of decision rules i.e. $F := \{f : I \to \bigcup_{i \in I} A(i) \mid f(i) \in A(i)$ for all $i \in I\}$. Then each decision rule $f \in F$ defines a Markov chain (M.c.) $((P_{i,f})_{i \in I}, \{X_n\}_{n \in \mathbb{N}_0})$ on the infinite product space over I with transition probabilities $p_f(i,j) = p(i,f(i),j)$, $i, j \in I$, where X_n denotes the state of the system, when transition occurs for the n-th time. Thus the essential feature of our decision model is the family $(p_f, t_f, r_f)_{f \in F}$ with $t_f(i) := t(i,f(i))$ and $r_f(i) := r(i,f(i))$, $i \in I$. Generally, any (parametrized) family $(p_\lambda, t_\lambda, r_\lambda)_{\lambda \in \Lambda}$ of transition probabilities p_λ, non-negative "transition functions" t_λ and real valued "reward functions" r_λ on a denumerable set S is called a stationary semi-Markov decision model (SSMD) on S with parameter or strategy space Λ, cf. (Wijngaard, 1975).

To prove the validity of the functional equations (1) and (2), we follow an idea of Schweitzer (1982) and establish conditions for the gain rates and (special) relative values to depend continuously on f. Then there exist decision rules such that the policy iteration algorithm provides immediate convergence to a solution of the equations (1) and (2) if initialized with any of these decision rules.

Note that F is a compact metric space with respect to point-wise convergence.

2. Gain rates, relative values and embedded decision models

For any M.c. $\{X_n\}_{n\in\mathbb{N}_o}$ and any $L \subseteq I$, define

$$\tau_L := \min\{n \geqslant 1 \mid X_n \in L\}$$

and for any $f \in F$ and $L \subseteq I$, define for $i, j \in I$ and $n \geqslant 1$ the taboo probability

$$_L p_f^n(i,j) := P_{i,f}(X_n = j \text{ and } X_k \notin L \text{ for } 1 \leqslant k < n).$$

Throughout the paper we assume that for each $f \in F$ the induced Markov chain is non-dissipative, i.e. $\sum_{j\in I} \pi_f(i,j) = 1$ for all $i \in I$, where $\pi_f(i,j)$ is the Cesaro limit of the n-step transition probabilities $p_f^n(i,j)$, and that the family of induced Markov chains is quasi-finite with respect to a finite set K, i.e. $P_{i,f}(\tau_K < \infty) = 1$ for all $i \in I, f \in F$ and a fixed finite set $K \subseteq I$.

Then for each $f \in F$ the state space of the induced M.c. can be divided into a finite number of positive recurrent classes and a (possibly void) set of transient states. Denote by n_f, $R_{f,k}$ and $\phi_{f,k}(i), 1 \leqslant k \leqslant n_f$, $i \in I$, the number of recurrent classes, the k-th recurrent class and the probability of eventual absorption in $R_{f,k}$ starting in i, respectively.

The quasi-finiteness of the family of induced M.c.'s enables us to consider for each decision rule f the "embedded Markov chain on K" with transition probabilities $\bar{p}_f(i,j) := \sum_{n=1}^{\infty} {}_K p_f^n(i,j)$ ($= P_{i,f}(X_{\tau_K} = j)$), $i, j \in K$, which describes the states of the system, whenever transition to the set K occurs (cf. Çinlar, 1974; Federgruen, Hordijk and Tijms, 1979; Kolonko, 1980b; Wijngaard, 1975). In the next lemmata we shall summarize some results relating the family $(\bar{p}_f)_{f\in F}$ to the family $(p_f)_{f\in F}$.

Lemma 1

For each $f \in F$ the number $\bar{n}_f$ of recurrent classes of the embedded M.c. on K is equal to the number of recurrent classes of the original M.c. on I and the sets $\bar{R}_{f,k} := R_{f,k} \cap K$, $1 \leqslant k \leqslant n_f$, are the recurrent classes of the embedded M.c.; for starting states $i \in K$, the probabilities $\bar{\phi}_{f,k}(i)$, $\phi_{f,k}(i)$ of eventual absorption into corresponding recurrent classes with respect to the embedded and the original M.c., respectively, coincide.

Lemma 2

If the transition probabilities $p_f(i,j)$ depend continuously on f for all states $i, j \in I$, then the same is true for the transition probabilities $\bar{p}_f(i,j)$ of the embedded M.c. for all $i, j \in K$. Furthermore we have that $f \to \bar{p}_f(i,j)$ is continuous for all $i \in I$, $j \in K$, where the definition of $\bar{p}_f$ is extended to the set $\{(i,j), i \in I, j \in K\}$ in an obvious way. (Deppe, 1985, Theorem 5.2).

Lemma 3

If u is a non-negative real valued function on I such that $E_{i,f} \sum_{n=0}^{\tau_K - 1} u(X_n) < \infty$ for all $i \in I$, then $E_{i,f} \sum_{n=0}^{\tau_L - 1} u(X_n) < \infty$ for all $i \in I$ and any set $L \subseteq K$, that contains at least one representative of each recurrent class; L can be chosen as a singleton in the case where the considered initial states belong to one fixed recurrent class. (cf. Kolonko, 1980a, Lemma 6.2 for the unichained case).

In connection with the family $(\bar{p}_f)_{f \in F}$ of embedded M.c.'s, we shall consider decision models with finite state space K, which are intimately related to our original model with denumerable state space. Thus it will be possible to fall back upon results about decision models with finite state space to obtain corresponding results about decision models with denumerable state space.

The following condition will be needed in the sequel

$$(L_o) : \quad E_{i,f} \sum_{n=0}^{\tau_K-1} |r_f(X_n)| < \infty \quad \text{and} \quad E_{i,f} \sum_{n=0}^{\tau_K-1} t_f(X_n) < \infty$$

$$\text{for all } i \in I \text{ and all } f \in F$$

Remark

Under the non-dissipativity assumption we have that, if $L \subseteq I$ is such $E_{i,f} \sum_{n=0}^{\tau_L-1} t_f(X_n) < \infty$, then $P_{i,f}(\tau_L < \infty) = 1$.

Under (L_o) we can consider the so-called "embedded semi-Markov decision model" $(\bar{p}_f, \bar{t}_f, \bar{r}_f)_{f \in F}$, where for $f \in F$,

$\bar{p}_f(i,j)$, i, $j \in K$, are the transition probabilities of the embedded M.c.,

$\bar{t}_f(i)$, $i \in I$, is the expected first passage time from i to K and

$\bar{r}_f(i)$, $i \in K$, is the expected total reward until the first return to K starting in i.

Furthermore, under (L_o), for each $f \in F$ the <u>gain rate function</u> g_f given by

$$g_f(i) := \sum_{k=1}^{n_f} \phi_{f,k}(i) g_{f,k}, \quad \text{where for } 1 \leq k \leq n_f$$

$$g_{f,k} := \frac{\sum_{j \in R_{f,k}} \pi_{f,k}(j) r_f(j)}{\sum_{j \in R_{f,k}} \pi_{f,k}(j) t_f(j)}$$

and $\pi_{f,k}$ is the unique stationary distribution on the k-th recurrent class, is well-defined and finite.

Lemma 4

Under (L_o) for each $f \in F$

(a) $g_f : I \to \mathbb{R}$ is bounded, (b) $g_f(i) = \sum_{j \in I} p_f(i,j) g_f(j)$ for all $i \in I$ and (c) there exists a real valued function v_f on I such that

$$v_f(i) = r_f(i) - t_f(i)g_f(i) + \sum_{j \in I} p_f(i,j)v_f(j) \quad \text{for all} \quad i \in I.$$

A real valued function v_f satisfying the above equations is called a <u>relative value function</u>. Compare (Wijngaard, 1975) for more general results concerning the existence of relative value functions in stationary Markov decision models with general state space.

Remark

The gain rate functions and relative value functions are explained in terms of the SSMD $(p_f, t_f, r_f)_{f \in F}$. Gain rate functions and relative value functions in arbitrary SSMD's with denumerable state space are defined in an analogous way.

Proof of Lemma 4

(a) As $1 \leqslant n_f \leqslant |K|$ it remains to show that for each $f \in F$ the gain rates $g_{f,k}$, $1 \leqslant k \leqslant n_f$, are well-defined and finite. By Lemma 3 there exist for each $f \in F$ states $s_k = s_{f,k} \in R_{f,k}$, $1 \leqslant k \leqslant n_f$, such that

$$E_{s_k,f} \sum_{n=0}^{\tau_{s_k}-1} |r_f(X_n)| < \infty. \quad \text{Thus} \quad \sum_{j \in R_{f,k}} \pi_{f,k}(j)r_f(j) = \pi_{f,k}(s_k).$$

$E_{s_k,f} \sum_{n=0}^{\tau_{s_k}-1} r_f(X_n)$ is finite for all $f \in F$, $1 \leqslant k \leqslant n_f$ and the same is true if r_f is replaced by t_f. As $\sum_{j \in I} \pi_{f,k}(j)t_f(j) > 0$ for all $f \in F$, $1 \leqslant k \leqslant n_f$, we have the desired result.

The assertion in (b) follows immediately since $\phi_{f,k}(i) = \sum_{j \in I} p_f(i,j)\phi_{f,k}(j)$ for all $f \in F$, $1 \leqslant k \leqslant n_f$.

(c) Consider the gain rates $\bar{g}_{f,k}$, $f \in F$, $1 \leqslant k \leqslant \bar{n}_f$ of the embedded semi-Markov decision model $(\bar{p}_f, \bar{t}_f, \bar{r}_f)_{f \in F}$ and define for $i \in I$

$$\mathfrak{g}_f(i) := \sum_{k=1}^{n_f} \phi_{f,k}(i)\bar{g}_{f,k} \quad - \text{ observe that } \bar{n}_f = n_f - \text{ and for}$$

$i \in K$

$$r_f'(i) := E_{i,f} \sum_{n=0}^{\tau_K-1} (r_f(X_n) - t_f(X_n)\mathfrak{g}_f(X_n)).$$

Because of the boundedness of $\mathfrak{g}_f$ the function r_f' on K is well-defined and finite. Since $\mathfrak{g}_f(i) = \bar{g}_{f,k}$ for all $i \in R_{f,k}$ and only states $j \in R_{f,k}$ are accessible from states $i \in \bar{R}_{f,k} = R_{f,k} \cap K$ one gets

$$r_f'(i) = E_{i,f} \sum_{n=0}^{\tau_K-1} r_f(X_n) - \bar{g}_{f,k} E_{i,f} \sum_{n=0}^{\tau_K-1} t_f(X_n)$$

$$\text{for all } i \in \bar{R}_{f,k}.$$

Hence, integrating both sides of the above equalities by $\bar{\pi}_{f,k}$, $1 \leq k \leq n_f$, and observing that by definition

$$\bar{g}_{f,k} = \frac{\displaystyle\sum_{j \in \bar{R}_{f,k}} \bar{\pi}_{f,k}(j)\bar{r}_f(j)}{\displaystyle\sum_{j \in \bar{R}_{f,k}} \bar{\pi}_{f,k}(i)\bar{t}_f(i)} \quad , \text{ one gets}$$

$$\sum_{j \in K} \bar{\pi}_f(i,j)r_f'(j) = \sum_{k=1}^{n_f} \bar{\phi}_{f,k}(i) \sum_{j \in \bar{R}_{f,k}} \bar{\pi}_{f,k}(j)r_f'(j) = 0$$

$$\text{for all } i \in K$$

and thus $v_f'(i) = r_f'(i) + \sum_{j \in K} \bar{p}_f(i,j)v_f'(j)$ for all $i \in K$, where v_f' is the bias function, i.e. the unique relative value function such that $\sum_{j \in K} \bar{\pi}_f(i,j)v_f'(j) = 0$ for all $i \in K$, of the finite M.c. with rewards $(\bar{p}_f, \bar{r}_f')$. Considering the extension of $\bar{p}_f$ to the set of pairs (i,j), $i \in I$, $j \in K$, v_f' can be extended to a function v_f on I by defining for $i \in I$

$$v_f(i) = E_{i,f} \sum_{n=0}^{\tau_K - 1} (r_f(X_n) - t_f(X_n)g_f(X_n)) + \sum_{j \in K} \bar{p}_f(i,j)v_f'(j) \; .$$

Observing that the first condition in (L_o) is formulated with respect to the absolute values of the rewards, it is easy to check in the same way as Wijngaard (1975, page 55) did for Markov decision models that

$$v_f(i) = r_f(i) - t_f(i)g_f(i) + \sum_{j \in I} p_f(i,j)v_f(j) \quad \text{for all} \quad i \in I$$

and it is immediate (Wijngaard, 1975, Lemma 3.6) that

$$\sum_{j \in I} p_f^n(i,j)|v_f(j)| < \infty \quad \text{for all} \quad i \in I, \; n \geq 1 \quad \text{and}$$

$$\lim_{n \to \infty} \frac{1}{n} \sum_{j \in I} p_f^n(i,j)v_f(j) = 0 \quad \text{for all} \quad i \in I.$$

This implies, using a standard ergodic theorem (Chung, 1967, Th. 15.3), that $g_{f,k} = \bar{g}_{f,k}$ for all $1 \leq k \leq n_f$ and this completes the proof. $\square$

The next lemma shows that the gain rate functions g_f of our decision model can be expressed by the gain rate functions $\bar{g}_f$ of the embedded decision model $(\bar{p}_f, \bar{t}_f, \bar{r}_f)_{f \in F}$.

<u>Lemma 5</u>

Under (L_o)

$$g_f(i) = \sum_{j \in K} \bar{p}_f(i,j)\bar{g}_f(j) \quad \text{for all} \quad f \in F, \; i \in I.$$

<u>Proof</u>

By Lemma 4 we have that g_f is bounded and $g_f(i) = \sum_{j \in I} p_f(i,j)g_f(j)$ for all $i \in I$, which implies (Wijngaard, 1975, Lemma 2.21) that $g_f(i) = \sum_{j \in K} \bar{p}_f(i,j)g_f(j)$ for all $i \in I$. It remains to show that $g_f(i) = \bar{g}_f(i)$ for $i \in K$. We know by Lemma 1 that for $i \in K$

$$g_f(i) := \sum_{n=1}^{n_f} \phi_{f,k}(i)g_{f,k} = \sum_{n=1}^{\bar{n}_f} \bar{\phi}_{f,k}(i) \; g_{f,k}.$$ But in the proof of Lemma 4 the gain rates on corresponding recurrent classes of the original and the embedded model turned out to be equal and thus we have that g_f equals $\bar{g}_f$ on K. $\square$

3. Finite state SSMD's

Let $(p_\lambda, t_\lambda, r_\lambda)_{\lambda \in \Lambda}$ be any SSMD with finite state space $S = \{s_1, \ldots, s_n\}$ and metric parameter space Λ such that the functions $\lambda \to p_\lambda(s,s')$, $\lambda \to t_\lambda(s)$ and $\lambda \to r_\lambda(s)$ are continuous for all $s, s' \in S$ and $0 < t_\lambda(s) < \infty$ for all $s \in S$. For each $\lambda \in \Lambda$ let n_λ denote the number of recurrent classes of the associated M.c. and assume throughout the section that $\lambda \to n_\lambda$ is continuous. Then it is well-known (cf. Schweitzer, 1968) that the Cesaro limits $\pi_\lambda(s,s')$ of the n-step transition probabilities $p_\lambda^n(s,s')$ depend continuously on λ for all $s, s' \in S$; denote by Π_λ and P_λ the matrices that are defined in an obvious way by $\pi_\lambda(s,s')$ and $p_\lambda(s,s')$, $s, s' \in S$, and denote by E the $n \times n$ identity matrix. As for the gain rate functions g_λ, $\lambda \in \Lambda$,

$$g_\lambda(s) = \sum_{s' \in S} \pi_\lambda(s,s') \frac{\sum\limits_{s'' \in S} \pi_\lambda(s',s'') r_\lambda(s'')}{\sum\limits_{s'' \in S} \pi_\lambda(s',s'') t_\lambda(s'')} \quad \text{for all } s \in S,$$

this yields the following result.

Lemma 6
$\lambda \to g_\lambda(s)$ is continuous on Λ for all $s \in S$.

The bias functions v_λ, $\lambda \in \Lambda$ – regarded as vectors in $\mathbb{R}^n$ – of the (associated) stationary Markov decision model $(p_\lambda, r_\lambda)_{\lambda \in \Lambda}$ admit the following representation (cf. Blackwell, 1962) :
$v_\lambda = (E - P_\lambda + \Pi_\lambda)^{-1} r_\lambda - \Pi_\lambda r_\lambda$, and thus the following statement is evident.

Lemma 7
$\lambda \to v_\lambda(s)$ is continuous on Λ for all $s \in S$.

In the next section we shall apply these results to the embedded decision models $(\bar{p}_f, \bar{t}_f, \bar{r}_f)_{f \in F}$ and $(\bar{p}_f, r'_f)_{f \in F}$ with

$$r'_f(i) = E_{i,f} \sum_{n=0}^{\tau_K - 1} (r_f(X_n) - t_f(X_n)g_f(X_n)), \quad i \in K, \text{ to obtain corres-}$$

ponding results for our original model.

4. Main results

In the sequel, we will use usual continuity conditions concerning the law of motion, the reward function and the holding times, summarized in condition (C) below.

(C) : $a \to p(i,a,j)$, $a \to t(i,a)$ and $a \to r(i,a)$ are continuous
        on $A(i)$ for all $i, j \in I$

Furthermore, the following condition will be needed.

(C*) : $f \to n_f$ is continuous on F

We shall use the representation of the gain rate functions given in Lemma 5 to show that the gain rates $g_f(i)$ depend continuously on f for all starting states i under the following additional condition of Liapunov function type.

(L_1) : There exists a non-negative real valued function y_1 on I such that

(i) $|r_f(i)| + t_f(i) + \sum_{j \notin K} p_f(i,j)y_1(j) \leq y_1(i)$

        for all $i \in I$, $f \in F$

(ii) $\lim_{n \to \infty} \sum_{j \notin K} {}_K p_f^n(i,j)y_1(j) = 0$

        for all $i \in K$, $f \in F$

(iii) $f \to \sum_{j \notin K} p_f(i,j)y_1(j)$ is continuous on F

        for all $i \in I$

Hordijk (1974) and Hordijk and Sladky (1977) introduced conditions of Liapunov function type and showed the existence of finite solutions of the functional equations in the strong unichain case.

<u>Remark</u>

$(L_1(i))$ implies (L_o).

<u>Theorem 1</u>

Under (C), (C^*) and (L_1)

$$f \rightarrow g_f(i) \quad \text{is continuous on} \quad F \quad \text{for all} \quad i \in I.$$

<u>Proof</u>

The conditions (C) and (L_1) imply the continuity of the functions
$f \rightarrow \bar{r}_f(i)$ and $f \rightarrow \bar{t}_f(i)$ for all $i \in K$ (cf. Kolonko, 1980b,
Lemma 13). By Lemma 1 and 2 respectively, we have that $\bar{n}_f = n_f$
and $f \rightarrow \bar{p}_f(i,j)$ is continuous for all $i \in I, j \in K$. Thus
$(\bar{p}_f, \bar{t}_f, \bar{r}_f)_{f \in F}$ is a SSMD as considered in section 3 and the desired
result follows from Lemma 6 and Lemma 5. $\square$

As an immediate consequence we get :

<u>Corollary</u>

Under (C), (C^*) and (L_1) for each initial state $i \in I$, there
exists a decision rule $f_i \in F$ such that $g_{f_i}(i) = \max_{f \in F} g_f(i)$.

Applying the result of Lemma 7 to the SMD $(\bar{p}_f, r'_f)_{f \in F}$, we
obtain corresponding continuity results for a family of relative
value functions under the following second condition of Liapunov
function type.

(L_2) : There exist non-negative real valued functions y_1, y_2
on I such that

(i) $\quad |r_f(i)| + t_f(i) + \sum\limits_{j \notin K} p_f(i,j)y_1(j) \leqslant y_1(i)$

$\quad$ for all $i \in I, f \in F$

(ii) $\quad y_1(i) + \sum\limits_{j \notin K} p_f(i,j)y_2(j) \leqslant y_2(i)$

$\quad$ for all $i \in I, f \in F$

(iii) $f \to \sum\limits_{j \notin K} p_f(i,j)y_1(j)$ is continuous on F for all $i \in I$

<u>Remark</u>

$(L_2(ii))$ implies $(L_1(ii))$ (then valid for all $i \in I$).

Let $(H(I), \geqslant)$ denote the partially ordered set of real valued functions on I, where the order "$\geqslant$" corresponds to pointwise order, i.e. $h_1 \geqslant h_2$ iff $h_1(i) \geqslant h_2(i)$ for all $i \in I$.

<u>Proposition</u>

Under (C), (C^*) and (L_2) there exists a family $(v_f)_{f \in F}$ of real valued relative value functions such that

(a) $f \to v_f(i)$ is continuous on F for all $i \in I$ and

(b) for each $f \in F$, policy iteration based on the relative value function v_f can only yield successor decision rules f' such that

$$g_{f'} \geqslant g_f \quad \text{and} \quad g_{f'} = g_f \quad \text{implies} \quad v_{f'} \geqslant v_f.$$

Moreover, if $f' \neq f$ then there is a strict improvement in the sense that $(g_{f'}, v_{f'}) \neq (g_f, v_f)$.

<u>Proof</u>

(a) As (L_2) implies (L_o), we know by the proof of Lemma 4 that there exist relative value functions v_f, $f \in F$, which admit the following representation :

$$v_f(i) = E_{i,f} \sum_{n=0}^{\tau_K - 1} ((r_f(X_n) - t_f(X_n)g_f(X_n)) + \sum_{j \in K} \bar{p}_f(i,j)v_f'(j), \quad i \in I,$$

the v_f', $f \in F$, being the bias functions of the SMD $(\bar{p}_f, r_f')_{f \in F}$. By Lemma 5 and Lemma 6 the family $(g_f)_{f \in F}$ of gain rate functions is uniformly bounded. Hence there exists a constant M such that the function $y = My_1$ satisfies

$$|r_f(i)| + |g_f(i)|t_f(i) + \sum_{j \notin K} p_f(i,j)y(j) \leq y(i) \quad \text{for all} \quad i \in I.$$

Furthermore we know by Theorem 1 that the gain rates $g_f(i)$ depend continuously on f for all $i \in I$ and thus, by the same reasoning as in the proof of Theorem 1, the first term of the sum defining $v_f(i)$ depends continuously on f for all states $i \in I$ and $(p_f, r_f')_{f \in F}$ is a SMD as considered in section 3. The continuity of $f \to v_f(i)$, $i \in I$, then follows from Lemma 2 and Lemma 7.

(b) As the family $(v_f')_{f \in F}$ is uniformly bounded and for non-negative functions $u_1, u_2 \in H(I)$ the inequality $u_1 + \sum_{j \notin K} p_f(.,j)u_2(j)$

$\leq u_2$ implies that $E_{i,f} \sum_{n=0}^{\tau_K - 1} u_1(X_n) \leq u_2(i)$ for all $i \in I$, there exist constants M_1, M_2 such that $|v_f| \leq M_1 y_1 + M_2$.

Thus, for each $f \in F$, the action sets

$$A_f(i) = \{a \in A(i) \mid \sum_{j \in I} p(i,a,j)g_f(j) = \max!\} \quad \text{and}$$

$$B_f(i) = \{a \in A_f(i) \mid r(i,a) - t(i,a)g_f(i)$$

$$+ \sum_{j \in I} p(i,a,j)v_f(j) = \max_{A_f(i)}!\} \ , \ i \in I,$$

are non-empty.

To prove the desired result, we take a pattern from the reasoning used by Veinott (1966, section 4) to obtain similar results for SMD's with finite state space. Let for any $f \in F$, R_f denote the set of positive recurrent states of the associated M.c. . Now fix some $f_o \in F$ and consider any $f_1 \in F_o := \{f \in F \mid f(i) \in B_{f_o}(i)$ and $f(i) = f_o(i)$ if $f_o(i) \in B_{f_o}(i)$, $i \in I\}$.

Then $f_1(i) \in A_{f_o}(i)$ for all $i \in I$ implies that

$$\sum_{j \in I} p_{f_1}(i,j)g_{f_o}(j) \geq \sum_{j \in I} p_{f_o}(i,j)g_{f_o}(j) = g_{f_o}(i) \quad \text{for all} \quad i \in I.$$

As $\sum_{j \in I} \pi_{f_1}(i,j)(\sum_{\ell \in I} p_{f_1}(j,\ell)g_{f_o}(\ell) - g_{f_o}(j)) = 0$ by the invariance of π_{f_1}, one gets that $g_{f_0}(i) = \sum_{j \in I} p_{f_1}(i,j) g_{f_0}(j)$ for all $i \in R_{f_1}$ and thus $f_0(i) \in A_{f_0}(i)$ for all $i \in R_{f_1}$, g_{f_0} has identical components on each recurrent class $R_{f_1,k}$, $1 \le k \le n_f$, and

$$r_{f_1}(i) - t_{f_1}(i) g_{f_0}(i) + \sum_{j \in I} p_{f_1}(i,j) v_{f_0}(j) \ge v_{f_0}(i)$$

for all $i \in R_{f_1}$.

In the sequel, we use that condition $(L_2(ii))$ implies that

$$E_{i,f} \sum_{n=0}^{\tau_K - 1} y_1(X_n) < \infty \quad \text{for all} \quad i \in I, \; f \in F \quad \text{and therefore, by}$$

Lemma 4

$$\sum_{j \in I} \pi_f(i,j) y_1(j) < \infty \quad \text{and}$$

$$\sum_{j \in I} \pi_f(i,j) y_1(j) = \lim_{n \to \infty} \frac{1}{n} \sum_{k=1}^{n} \sum_{j \in I} p_f^k(i,j) y_1(j)$$

for all $i \in I$, $f \in F$. The last assertion follows from an ergodic theorem formulated in (Deppe, 1981, Theorem 7.1) for semiregenerative processes. Then it is easy to see that $g_{f_0}(i) \le g_{f_1}(i)$ for all $i \in R_{f_1,k}$, $1 \le k \le n_{f_1}$ and with $g_{f_1}(i) - g_{f_0}(i) \ge \sum_{j \in I} \pi_{f_1}(i,j)(g_{f_1}(j) - g_{f_0}(j))$ we have that $g_{f_1} \ge g_{f_0}$.

Next assume that $g_{f_1} = g_{f_0}$. Then it follows from the above assertions that $r_{f_1}(i) - t_{f_1}(i) g_{f_1}(i) + \sum_{j \in I} p_{f_1}(i,j) v_{f_0}(j) \ge v_{f_0}(i)$ for all $i \in I$ with equality for $i \in R_{f_1}$, such that $f_1(i) = f_0(i)$ for $i \in R_{f_1}$ and thus, by the definition of v_f, $f \in F$, $v_{f_1}(i) = v_{f_0}(i)$ for all $i \in R_{f_1}$.

As $v_{f_1}(i) - v_{f_0}(i) \ge \sum_{j \in I} p_{f_1}(i,j)(v_{f_1}(j) - v_{f_0}(j))$ and therefore under $(L_2(ii))$

$$v_{f_1}(i) - v_{f_0}(i) \geq \sum_{j \in I} \pi_{f_1}(i,j)(v_{f_1}(j) - v_{f_0}(j))$$

$$\text{for all } i \in I,$$

we have that $v_{f_1} \geq v_{f_0}$.

If $(g_{f_1}, v_{f_1}) = (g_{f_0}, v_{f_0})$, then

$$\sum_{j \in I} p_{f_0}(i,j) g_{f_0}(j) = g_{f_0}(i) = g_{f_1}(i)$$

$$= \sum_{j \in I} p_{f_1}(i,j) g_{f_1}(j) = \sum_{j \in I} p_{f_1}(i,j) g_{f_0}(j) = \max!$$

and analogously

$$r_{f_0}(i) - t_{f_0}(i) g_{f_0}(i) + \sum_{j \in I} p_{f_0}(i,j) v_{f_0}(j) = v_{f_0}(i) = v_{f_1}(i)$$

$$= r_{f_1}(i) - t_{f_1}(i) g_{f_0}(i) + \sum_{j \in I} p_{f_1}(i,j) v_{f_0}(j) = \max_{A_{f_0}(i)}!$$

for all $i \in I$ and thus $f_1 = f_0$ and this completes the proof. $\square$

For the remainder of the section we assume that the underlying family of relative value functions $(v_f)_{f \in F}$ is the family of relative value functions as used in the proof of the preceding proposition.

Then, if there exists a decision rule $f_0 \in F$ which cannot be improved by one policy iteration step (with respect to the gain rates and the relative values), the corresponding gain rate function g_{f_0} together with relative value function v_{f_0} constitute a solution of the functional equations (1) and (2), the so-called <u>average optimality equations</u>.

By the results obtained so far, we can now show the existence of such decision rules, following an idea of Schweitzer (1982).

Theorem 2

Under (C), (C*) and (L$_2$) there exists a finite solution (g,v) of the average optimality equations (1) and (2) such that g is bounded and $|v(i)| \leqslant M_1 y_1(i) + M_2$ for some constants M_1, M_2 independent of $i \in I$. For each such solution (g,v) of (1) and (2) we have that $g = \sup_{f \in F} g_f$ and furthermore, $g_{f^*} = g$ for each $f^* \in F$ such that $f^*(i)$ maximizes the right hand sides of (1) and (2) for all $i \in I$.

Proof

By Theorem 1 the function $f \to g_f$ from F to $(H(I), \geqslant)$ is (upper semi-) continuous and thus, by the compactness of F, there exists a decision rule f_1 such that g_{f_1} is a maximal element in $G := \{g_f \mid f \in F\}$, (Birkhoff, 1948, Theorem 4.16), i.e. there exists a decision rule f_1 which cannot be improved uniformly with respect to the gain rates.

Now, define $F_1 := \{f \in F \mid g_f = g_{f_1}\}$ and $V_1 := \{v_f \mid f \in F_1\}$. As F_1 is a compact subset of F and the function $f \to v_f$ from F_1 to $(H(I), \geqslant)$ is continuous, a further application of the above mentioned result in (Birkhoff, 1948) provides the existence of a decision rule $f_2 \in F_1$ such that v_{f_2} is a maximal element in V_1, i.e. f_2 cannot be improved uniformly with respect to the relative values within the set of decision rules F_1. Hence, f_2 is a decision rule that cannot be improved by one policy iteration step and thus (g_{f_2}, v_{f_2}) is a finite solution of (1) and (2).

The proof of the remaining assertions is standard, observing that $\sum_{j \in I} \pi_f(i,j) v(j)$ is finite for all $i \in I, f \in F$. $\square$

Recently, similar results were formulated by Zijm (1984) for Markov decision models. The communicating case is investigated by Hordijk (1974) and Federgruen, Schweitzer and Tijms (1983). Hordijk and Dekker (1983), Hordijk and Sladky (1977), Mann (1983a

and 1983b) establish conditions for the existence of a solution
of the optimality equations that arise in Markov decision processes
in connection with sensitive criteria.

References

Birkhoff, G. (1948). Lattice theory. American Math.Society
    Colloquium, Volume 25.

Blackwell, D. (1962). Discrete dynamic programming. Ann.Math.Stat.
    33, 719-726.

Chung, K.L. (1967). Markov chains with stationary transition
    probabilities. Springer, Berlin.

Çinlar, E. (1974). Periodicity in Markov renewal theory. Advances
    in Appl.Probability 6, 61-78.

Deppe, H. (1985). Continuity of mean recurrence times in denumerable
    semi Markov processes. Z. Wahrscheinlichkeitstheorie verw.
    Gebiete 69.

Deppe, H. (1981). Semiregenerative processes with costs. Preprint
    475, SFB 72, University of Bonn.

Federgruen, A., A.Hordijk and H.C.Tijms (1979). Denumerable state
    semi Markov decision processes with unbounded costs, average
    reward criterion. Stoc.Proc.Appl. 9, 223-225.

Federgruen, A., P.J.Schweitzer and H.C.Tijms (1983). Denumerable
    undiscounted semi-Markov decision processes with unbounded
    rewards. Math.of Oper.Res. 8, 298-313.

Hordijk, A. (1974). Dynamic programming and Markov potential theory.
    Math.Centre Tract 51, Mathematisch Centrum, Amsterdam.

Hordijk, A. and R.Dekker (1983). Average, sensitive and Blackwell
    optimal policies in denumerable Markov decision chains with
    unbounded rewards. Report 83-36, Institute of Appl.Math. and
    Comp.Science, University of Leiden.

Hordijk, A. and K.Sladky (1977). Sensitive optimality criteria in
    countable state dynamic programming. Math.of Oper.Res. 2, 1-14.

Kolonko, M. (1980a). Dynamische Optimierung unter Unsicherheit in
einem Semi-Markoff Modell mit abzählbarem Zustandsraum.
Dissertation, Institut für Angewandte Mathematik, Universität
Bonn.

Kolonko, M. (1980b). A countable Markov chain with reward structure-
continuity of the average reward. Preprint 415, SFB 72,
University of Bonn.

Mann, E. (1983a). Optimality equations and bias-optimality in
bounded Markov decision processes. Preprint 574, SFB 72,
University of Bonn. To appear in Optimization 16.

Mann, E. (1983b). Optimalitätsgleichungen und optimale Politiken
für sensitive Kriterien. Operations Research Proceedings 1983,
Springer Berlin Heidelberg.

Schweitzer, P.J. (1968). Perturbation theory and finite Markov
chains. J.Appl.Prob. 5, 401-413.

Schweitzer, P.J. (1982). Solving MDP functional equations by
lexicographic optimization. RAIRO 16, 91-98.

Veinott, A.F. (1966). On finding optimal policies in discrete
dynamic programming. Ann.Math.Stat. 37, 1284-1294.

Wijngaard, J. (1975). Stationary Markov decision problems, discrete
time, general state space. Proefschrift, Technische Hogeschool
Eindhoven.

Zijm, H. (1984). The optimality equations in multichain denumerable
state Markov decision processes with the average cost
criterion : the unbounded cost case. CQM-Note 022, Centre of
Quantitative Methods, Eindhoven.

SECTION III. ALGORITHMIC AND COMPUTER-ORIENTED APPROACH

RECURSIVE MOMENT FORMULAS FOR REGENERATIVE SIMULATION

Peter W. Glynn and Donald L. Iglehart*

University of Wisconsin

Stanford University*

1. Introduction

Let f be a real-valued function defined on the state space of a regenerative process $\underset{\sim}{X} = \{X(t) : t \geq 0\}$ with regeneration times $0 = T_0 < T_1 < \ldots$, and suppose that

$$r_t = \frac{1}{t} \int_0^t f(X(s))ds \to r \qquad \text{a.s.} \tag{1.1}$$

as $t \to \infty$. The problem of estimating r via a simulation of $\underset{\sim}{X}$ is called the steady state simulation problem.

Relation (1.1) implies that r_t is a strongly consistent point estimator for r. To obtain confidence intervals for r, set (for $k \geq 1$)

$$Y_k(f) = \int_{T_{k-1}}^{T_k} f(X(s))ds$$

$$\tau_k = T_k - T_{k-1}$$

$$Z_k = Y_k(f) - r\tau_k.$$

The regenerative structure of $\underset{\sim}{X}$ guarantees that $r = E\{Y_1(f)\}/E\{\tau_1\}$ and that $\{(Y_k(f), \tau_k) : k \geqslant 1\}$ is a sequence of i.i.d. random vectors. Standard arguments (see Crane and Iglehart, 1975) show that if $E\{Y_1^2(f) + \tau_1^2\} < \infty$, then

$$\sqrt{t}(r_t - r) \Rightarrow \sigma N(0,1) \tag{1.2}$$

as $t \to \infty$, where $\sigma^2 = \sigma^2\{Z_1\}/E\{\tau_1\}$. To use (1.2) for confidence intervals, the regenerative cycle structure of $\underset{\sim}{X}$ is exploited to obtain a strongly consistent estimator v_t for σ^2.

These confidence intervals, while asymptotically correct, often have poor small-sample behavior. For example, such confidence intervals often tend to significantly undercover the parameter r. Several recent studies have examined this problem. Glynn (1982), in considering regenerative confidence intervals on the time scale of regenerative cycles, obtained asymptotic expansions for the coverage error which indicated that skewness/kurtosis effects play a significant role in determining quality of the confidence interval. To be more precise, the error, to a first approximation, is determined by the magnitude of quantities of the form $E\{Z_1^m \tau_1^n\}$ for $m + n \leqslant 4$. Glynn and Iglehart (1984) obtain expressions for the asymptotic covariance between r_t and v_t, and the variance of v_t; these expressions also involve mixed moments of the form $E\{Z_1^m \tau_1^n\}$. Consequently, in studying small-sample behavior of regenerative confidence intervals, it is of some interest to be able to calculate the exact values of the mixed moments for some "test case" stochastic models. Hordijk, Iglehart and Schassberger (1976) showed how to do this for $m + n \leqslant 2$, when X is a discrete or continuous time Markov chain with countably many states. In this note, we show how to calculate such quantities when X is a semi-Markov process with countably many states; discrete and continuous time Markov chain results follow as special cases.

2. Statement of the Recursive Moment Formulas

Let $X = \{X(t) : t \geq 0\}$ be an irreducible non-explosive regenerative semi-Markov process on countable state space E. Thus, $X(t)$ may be represented as

$$X(t) = \sum_{k=0}^{\infty} R_k \, I(S_k \leq t < S_{k+1}) \, ,$$

where :

(i) $R = \{R_n : n \geq 0\}$ is a discrete-time Markov chain on E with transition matrix $P = (p_{xy} : x, y \in E)$

(ii) $S = \{S_n : n \geq 0\}$ is an increasing sequence of jump times with $S_0 = 0$ and differences $\alpha_n = S_{n+1} - S_n$ which are conditionally independent r.v.'s given R.

The conditional distribution of α_n is given by $F(R_n, R_{n+1}, dt) = P\{\alpha_n \in dt \mid R\}$, where $F(x,y,0) = 0$ for all $x, y \in E$. Note that $S_n \to \infty$ a.s., since X is non-explosive by assumption. Fix $z \in E$ as the regenerative state; let $T(z) = \inf\{t > 0 : X(t-) \neq z, X(t) = z\}$ and set

$$Y(u) = \int_0^{T(z)} u(X(t)) dt$$

where $u : E \to R$ is an arbitrary function. We wish to study mixed moments of the form

$$a_{ij}(x) = E_x\{Y(g)^i \, Y(h)^j\}$$

for $x \in E$, $0 \leq i \leq m$, and $0 \leq j \leq n$, when g and h are fixed functions, and m and n are non-negative integers. Throughout the paper we shall use $P_x\{\cdot\}$ and $E_x\{\cdot\}$ to denote conditional probabilities and expectations, given $X(0) = R_0 = x$. Note that by choosing $g(\cdot) = f(\cdot) - r$

and $h(\cdot) = 1$, $a_{ij}(z)$ yields $E_z\{Z_1^i \ \tau_1^j\}$.

To state our result, let $a \vee b$ denote $\max(a,b)$ and set $b_{mn}(x) = E_x\{Y(|g|) \vee 1)^m \ (Y(|h|) \vee 1)^n\}$. Let G_n and β_n be the matrix and function, respectively, defined by

$$G_n(x,y) = \begin{cases} P_{xy} \ \mu(x,y) \ ; & y \neq z \\ \\ 0 & ; & y = z \end{cases}$$

$$\beta_n(x) \ = \ \sum_{y \in E} \mu_n(x,y) \ P_{xy}$$

where $\mu_n(x,y) = \int_0^\infty t^n \ F(x,y,dt)$. Also, we shall identify real-valued functions $u(\cdot)$ on E with column vectors $\underset{\sim}{u}$, and shall use the notation $\underset{\sim}{u} \circ \underset{\sim}{v}$ to denote the vector with x^{th} component $(\underset{\sim}{u} \circ \underset{\sim}{v})(x) = u(x)v(x)$. Set $\underset{\sim}{u}^0(\cdot) = 1$ and $\underset{\sim}{u}^{n+1} = (\underset{\sim}{u} \circ \underset{\sim}{u}^n)$ for $n \geq 0$.

Let C denote the class of all $m \times n$ matrix-valued functions, $\underset{\sim}{C}$, on E. Then set

$$C_{mn} = \{\underset{\sim}{C} \in C \ : \ \sum_{k=0}^i \sum_{\ell=0}^j |g(x)|^{i-k} \cdot |h(x)|^{j-\ell}$$

$$\cdot \sum_{y \in E} G_{i+j-k-\ell}(x,y)|c_{k\ell}(y)| < \infty$$

$$\text{and } (G_0^k \ c_{ij})(x) \rightarrow 0 \text{ as } k \rightarrow \infty ,$$

$$\text{for all } x \in E, \ 0 \leq i \leq m, \text{ and } 0 \leq j \leq n\}.$$

2.1. Theorem

If $b_{mn}(z) < \infty$, then $\underset{\sim}{A} = \{\underset{\sim}{a}_{ij} : 0 \leq i \leq m, \ 0 \leq j \leq n\}$ is the unique solution in C_{mn} to the system :

$$\underset{\sim}{c}_{ij} = \underset{\sim}{g}^j \circ \underset{\sim}{h}^j \circ \underset{\sim}{\beta}_{i+j} + \sum_{(k,\ell) \in B_{ij}} \binom{i}{k}\binom{j}{\ell} (\underset{\sim}{g}^{i-k} \circ \underset{\sim}{h}^{j-\ell} \circ \underset{\sim}{G}^{i+j-k-\ell} \underset{\sim}{c}_{k\ell})$$

$$(2.2)$$

where $0 \leqslant i \leqslant m$, $0 \leqslant j \leqslant n$, and $B_{ij} = \{(k,\ell) : 0 \leqslant k \leqslant i, 0 \leqslant \ell \leqslant j, k+\ell > 0\}$.

Set $\tau = \inf\{n \geqslant 1 : R_n = z\}$ and observe that

$$G_0^k(x,y) = P_x\{R_k = y, \tau > k\}.$$

Hence, $\underset{\sim}{G}_0^k \to 0$ as $k \to \infty$. It follows that if E has a finite number of elements, $C_{mn} \equiv C$, so that the $\underset{\sim}{a}_{ij}$'s are unique in the class of all possible solutions to (2.2). Also, in the presence of a finite state space, it is well known that $(\underset{\sim}{I} - \underset{\sim}{G}_0)^{-1}$ exists, so that (2.2) may be re-written as

$$\underset{\sim}{c}_{ij} = (\underset{\sim}{I} - \underset{\sim}{G}_0)^{-1}\{\underset{\sim}{g}^i \circ \underset{\sim}{h}^i \circ \underset{\sim}{\beta}_{i+j}$$

$$+ \sum_{(k,\ell) \in A_{ij}} \binom{i}{k}\binom{j}{\ell} (\underset{\sim}{g}^{i-k} \circ \underset{\sim}{h}^{j-\ell} \circ (\underset{\sim}{G}_{i+j-k-\ell}\underset{\sim}{c}_{k\ell}))\} ,$$

$$(2.3)$$

where $0 \leqslant i \leqslant m$, $0 \leqslant j \leqslant n$, $i + j > 0$, and $A_{ij} = \{(k,\ell) : 0 \leqslant k \leqslant i, 0 \leqslant \ell \leqslant j, 0 < k+\ell < i+j\}$. Also observe that $\underset{\sim}{c}_{00} \equiv 1$. Note that the system of equations (2.3) is recursive in $i+j$, in the sense that the $\underset{\sim}{c}_{ij}$'s may be solved in terms of the $\underset{\sim}{c}_{k\ell}$'s, where $k+\ell < i+j$. By successively solving for the $\underset{\sim}{c}_{k\ell}$'s with fixed $k+\ell$ on each iteration, one eventually obtains $\underset{\sim}{c}_{mn}$.

Formula (2.3) can be further simplified when X has special structure. Note that if X is a continuous time Markov chain, then

$$F(x,y,dt) = \lambda(x) \exp(-\lambda(x)t)dt$$

for $t > 0$, so that

$$\mu_n(x,y) = n!/\lambda(x)^n \equiv \eta_n(x) .$$

We find that (2.3) can be re-written as

$$\underset{\sim}{c}_{ij} = (I - \underset{\sim}{G}_0)^{-1}\{g^i \circ h^j \circ \underset{\sim}{\eta}_{i+j}$$

$$+ \sum_{(k,\ell) \in A_{ij}} \binom{i}{k}\binom{j}{\ell}(g^{i-k} \circ h^{j-\ell} \circ \underset{\sim}{\eta}_{i+j-k-\ell} \circ (\underset{\sim}{G}_0 \underset{\sim}{c}_{k\ell}))\} ,$$

where $0 \leqslant i \leqslant m$, $0 \leqslant j \leqslant n$, and $i+j > 0$.

For discrete time Markov chains, $\underset{\sim}{\beta}_n \equiv 1$ and $\underset{\sim}{G}_n = \underset{\sim}{G}_0$, so (2.3) takes the form

$$\underset{\sim}{c}_{ij} = (I - \underset{\sim}{G}_0)^{-1}\{g^i \circ h^j$$

$$+ \sum_{(k,\ell) \in A_{ij}} \binom{i}{k}\binom{j}{\ell}(g^{i-k} \circ h^{k-\ell} \circ (\underset{\sim}{G}_0 \underset{\sim}{c}_{k\ell}))\} ,$$

where $0 \leqslant i \leqslant m$, $0 \leqslant j \leqslant m$, and $i+j > 0$.

Relation (2.4) expresses $\underset{\sim}{c}_{ij}$ in terms of $\underset{\sim}{G}_0 \underset{\sim}{c}_{k\ell}$, where $(k,\ell) \in A_{ij}$. Equation (2.4) can be re-written, when $\underset{\sim}{g} = \underset{\sim}{h}$, so that the $\underset{\sim}{c}_{ij}$'s are written directly in terms of the $\underset{\sim}{c}_{k\ell}$'s. If $\underset{\sim}{g} = \underset{\sim}{h}$, we write $\underset{\sim}{c}_{ij}$ as $\underset{\sim}{c}_{i+j}$, and observe that (2.4) takes the form

$$\underset{\sim}{c}_i = (I - \underset{\sim}{G}_0)^{-1}\{g^i + \sum_{k=1}^{i-1} \binom{i}{k}(g^{i-k} \circ \underset{\sim}{G}_0 \underset{\sim}{c}_k)\}, \quad 1 \leqslant i \leqslant n. \quad (2.5)$$

Recall also that from (2.2), $\underset{\sim}{c}_0 \equiv 1$. We claim that the system (2.5) can be re-written as

$$\underset{\sim}{c}_i = (\underset{\sim}{I} - \underset{\sim}{G}_0)^{-1}\{ \sum_{k=1}^{i} (-1)^{k+1} \binom{i}{k} \underset{\sim}{g}^k \circ \underset{\sim}{c}_{i-k}\}, \quad 1 \leq i \leq n. \quad (2.6)$$

The proof is by induction. For $n = 1$, the result is obvious, so suppose (2.5) and (2.6) are equivalent systems for $n = m$. To check the $(m+1)$'st equation in the $(m+1)$'st system, observe that a solution of (2.5) satisfies

$$\underset{\sim}{c}_{m+1} = (\underset{\sim}{I} - \underset{\sim}{G}_0)^{-1}\{\underset{\sim}{g}^{m+1} + \sum_{i=1}^{m} \binom{m+1}{i} (\underset{\sim}{g}^{m+1-i} \circ \underset{\sim}{G}_0 \underset{\sim}{c}_i)\}. \quad (2.7)$$

By the inductive hypothesis, (2.6) shows that

$$\underset{\sim}{G}_0 \underset{\sim}{c}_i = - \sum_{k=0}^{i} (-1)^{k+1} \binom{i}{k} \underset{\sim}{g}^k \circ \underset{\sim}{c}_{i-k} \quad (2.8)$$

for $i \leq m$. Substituting (2.8) into (2.7), we get that $(\underset{\sim}{I} - \underset{\sim}{G}_0)\underset{\sim}{c}_{m+1}$ equals

$$\underset{\sim}{g}^{m+1} + \sum_{i=1}^{m} \sum_{k=0}^{i} \binom{m+1}{i} \binom{i}{k} (-1)^{i-k} \underset{\sim}{g}^{m+1-k} \circ \underset{\sim}{c}_k$$

$$= \sum_{k=0}^{m} \binom{m+1}{k} (\underset{\sim}{g}^{m+1-k} \circ \underset{\sim}{c}_k) \sum_{j=1}^{m+1-k} \binom{m+1-k}{j} (-1)^{m+1-k-j}$$

$$= \sum_{k=0}^{m} \binom{m+1}{k} (\underset{\sim}{g}^{m+1-k} \circ \underset{\sim}{c}_k) (-1)^{m+2-k}$$

$$= \sum_{\ell=1}^{m+1} \binom{m+1}{\ell} (\underset{\sim}{g}^{\ell} \circ \underset{\sim}{c}_{m+1-\ell}) (-1)^{\ell+1},$$

which is equivalent to the $(m+1)$'st relation of (2.6) (the binomial identity was used for the third equality). The steps being reversible, this proves the claimed result. We remark that (2.6) yields the equations of A. Hordijk, D.L. Iglehart and R. Schassberger, Discrete-Time Methods of Stimulating Continuous-Time Markov Chains, for $i = 1,2$.

3. Proof of the Theorem

We proceed via a series of lemmas.

3.1. Lemma

If $b_{mn}(z) < \infty$, then the r.v.'s $Y(g)^i \, Y(h)^j$ are integrable under the probability distribution P_x for $0 \leq i \leq m$, $0 \leq j \leq n$, $x \in E$.

Proof.

Note that

$$(Y(|g|) \vee 1)^i \, (Y(|h|) \vee 1)^j \leq (Y(|g|) \vee 1)^m \, (Y(|h|) \vee 1)^n$$

so that $b_{ij}(z) < \infty$ for $0 \leq i \leq m$, $0 \leq j \leq n$. Since $\underset{\sim}{X}$ is irreducible, it follows that $P_z\{T(y) < T(z)\} > 0$ for all $y \in E$, for otherwise the regenerative property guarantees that $P_z\{T(y) = \infty\} = 1$, which violates our irreducibility assumption.

Then, by the strong Markov property applied at time $T(y)$,

$$E_z\{ (\int_{T(y)}^{T(z)} |g(X(s))| ds \vee 1)^i \, (\int_{T(y)}^{T(z)} |h(X(s))| ds \vee 1)^j ; \, T(y) < T(z)\}$$

$$= b_{ij}(y) \, P\{T(y) < T(z) | X(0) = z\} \leq b_{ij}(z)$$

so that $b_{ij}(y) < \infty$ for all $y \in E$, which proves the result. $\square$

3.2 Proposition

If $b_{mn}(z) < \infty$, then $a_{ij}(x)$ exists and is finite for $0 \leq i \leq m$, $0 \leq j \leq n$, $x \in E$. Furthermore, $\underset{\sim}{A}$ solves (2.2).

Proof

The first part follows immediately from lemma 3.1. For the second part, the integrability of $Y(g)^i \, Y(h)^j$ ensures that the following manipulations of conditional expectations are valid :

$$a_{ij}(x) = E_x\{Y(g)^i \, Y(h)^j\}$$

$$= E_x\{Y(g)^i \, Y(h)^j; \ \tau = 1\} + E_x\{Y(g)^i \, Y(h)^j; \ \tau > 1\}$$

$$= g^i(x) \, h^j(x) \, \mu_{i+j}(x,z) \, P_{xz}$$

$$+ E_x\{(g(R_0) \, \alpha_0 + \int_{S_1}^{T(z)} g(X(t)dt)^i \, (h(R_0) \, \alpha_0$$

$$+ \int_{S_1}^{T(z)} h(X(t))dt)^j; \ \tau > 1\}$$

$$= \sum_{k=0}^{i} \sum_{\ell=0}^{j} \binom{i}{k} \binom{j}{\ell} E_x\{g(R_0)^{i-k} \, h(R_0)^{j-\ell} (\int_{S_1}^{T(z)} g(X(t))dt)^k$$

$$\cdot (\int_{S_1}^{T} h(X(t))dt)^\ell; \ \tau > 1\}$$

$$= g^i(x) \, h^j(x) \, \mu_{i+j}(x,z) \, P_{xz}$$

$$+ \sum_{k=0}^{i} \sum_{\ell=0}^{j} \binom{i}{k} \binom{j}{\ell} E_x\{g(R_0)^{i-k} \, h(R_0)^{j-\ell} \alpha_0^{i+j-k-\ell}$$

$$(\int_{S_1}^{T(z)} g(X(t))dt)^k \cdot (\int_{S_1}^{T} h(X(t))dt)^\ell; \ \tau > 1\}$$

$$= g^i(x) \, h^j(x) \, \mu_{i+j}(x,z) \, P_{xz}$$

$$+ \sum_{k=0}^{i} \sum_{\ell=0}^{j} \binom{i}{k} \binom{j}{\ell} g^{i-k}(x) h(x)^{j-\ell} \sum_{y \neq z} P_{xy} \, \mu_{i+j-k-\ell}(x,y) \, a_{k\ell}(y)$$

$$= (\underset{\sim}{g}^i \circ \underset{\sim}{h}^i \circ \underset{\sim}{\beta}_{i+j})(x) + \sum_{(k,\ell) \in B_{ij}} \binom{i}{k} \binom{j}{\ell} (\underset{\sim}{g}^{i-k} \circ \underset{\sim}{h}^{j-\ell}$$

$$\circ (\underset{\sim}{G}_{i+j-k-\ell} \, \underset{\sim}{a}_{k\ell})(x)) ;$$

the strong Markov property at time S_1 was used to obtain the second last equality. $\square$

3.3. Lemma

If $b_{mn}(z) < \infty$, then $\underset{\sim}{A} \in C_{mn}$.

Proof

For the absolute summability observe that $d_{ij}(x) = E_x\{Y(|g|)^i\, Y(|h|)^j\}$ satisfies

$$\underset{\sim}{d}_{ij} = |\underset{\sim}{g}|^i \circ |\underset{\sim}{h}|^j \circ \underset{\sim}{\beta}_{i+j}$$

$$+ \sum_{(k,\ell)\,\in\, B_{ij}} \binom{i}{k}\binom{j}{\ell} \left(|\underset{\sim}{g}|^{i-k} \circ |\underset{\sim}{h}|^{j-\ell} \circ (\underset{\sim}{G}_{i+j-k-\ell}\, \underset{\sim}{d}_{k\ell})\right).$$

But $d_{ij}(x) \leqslant b_{ij}(x) < \infty$, so $|\underset{\sim}{g}|^{i-k} \circ |\underset{\sim}{h}|^{j-\ell} \circ (\underset{\sim}{G}_{i+j-k-\ell}\underset{\sim}{d}_{k\ell})$ is finite, proving the first part, since $|\underset{\sim}{a}_{ij}| \leqslant \underset{\sim}{d}_{ij}$. For the second,

$$(\underset{\sim}{G}_0^k\, \underset{\sim}{a}_{ij})(x) = E_x\{(\int_{S_k}^{T(z)} Y(g))^i\, (\int_{S_k}^{T(z)} Y(h))^j;\ \tau > k\}\,,$$

which tends to zero by integrability of $Y(|g|)^i\, Y(|h|)^j$. $\square$

3.4. Lemma

If $b_{mn}(z) < \infty$, then $\underset{\sim}{A}$ is the unique solution to (2.2) in C_{mn}.

Proof

Suppose $\{\underset{\sim}{a}_{rs} : 0 \leqslant r \leqslant k,\ 0 \leqslant s \leqslant \ell\}$ is unique in $C_{k\ell}$ for $k+\ell < i+j$, where $0 \leqslant i \leqslant m$, $0 \leqslant j \leqslant n$. We shall prove uniqueness in C_{ij}; "bootstrapping" this result yields the lemma. Since the $\underset{\sim}{a}_{rs}$'s are unique in $C_{k\ell}$, any solution to the (i,j) equation must satisfy

$$\underset{\sim}{c}_{ij} = \underset{\sim}{g}^i \circ \underset{\sim}{h}^j \circ \underset{\sim}{\beta}_{i+j} + \sum_{(k,\ell) \in A_{ij}} \binom{i}{k}\binom{j}{\ell} (\underset{\sim}{g}^{i-k} \circ \underset{\sim}{h}^{j-\ell}$$

$$\circ \underset{\sim}{G}_{i+j-k-\ell} \underset{\sim}{a}_{k\ell}) + \underset{\sim}{G}_0 \underset{\sim}{c}_{ij}. \tag{3.5}$$

By the absolute summability, any two solutions $\underset{\sim}{c}_{ij}$, $\underset{\sim}{c}'_{ij}$ of (3.5) must satisfy $(\underset{\sim}{c}_{ij} - \underset{\sim}{c}'_{ij}) = \underset{\sim}{G}_0(\underset{\sim}{c}_{ij} - \underset{\sim}{c}'_{ij})$. Since $\underset{\sim}{G}_0^k(\underset{\sim}{c}_{ij} - \underset{\sim}{c}'_{ij}) \to 0$ as $k \to \infty$, it follows that $\underset{\sim}{c}_{ij} = \underset{\sim}{c}'_{ij}$, proving uniqueness in C_{ij}. $\square$

These above results prove all the assertions of the theorem.

Acknowledgment

Both authors gratefully acknowledge support by U.S. Army Research Office Contract DAAG29-84-K-0030. The second author was also partially supported by National Science Fondation Grant MCS-8203483.

References

M.A. Crane, D.L. Iglehart (1975), "Stimulating Stable Stochastic Systems, III : Regenerative Processes and Discrete-Event Stimulations", Oper. Res. <u>23</u>, 33-45.

P.W. Glynn (1982), "Asymptotic Theory for Nonparametric Confidence Intervals", Technical Report n° <u>19</u> Dept. of Operations Research, Stanford University, Stanford, CA.

P.W. Glynn and D.L. Iglehart (1984), "The Joint Limit Distribution of Sample Mean and the Regenerative Variance Estimator". Forthcoming technical report, Dept. of Operations Research, Stanford University, Stanford, CA.

A. Hordijk, D.L. Iglehart and R. Schassberger (1976), "Discrete-Time Methods of Stimulating Continuous-Time Markov Chains", Advances in Appl. Probability <u>8</u>, 772-788.

COMPUTATION OF THE STATE PROBABILITIES IN A CLASS OF SEMI-REGENERA-
TIVE QUEUEING MODELS

Helmut Schellhaas

Fachbereich Mathematik, Technische Hichschule Darmstadt
Schlossgartenstrasse, 7, D-6100 Darmstadt, Germany

1. Introduction

One of the standard problems in queueing analysis concerns the
computation of the steady state probabilities. In recent years significant contributions to this problem have been made. Important research papers by M.F. Neuts in the context of phase type distributions and by H.C. Tijms and M.H. Van Hoorn in the context of regenerative processes were very stimulating for our investigations. We
refer to Neuts (1981), (1984) and the literature cited there and to
Tijms and Van Hoorn (1981) and Van Hoorn (1983). In thie paper we
consider the unifying queueing model including several well known
models as special cases. Using a semi-regenerative analysis we develop
an efficient and stable algorithm for the computation of the steady
state probabilities. The usefulness of the results is demonstrated
by applications to several modifications of state dependent M/G/1
queues.

In section 2 we describe our model, a generalized state dependent
M/G/1 queue with vacations. Section 3 is devoted to the semi-regenerative analysis of the model leading to a stable and efficient algorithm for the steady-state probabilities. Explicit representations

of the coefficients in the algorithm are given. Concerning the numerical effort considerable simplifications are possible in the case of
phase type distributions. The usefulness of the model and the algorithm is demonstrated in section 4 by investigating several modifications of generalized M/G/1 queues.

Notation . We put $\mathbb{N} = \{1,2,\ldots\}$, $\mathbb{N}_o = \mathbb{N} \cup \{0\}$, $\mathbb{R}_+ = [0,\infty)$.

2. The model

We consider a single server queueing system with finite or infinite waiting room. Customers are served individually, the service
times of the customers are independent random variables having probability distribution function B_i when i customers are in the waiting
room at the beginning of the service. During the service of a customer new customers arrive singly according to a state dependent Markovian input process independent of the service process. The rate of
the input process is λ_j when j customers are in the waiting room.
The arriving customers join the queue in the waiting room. As soon
as a customer is served the server takes a vacation. The vacation
times are independent random variables having probability distribution C_1 when 1 customers are in the waiting room at the beginning of
the vacation. We distinguish the cases 1 > 0 and 1 = 0. If 1 > 0
then new customers are not allowed to arrive before the termination
of the vacation, but the termination of the vacation may be caused by
an arriving customer. Then the time epoch of arrival of this customer equals the time epoch of termination of the vacation. Just after
a vacation the service of a customer is started leaving 1 customers
(if a customer arrived at the termination of the vacation) or 1 - 1
customers (if the termination of the vacation has not been initiated
by an arriving customer) in the waiting room. If 1 = 0 then new customers arrive singly during the vacation according to a state dependent Markovian input process with rate κ_j when j customers are in the
waiting room.

This input process may be either independent of the vacation process
or the arrival of the m-th customer, $m \in \mathbb{N}$, (counted from the beginning of the vacation) may cause termination of the vacation. Then the
time epoch of arrival of the m-th customer equals the time epoch of
termination of the vacation. If at the end of a vacation customers
are present in the waiting room the service of one of the customers
is started immediatly. (This is always the case when the arrival of
the m-th customer causes termination of the vacation. Then m-1 customers are left in the waiting room). If at the end of a vacation no
customers are present then the server takes another vacation and the
described procedure starts anew. We note that C_1 may be degenerate,
particularly for $1 > 0$ the vacation times may be zero with probability one.

This unifying queueing model includes several well known models
as special cases (cf. section 4). Examples are the M/G/1 queue, the
repairman problem, the M/G/1 queue under a T-policy, the M/G/1 queue
under a k-policy, the M/G/1 queue with repeated attempts. Generalizing
the usual assumptions for these models we allow the input and the service to be state dependent, furthermore we treat models with finite
and infinite waiting room simultaneously.

For models with $K < \infty$ waiting places we have $0 < \lambda_j < \infty$ for
$0 \leqslant j < K$ and $\lambda_j = 0$ for $j \geqslant K$. To avoid trivialities we assume
$K \geqslant 2$. In this case we put $J = \{0,1,\ldots,K\}$. For models with an unli-
mited number of waiting places we have $0 < \lambda_j < \infty$ for $0 \leqslant j < \infty$. In
this case we put $J = \mathbb{N}_o$. Similar conclusions are valid for the κ_j.

3. Semi-regenerative analysis of the model

Let $Z = \{Z(t), t \in \mathbb{R}_+\} = \{(Y(t), L(t)), t \in \mathbb{R}_+\}$ be the stochastic process with state space $J \times \{0,1\}$ where $Y(t)$ denotes the number
of customers in the waiting room and $L(t) = 0$ [$=1$] if the server is
in vacation [is active] at time t. We assume the trajectories of the

Z-process to be right-continuous. Let T_n, $n \in \mathbb{N}_0$, be the service completion epochs and the vacation completion epochs respectively, ordered with respect to magnitude. Let the time origin be such an epoch. Define $X = \{X_n, n \in \mathbb{N}_0\} = \{(Y_n, L_n), n \in \mathbb{N}_0\}$ as follows. If T_n is a service completion epoch leaving $k \geq 0$ customers in the waiting room then $Y_n = k$, $L_n = 0$. If T_n is a vacation completion epoch with $k \geq 1$ [$k = 0$] customers present at this epoch (possibly including a customer arriving at T_n) then $Y_n = k - 1$, $L_n = 1$ [$Y_n = 0$, $L_n = 0$]. The assumptions for the queueing model are assumed such that $(X,T) = \{(X_n, T_n), n \in \mathbb{N}_0\}$ is a Markov renewal process with state space $J \times \{0,1\}$ and that Z is a semi-regenerative process with embedded Markov renewal process (X,T). For the definition of a semi-regenerative process we refer to Çinlar (1975).

We first consider the embedded Markov chain X. The structure of the transition matrix

$$\mathbb{P} = (p_{(i,k)(j,1)}, \ (i,k),(j,1) \in J \times \{0,1\}) \tag{3.1}$$

with

$$p_{(i,k)(j,1)} = P(Y_{n+1}=j, L_{n+1} = 1 \mid Y_n = i, L_n = k), \ n \in \mathbb{N}_0 \tag{3.2}$$

is given by

$$
p_{(i,k),(j,1)} =
\begin{cases}
q_{ij}, & i \geq 0, \quad j \geq i, \quad k = 1, \quad 1 = 0 \\
r_o, & i = 0, \quad j = 0, \quad k = 0, \quad 1 = 0 \\
d_j, & i = 0, \quad j \geq 0, \quad k = 0, \quad 1 = 1 \\
g_i, & i \geq 1, \quad j = i, \quad k = 0, \quad 1 = 1 \\
h_i, & i \geq 1, \quad j = i-1, k = 0, \quad 1 = 1 \\
0 & \text{else}
\end{cases}
\tag{3.3}
$$

Observing that if $(Y_n, L_n) = (i, 1)$ the Y-process between T_n and T_{n+1} is a pure birth process with birth rates λ_j, $j \geq i$, one readily verifies the following representation of the q_{ij}

$$q_{ij} = \int_0^\infty z_{ij}(u) B_i(du) \tag{3.4}$$

where the z_{ik}, $i, k \in J$, are the solutions of the Kolmogorov differential equations for the birth process

$$\begin{cases} \dot{z}_{ii}(t) = -\lambda_i z_{ii}(t) \\ \dot{z}_{ik}(t) = -\lambda_k z_{ik}(t) + \lambda_{k-1} z_{i,k-1}(t), \; k > i \\ z_{ik}(0) = \delta_{ik} \end{cases} \tag{3.5}$$

The representation of r_o, d_j, g_i, h_i depends on the model specific rules for terminating a vacation (cf. section 2). Explicit representations of these coefficients for special models are given in section 4.

Denoting by

$$b_i = \int_0^\infty [1 - B_i(t)] dt, \; c_i = \int_0^\infty [1 - C_i(t)] dt, \; i \in J \tag{3.6}$$

the state dependent expected service times and expected vacation times respectively, we state the following conditions.

(A1) $0 \leq r_o < 1$, $0 \leq d_i \leq 1$, $0 \leq g_i < 1$, $0 < h_i \leq 1$ for $i \in J$

(A2) $\sup_{i \in J} b_i < \infty$, $\sup_{i \in J} c_i < \infty$
and there exist $\lambda^\star < \infty$, $\kappa^\star < \infty$ and $m \in \mathbb{N}$ such that $\sup_{i \in J} \kappa_i \leq \kappa^\star$,

$\sup_{i \geq m} \lambda_i \leq \lambda^\star$, $\sup_{i \geq m} \lambda^\star b_i = : \rho < 1$, $\inf_{i \geq m} h_i > \rho$

(A3) There exists some $\tau > 0$ with $\sup_{i \in J} B_i(\tau) < 1$

(A4) At least one of the distribution functions B_i, C_i, $i \in J$, is non-
   arithmetic.

Remark.   (A3) seems to be a natural condition for applications.   It
may be replaced by any other condition ensuring infinite lifetime of
the (X,T)-process.   For example (A3) may be replaced by $B_i(0)<1, i \in J$,
in the case of a finite waiting room.

   The Markov chain X is aperiodic and irreducible.   Furthermore
applying a generalized Foster criterion, see Pakes (1969), one easily
verifies by (A2) that X is positive recurrent.

Let $\pi = (\pi_{(i,k)}, (i,k) \in J \times \{0,1\})$ be its stationary distribution, π
is the unique solution of

$$
\begin{cases}
\pi_{(j,1)} = \sum_{i \in J} \sum_{k \in \{0,1\}} \pi_{(i,k)} P_{(i,k),(j,1)}, & (j,1) \in J \times \{0,1\} \\
\sum_{j \in J} \sum_{l \in \{0,1\}} \pi_{(j,1)} = 1.
\end{cases}
\tag{3.7}
$$

Our aim is the computation of the limiting state probabilities

$$
P^\star(j,1) := \lim_{t \to \infty} P(Y(t) = j, L(t) = 1 \mid Y_0 = i, L_0 = k)
\tag{3.8}
$$

for $(j,1),(i,k) \in J \times \{0,1\}$.   A well known limit theorem for semi-regenerative processes gives a representation of these state probabilities in terms of the stationary distribution π of the Markov chain X.
Explicitely, this limit theorem states (see Schäl, 1970, Corollary
9.1, Çinlar, 1975, p. 347), that the limits $P^\star(j,1)$ exist with

$$
P^\star(j,1) = \sum_{i \in J} \sum_{k \in \{0,1\}} \pi_{(i,k)} \int_0^\infty \psi_{(i,k)}(t,j,1)\,dt/D
\tag{3.9}
$$

provided that X is irreducible and recurrent, (X,T) is aperiodic with
infinite lifetime,

$$
D = \sum_{i \in J} \pi_{(i,0)} c_i + \sum_{i \in J} \pi_{(i,1)} b_i < \infty \;\; ,
\tag{3.10}
$$

$t \to \psi_{(i,k)}(t,j,1)$ is Riemann integrable for each $(i,k) \in J \times \{0,1\}$ where $\psi_{(i,k)}(t,j,1) = P(Y(t) = j, L(t) = 1, T_1 > t | Y_o = i, L_o = k)$

$$(3.11)$$

Using this result we will derive a recursive scheme allowing the computation of $p\star(j,1)$ without knowing the $\pi_{(i,k)}$.

Theorem 1. The limiting state probabilities

$$P\star(j,1) = \lim_{t \to \infty} P(Y(t) = j, L(t) = 1 | Y_o = i, L_o = k), \quad j \in J$$

$$(3.12)$$

exist independently of $(i,k) \in J \times \{0,1\}$. For $j \in J$

$$P\star(j,1) = \{[f_j + \sum_{i=1}^{j} g_i e_{i-1} a_{ij} / h_i + \sum_{i=0}^{j} e_i a_{ij}] \cdot p_o$$
$$+ \sum_{i=0}^{j-1} [g_{i+1} a_{i+1,j} / h_{i+1} + a_{ij}] \lambda_i P\star(i,1) \} / \delta_j \qquad (3.13)$$

with

$$a_{ij} = \int_0^{\infty} z_{ij}(t) [1 - B_i(t)] dt \qquad (3.14)$$

$$e_j = 1 - r_o - \sum_{i=0}^{j} d_i \qquad (3.15)$$

$$f_j = \sum_{i=0}^{j} d_i a_{ij} \qquad (3.16)$$

$$\delta_j = \int_0^{\infty} e^{-\lambda_j t} B_j(dt) \qquad (3.17)$$

$$p_o = \pi_{(0,0)} / D \qquad (3.18)$$

Proof

First we verify the conditions of the limit theorem cited above. As mentioned X is irreducible and positive recurrent. Aperiodicity of (X,T) follows by (A4), infinite lifetime of (X,T) is ensured by (A3), $D < \infty$ is implied by (A2).

We have for $(i,k),(j,l) \in J \times \{0,1\}$, $t \in \mathbb{R}_+$

$$
\psi_{(i,k)}(t,j,l) = \begin{cases}
z_{ij}(t)[1-B_i(t)] & i \geq 0, j \geq i, k = 1, l = 1 \\
1 - C_i(t) & i \geq 1, j = i, k = 0, l = 0 \\
w_j(t) & i = 0, j \geq 0, k = 0, l = 0 \\
0 & \text{else}
\end{cases} \tag{3.19}
$$

where the representation of the $w_j(t)$ depends on the assumptions concerning the vacation when no customers are present at the beginning
of the vacation (cf. chapter 2). If the state dependent Markovian
input process during vacation is independent of the vacation process,
we have

$$
w_j(t) = y_j(t)[1 - C_o(t)], \quad j \in J \tag{3.20}
$$

where the $y_j(t)$, $j \in J$, are the solution of the Kolmogorov differential equations for the birth process

$$
\begin{cases}
\dot{y}_o(t) = -\kappa_o y_o(t) \\
\dot{y}_j(t) = -\kappa_j y_j(t) + \kappa_{j-1} y_{j-1}(t), \quad j > 0 \\
y_j(0) = \delta_{oj} \,.
\end{cases} \tag{3.21}
$$

If on the other hand the arrival of the m-th customer, $m \in \mathbb{N}$, causes
the termination of the vacation, we have with (3.21)

$$
w_j(t) = \begin{cases}
y_j(t) & j < m \\
0 & j \geq m
\end{cases} \tag{3.22}
$$

(3.19), (3.20), (3.22) ensure Riemann integrability of ψ. Thus the
conditions of the limit theorem are fulfilled.

Defining $g_o = 0$ and $h_j = 0$ for $j \notin J$ (in case of finite J) we
have by (3.7) and (3.3) for $i \in J$

$$\pi_{(i,1)} = d_i \cdot \pi_{(0.0)} + g_i \pi_{(i,0)} + h_{i+1} \pi_{(i+1,0)} \tag{3.23}$$

$$\pi_{(i,0)} = r_0 \cdot \pi_{(0,0)} \delta_{0,i} + \sum_{k=0}^{i} q_{ki} \pi_{(k,1)} \tag{3.24}$$

Summing (3.23) we get gor $j \in J$ using $g_i + h_i = 1$

$$\sum_{i=0}^{j} \pi_{(i,1)} - \sum_{i=0}^{j} \pi_{(i,0)} = (\sum_{i=0}^{j} d_i - 1) \pi_{(0,0)} + h_{j+1} \pi_{(j+1,0)}$$

$$\tag{3.25}$$

By (3.9) and (3.19) we have for $j \in J$

$$D \cdot P\star(j,1) = \sum_{i=0}^{j} \pi_{(i,1)} \int_{0}^{\infty} \psi_{(i,1)}(t,j,1)dt$$

$$= \sum_{i=0}^{j} \pi_{(i,1)} \int_{0}^{\infty} z_{ij}(t) [1-B_i(t)] dt \tag{3.26}$$

$$= \sum_{i=0}^{j} \pi_{(i,1)} \int_{t=0}^{\infty} z_{ij}(t) \int_{u=t+}^{\infty} B_i(du)dt$$

$$= \sum_{i=0}^{j} \pi_{(i,1)} \int_{u=0}^{\infty} B_i(du) \int_{t=0}^{u} z_{ij}(t)dt \tag{3.27}$$

Using the relation (cf. Schellhaas, 1983)

$$\int_{0}^{u} z_{ij}(t)dt = \frac{1}{\lambda_j} [1 - \sum_{k=i}^{j} z_{ik}(u)], \quad i, j \in J, \ j \geqslant i, \ u \in \mathbf{R}_+$$

$$\tag{3.28}$$

we have by (3.27) and (3.24)

$$\lambda_j D \cdot P\star(j,1) = \sum_{i=0}^{j} \pi_{(i,1)} [1 - \sum_{k=i}^{j} \int_{0}^{\infty} z_{ik}(u)B_i(du)]$$

$$= \sum_{i=0}^{j} \pi_{(i,1)} - \sum_{i=0}^{j} \sum_{k=i}^{j} \pi_{(i,1)} q_{ik}$$

$$= \sum_{i=0}^{j} \pi_{(i,1)} - \sum_{k=0}^{j} \sum_{i=0}^{k} \pi_{(i,1)} q_{ik}$$

$$= \sum_{i=0}^{j} \pi_{(i,1)} - \sum_{k=0}^{j} [\pi_{(k,0)} - r_o \pi_{(0,0)} \delta_{0,k}]$$

$$= \sum_{i=0}^{j} \pi_{(i,1)} - \sum_{i=0}^{j} \pi_{(i,0)} + r_o \pi_{(0,0)} \qquad (3.29)$$

Using (3.25) in (3.29) gives

$$\lambda_j \, D \cdot P^\star(j,1) = h_{j+1} \pi_{(j+1,0)} + (r_o - 1 + \sum_{i=0}^{j} d_i) \pi_{(0,0)}$$

$$(3.30)$$

By (3.26), (3.14), (3.23), (3.15), (3.30) we have

$$D \cdot P^\star(j,1) = \sum_{i=0}^{j} \pi_{(i,1)} a_{ij} = \sum_{i=0}^{j} [d_i \pi_{(0,0)} + g_i \pi_{(i,0)}$$

$$+ h_{i+1} \pi_{(i+1,0)}] a_{ij}$$

$$= \sum_{i=0}^{j} d_i a_{ij} \pi_{(0,0)} + \sum_{i=1}^{j} g_i a_{ij} [\lambda_{i-1} \, D \cdot P^\star(i-1,1)$$

$$+ e_{i-1} \pi_{(0,0)}]/h_i + \sum_{i=0}^{j} a_{ij} [\lambda_i \, D \cdot P^\star(i,1)$$

$$+ e_i \pi_{(0,0)}] \qquad (3.31)$$

Observing that

$$a_{jj} = \int_o^\infty e^{-\lambda_j t} [1 - B_j(t)] \, dt = \frac{1}{\lambda_j} [1 - \delta_j] \qquad (3.32)$$

by (3.17) we can write

$$\delta_j \, D \, P^\star(j,1) = [f_j + \sum_{i=1}^{j} g_i e_{i-1} a_{ij}/h_i + \sum_{i=0}^{j} e_i a_{ij}] \pi_{(0,0)}$$

$$+ \sum_{i=1}^{j} g_i a_{ij} \lambda_{i-1} \, D \cdot P^\star(i-1,1)/h_i$$

$$+ \sum_{i=0}^{j-1} a_{ij} \lambda_i \, D \cdot P^\star(i,1)$$

by (3.16). This proves the theorem by (3.18). $\square$

<u>Theorem 2</u>. The limiting state probabilities

$$P\star(j,0) = \lim_{t\to\infty} P(Y(t) = j, L(t) = 0 \mid Y_o = i, L_o = k), \quad j \in J$$

$$(3.33)$$

exist independent of $(i,k) \in J \times \{0,1\}$. For $j \in J$

$$P\star(j,0) = \begin{cases} s_o \cdot p_o & \text{for } j = 0 \\ \\ (s_j + c_j e_{j-1}/h_j)p_o + c_j \lambda_{j-1} \, P\star(j-1,1)/h_j & j \geq 1 \end{cases}$$

$$(3.34)$$

with

$$s_j = \int_0^\infty w_j(t)dt \tag{3.35}$$

and c_j, e_j, p_o by (3.6), (3.15), (3.18) respectively, $w_j(t)$ by (3.19) and (3.20) or (3.22).

<u>Proof</u>

Defining $\eta(j) = 1 \ [=0]$ if $j > 0 \ [\,j=0]$ we have for $j \in J$ by (3.9), (3.19), (3.6)

$$\begin{aligned} D \cdot P\star(j,0) &= \pi_{(0,0)} \int_0^\infty \psi_{(0,0)}(t,j,0)dt \\ &\quad + \pi_{(j,0)} \int_0^\infty \psi_{(j,0)}(t,j,0)dt \cdot \eta(j) \\ &= \pi_{(0,0)} \int_0^\infty w_j(t)dt + \pi_{(j,0)} \, c_j \eta(j) \end{aligned} \tag{3.36}$$

Using (3.35) and (3.30), (3.15) we have

$$D \cdot P^\star(j,0) = \pi_{(0,0)} s_j + c_j \eta(j) \cdot [e_{j-1} \pi_{(0,0)}$$

$$+ \lambda_{j-1} D \cdot P^\star(j-1,1)] / h_j$$

This proves the theorem by (3.18). $\square$

Theorem 1 and theorem 2 imply the following numerically stable algorithm for the computation of $P^\star(j,1)$ and $P^\star(j,0)$, $j \in J$

Step 1 : fix $\hat{p}_o > 0$ arbitrarily
Step 2 : for $j = 0, 1, \ldots, j \in J$, compute recursively

$$p_j = \{[f_j + \sum_{i=1}^{j} g_i e_{i-1} a_{ij}/h_i + \sum_{i=0}^{j} e_i a_{ij}] \hat{p}_o$$

$$+ \sum_{i=0}^{j-1} [g_{i+1} a_{i+1,j}/h_{i+1} + a_{ij}] \lambda_i p_i \}/\delta_j$$

Step 3 : Put $q_o = s_o \hat{p}_o$, for $j = 1, 2, \ldots, j \in J$ compute recursively

$$q_j = [s_j + c_j e_{j-1}/h_j] \hat{p}_o + c_j \lambda_{j-1} p_{j-1}/h_j$$

Step 4 : Obtain $P^\star(j,1)$ and $P^\star(j,0)$, $j \in J$, by normalization

$$P^\star(j,1) = p_j/S$$

$$P^\star(j,0) = q_j/S$$

with

$$S = \sum_{j \in J} (p_j + q_j) .$$

In general the greater part of the work in the algorithm is generated by the determination of the a_{ij} according to (3.14). In special ca-ses the z_{ij} in (3.14) are known, for example in the case of $|J| = \infty$, $\lambda_j = \lambda$ for $j \in J$, where we have

$$a_{ij} = \int_0^\infty \frac{(\lambda t)^{j-i}}{(j-i)!} e^{-\lambda t} [1 - B_i(t)] dt \qquad (3.37)$$

Another example of explicitely known z_{ij} is the well known repairman problem, see Schellhass (1983).

Occasionally the interpretation of the a_{ij} may be helpful : a_{ij} is the expected amount of time during which j customers are in the waiting room in a service time given that at the start of this service i are present.

In the general case the Kolmogorov differential equations (3.5) have to be solved. This can be done recursively for k = i, i+1, ... by using an appropriate implicit numerical procedure. There is an important special class of service time distributions where the computation of the a_{ij} becomes elementary. This is the case for distributions of phase type (PH-distributions). For the definition of PH-distributions we refer to Neuts (1981). Let B_i be a PH-distribution with representation (α_i, T_i) where α_i is a non-negative row vector of dimension m_i and T_i is a $m_i \times m_i$ matrix. For conditions on T_i we refer to Neuts (1981, p. 44). Assume T_i is nonsingular and $\alpha_i 1_i = 1$ where 1_i is a m_i-dimensional column vector with all its components equal to one. Denoting by I the identity matrix of appropriate dimension we have

<u>Theorem 3</u>. Let B_i, $i \in J$, be a PH-distribution with representation (α_i, T_i). Then the a_{ij} in (3.14) are

$$a_{ij} = (\prod_{s=i}^{j-1} \lambda_s) \alpha_i (\prod_{r=i}^{j} (\lambda_r I - T_i)^{-1}) 1_i \qquad (3.38)$$

A proof of this theorem can be found in Schellhaas (1983). It is seen by (3.38) that the a_{ij} are given explicitely by matrix-vector operations, the z_{ij} may be unknown, integrations are no longer needed. A recursive computational scheme for the a_{ij} in (3.38) circumventing inverse matrices is given in Schellhaas (1983).

4. Examples

We demonstrate the usefulness of the model and the algorithm by several queueing models which are generalizations (with respect to state dependency) of well known models. In all these models the waiting room may be finite or infinite.

4.1. The state dependent M/G/1 queue

We consider the M/G/1 queue with state dependent Markovian input with rate λ_j when j customers are in the waiting room. The service is also state dependent with service time distribution function B_i when i customers are in the waiting room at the beginning of the service.

One easily verifies that this model may be considered as a special case of our general model by putting

$$C_i(t) = \begin{cases} 1, & i \geq 1, \ t \geq 0 \\ 1-e^{-\kappa_o t}, & i = 0, \ t \geq 0 \end{cases} \tag{4.1}$$

That is, as long as customers are present the vacation times are zero (with probability one) and a new service starts immediately as soon as a service is finished. If no customers are present at the end of a service the vacation time is the exponentially distributed waiting time until the arrival of the next customer.

The coefficients in our general model are

$$r_o = 0$$

$$d_j = \begin{cases} 1, & j = 0 \\ 0, & j \neq 0 \end{cases} \tag{4.2}$$

$$g_i = 0, \ i \geqslant 1$$

$$h_i = 1, \ i \geqslant 1$$

$$w_j(t) = \begin{cases} e^{-\kappa_o t}, & j = 0 \\ \\ 0, & j \neq 0 \end{cases}$$

Observing $s_o = 1/\kappa_o$, $s_j = c_j = 0$ for $j \geqslant 1$ we have by (3.13) and (3.34)

$$P^\star(j,1) = [a_{oj} \cdot p_o + \sum_{i=0}^{j-1} a_{ij}\lambda_i \, P^\star(i,1)]/\delta_j, \qquad j \in J$$

$$P^\star(j,0) = \begin{cases} p_o/\kappa_o, & j = 0 \\ \\ 0, & j \geqslant 1 \end{cases} \qquad\qquad (4.3)$$

This algorithm (4.3) is equivalent to an algorithm given by Tijms and Van Hoorn (1981) for the case $B_i = B$, $i \in J$. These authors used a regenerative analysis. Our semi-regenerative analysis has the advantage of giving an explicit representation of the coefficients a_{ij} in (3.14) which was not the case in the Tijms/Van Hoorn paper. By a semi-regenerative analysis the algorithm (4.3) was first derived by Schellhaas (1983). The well known repairman problem is contained as a special case, see Schellhaas (1983).

4.2. The state dependent M/G/1 queue under a T-policy

We consider the M/G/1 queue with state dependent Markovian input during service. The input rate is λ_j when j customers are in the waiting room. The service time distribution is B_i when i customers are in the waiting room at the beginning of the service. At the end of a busy period the server leaves the system for a stochastic length of time T with distribution function C_o. During this vacation customers arrive according to a Markovian input process with rate κ_j when j

customers are in the waiting room. After his vacation time T the server returns to the system and scans the queue to determine if customers are present. If customers are found a busy period starts, if no
customers are found the server leaves for another vacation and the
procedure of taking a vacation and scanning is repeated until customers are found. This model, in the absence of state dependency, was
studied by Levy and Yechiali (1975). For non-stochastic vacation
times Heyman (1977) provided an optimization setting for a cost model.
These investigations were continued by Schellhaas (1978).

One easily verifies that this model is a special case of our
general model. We have

$$C_i(t) = \begin{cases} 1, & i \geqslant 1, \ t \geqslant 0 \\ \\ C_o(t), & i = 0, \ t \geqslant 0 \end{cases} \tag{4.4}$$

$$r_o = \int_o^\infty e^{-\kappa_o t} C_o(dt)$$

$$d_i = \int_o^\infty y_{i+1}(t) C_o(dt), \qquad i \geqslant 0 \tag{4.5}$$

$$g_i = 0, \ i \geqslant 1$$

$$h_i = 1, \ i \geqslant 1$$

$$w_j(t) = y_j(t) [1 - C_o(t)], \qquad j \geqslant 0$$

and y_j, $j \in J$ is the solution of (3.21). We have

$$P^\star(j,1) = \{[f_j + \sum_{i=0}^{j} e_i \, a_{ij}] p_o + \sum_{i=0}^{j-1} a_{ij} \lambda_i \, P^\star(i,1)\}/\delta_j, \qquad j \in J$$

$$P^\star(j,0) = s_j p_o, \qquad j \in J \tag{4.6}$$

For the standard case $\kappa_j = \kappa$, $j \in J$, we have

$$y_j(t) = (\kappa t)^j \, e^{-\kappa t} / j!, \qquad j \in J.$$

4.3. The state dependent M/G/1 queue under a k-policy

We consider the M/G/1 queue with state dependent Markovian input during service. The input rate is λ_j when j customers are in the waiting room. The service time distribution is B_i when i customers are in the waiting room at the beginning of the service. At the end of a busy period the server is turned off. During this vacation customers arrive according to a Markovian input process with rate κ_j when j customers are in the waiting room. As soon as $k \geqslant 1$ customers are present the server is turned on. This model, in the absence of state dependency, was originally investigated by Heyman (1968), and was later on treated by several authors, for a corresponding cost analysis see Schellhaas (1979).

This model is a special case of our general model, where we have used k instead of m (cf. chapter 2) to be in conformity with the usual nomenclature. We have

$$C_i(t) = \begin{cases} 1 & i \geqslant 1, \quad t \geqslant 0 \\ 1 - \sum_{j=0}^{k-1} y_j(t), & i = 0, \quad t \geqslant 0 \end{cases}$$

$$r_o = 0$$

$$d_i = \begin{cases} 1 & i = k - 1 \\ 0 & i \neq k - 1 \end{cases}$$

$$g_i = 0, \quad i \geqslant 1$$

$$h_i = 1, \quad i \geqslant 1$$

$$w_j(t) = \begin{cases} y_j(t), & 0 \leqslant j < k \\ 0, & j \geqslant k \end{cases}$$

where y_j, $j \in J$, is the solution of (3.21). We have

$$P^\star(j,1) = \sum_{i=0}^{\min\{j,k-1\}} a_{ij} P_0 + \sum_{i=0}^{j-1} a_{ij} \lambda_i P^\star(i,1)\}/\delta_j, \quad j \in J$$

$$P^\star(j,0) = \begin{cases} s_j P_0 & \text{for } 0 \leqslant j < k \\ \\ 0 & j \geqslant k \end{cases}$$

4.4. The state dependent M/G/1 queue with repeated attempts

We consider a state dependent M/G/1 queue where any customer finding the server busy upon arrival will make repeated attempts to enter service. Customers in the waiting room try independently of each other to enter service after an exponentially distributed time with mean $1/\nu$. The input process is as follows. During a service customers arrive according to a state dependent Markovian input process with rate λ_j when j customers are in the waiting room. During the server's vacations customers arrive according to a state dependent Markovian input process with rate σ_j when j customers are in the waiting room. An arriving customer who finds the server busy joins the queue in the waiting room, an arriving customer who finds the server free is served immediately. The service time distribution is B_i when i customers are in the waiting room at the beginning of the service. After a service the server takes a vacation until either one of the customers in the waiting room or an arriving customer calls for service. This model is a special case of our general model. We have with $\tau_i = i\nu + \sigma_i$, $i \in J$,

$$C_i(t) = 1 - e^{-\tau_i t}, \quad i \in J, \ t \geqslant 0$$

$$r_0 = 0$$

$$d_i = \begin{cases} 1 & i = 0 \\ \\ 0 & i \neq 0 \end{cases}$$

$$g_i = \sigma_i/\tau_i, \quad i \geq 1$$

$$h_i = i\nu/\tau_i, \quad i \geq 1$$

$$w_j(t) = \begin{cases} e^{-\sigma_o t}, & j = 0 \\ 0, & j \neq 0 \end{cases}$$

Observing $e_j = 0$, $f_j = a_{oj}$, $j \in J$, $s_o = 1/\sigma_o$, $s_j = 0$ for $j \geq 1$ we have

$$P\star(j,1) = \{a_{oj}p_o + \sum_{i=0}^{j-1} [\sigma_{i+1} a_{i+1,j}/[(i+1)\nu] + a_{ij}] \lambda_i P\star(i,1)\}/\delta_j$$

$$P\star(j,0) = \begin{cases} \cdot p_o/\sigma_o & \text{for } j = 0 \\ \\ \lambda_{j-1} P\star(j-1,1)/(i\nu) & \text{for } j \geq 1 \end{cases}$$

Using a regenerative analysis de Kok (1982) derived the algorithm for the case $B_i = B$, $i \in J$. Again our general approach has the advantage of giving an explicit representation of the coefficients a_{ij}.

Using a regenerative analysis de Kok (1982) derived the algorithm

References

E. Çinlar (1975), Introduction to stochastic processes. Prentice Hall, Englewood Cliffs.

D.P. Heyman (1968), Optimal operating policies for M/G/1 queueing systems. Operations Research 16, 362–382.

D.P. Heyman (1977), The T-policy for the M/G/1 queue. Management Science 23, 775–778.

A.G. de Kok (1982), Computational methods for single server queueing systems with repeated attemps. Report, Interfaculteit der Actuariele Wetenschappen en Econometrie, Vrije Universiteit, Amsterdam.

Y. Levy, U. Yechiali (1975), Utilization of idle time in an M/G/1
    queueing system. Management Science 22, 202-211.

M.F. Neuts (1981), Matrix-geometric solutions in stochastic models.
    An algorithmic approach. The John Hopkins University Press.
    Baltimore.

M.F. Neuts (1984), Matrix-analiytic methods in queueing theory.
    European Journal of Operational Research 15, 2-12.

A.G. Pakes (1969), Some conditions for ergodicity and recurrence of
    Markov chains. Operations Research 17, 1058-1061.

M. Schäl (1970), Markov renewal processes with auxiliary path. Annals
    of Mathematical Statistics 41, 1604-1623.

H. Schellhaas (1978), On the T-policy for an M/G/1 queue. Operations
    Research Verfahren 29, 750-763.

H. Schellhaas (1983), Computation of the state probabilities in
    M/G/1 queues with state dependent input and state dependent
    service. OR Spektrum 5, 223-228.

H.C. Tijms, M.H. Van Hoorn (1981), Algorithms for the state probabili-
    ties and waiting times in single queueing systems with random and
    quasi-random input and phase type service times. OR Spektrum 2,
    145-152. Erratum OR-Spektrum 3, 1982, 244.

M.H. Van Hoorn (1983), Algorithms and approximations for queueing
    systems Doctoral thesis. Mathematisch Centrum, Amsterdam.

THE SUPERPOSITION OF TWO PH-RENEWAL PROCESSES

Marcel Neuts and Guy Latouche*

University of Delaware

Université Libre de Bruxelles*

1. Introduction

A large variety of stochastic models may be viewed as systems
of two interacting renewal processes. Unless at least one of these
is a Poisson process, the analytic properties of the model are difficult to study and are often intractable. It is clear that the
greater analytic tractability of such models as the M/G/1 and GI/M/1
queues is due to the memory-less property of the exponential distribution. This tractability is gained, however, at the price of a
severe distributional assumption. A useful compromise is found in
the introduction of the probability distributions of _phase type_ (PH-
distributions). For these, a formalism of matrix manipulations has
been developed which extends the elementary properties of the exponential distribution, yet is general and versatile enough to cover
the probability distributions needed in a large variety of modelling
problems.

The formalism of the PH-distributions often permits a thorough
matrix-analytic treatment of the stochastic models in which they
arise. The end results of this analysis are then frequently in a

form which permits an immediate numerical implementation. This major
utility of the PH-distributions has been demonstrated for a number of
queueing and reliability models in the references G. Latouche,(1982),
G. Latouche,(198),M.F. Neuts,(1981),M.F. Neuts (1981).

In the present paper, we examine the superposition of two PH-
renewal processes and we obtain a number of matrix formulas which are
useful in the solution of several queueing models. A few of these are
discussed as illustrations. We shall assume familiarity with the
basic properties of PH-distributions and PH-renewal processes as they
are discussed in chapter 2 of M.F. Neuts,(1981).

2. The superposition of two PH-renewal processes

We consider two independent PH-renewal processes, whose underlying probability distributions $F_i(.)$ have the irreductible represen-
tations $\{\underline{\alpha}(i), T(i)\}$ with $m(i)$ phases, $i = 1, 2$. The vectors $\underline{\alpha}(1)$
and $\underline{\alpha}(2)$ are assumed to be proper probability vectors, so that in
both processes the times between renewals are positive with probability one. The probability distribution $F_i(.)$ is given by

$$F_i(x) = 1 - \underline{\alpha}(i) \exp[T(i)x]\underline{e}, \quad \text{for } x \geqslant 0, \tag{1}$$

where $\underline{e}$ is a column vector of appropriate dimension with all its components equal to one. Its moments $\lambda'_j(i)$ are given by

$$\lambda'_j(i) = (-1)^j \, j! \, \underline{\alpha}(i) \, T^{-j}(i)\underline{e}, \quad \text{for } j \geqslant 1. \tag{2}$$

With the vectors $\underline{T}^0(i), i = 1, 2$, defined by $\underline{T}^0(i) = -T(i)\underline{e}$, the Markov processes with the irreductible generators

$$Q^*(i) = T(i) + \underline{T}^0(i) \alpha(i), \quad i = 1, 2, \tag{3}$$

and respectively $m(1)$ and $m(2)$ states play an important role in the

PH-renewal processes with the underlying distributions $F_1(.)$ and $F_2(.)$. This role is discussed in detail in M.F. Neuts, (1981). The vector $\underline{\pi}(i)$, given by

$$\underline{\pi}(i) = -\frac{1}{\lambda_1'(i)}\,\underline{\alpha}(i)\,T^{-1}(i),\tag{4}$$

is clearly the stationary probability vector of the generator $Q^{\star}(i)$, $i = 1, 2$. The stationary version of the renewal process with life time distribution $F_i(.)$ is obtained by choosing the initial state of the Markov process $Q^{\star}(i)$ according to the vector $\underline{\pi}(i)$, $i = 1, 2$.

We shall now construct a Markov renewal process which permits us to study the simultaneous evolution of the two independent PH-renewal processes and, in particular, their superposition. For notational convenience only, we shall limit our attention to their stationary versions. Other initial conditions may be treated in the same manner, but require somewhat more involved notation.

In what follows, we shall make frequent use of the matrix $T(1,2)$, which is the <u>Kronecker sum</u> (in R. Bellman, 1960) of the matrices $T(1)$ and $T(2)$, i.e.

$$T(1,2) = T(1) \oplus T(2) = T(1) \otimes I + I \otimes T(2).\tag{5}$$

The matrix $T(1,2)$ of order $m(1)m(2)$ is nonsingular, has negative diagonal elements, nonnegative off-diagonal elements and its vector of row sums is given by

$$T(1,2)\,(\underline{e} \otimes \underline{e}) = -\underline{T}^{0}(1) \otimes \underline{e} - \underline{e} \otimes \underline{T}^{0}(2).\tag{6}$$

In all Kronecker products in this paper, the identity matrix I and the vector $\underline{e}$ are of dimension $m(1)$, resp. $m(2)$, when they occur on the left, resp. the right, of the symbol $\otimes$.

It is well-known that for all real x,

$$\exp[\,T(1,2)x\,] \;=\; \exp[\,T(1)x\,] \;\oplus\; \exp[\,T(2)x\,]\,. \tag{7}$$

The exponential matrix $\exp[\,T(1,2)x\,]$ has, for $x \geqslant 0$, the following probabilistic significance. Consider the product of the (independent) Markov processes with generators $Q^{\star}(1)$ and $Q^{\star}(2)$. The element $\{\exp[\,T(1,2)x\,]\}_{j'j;h'h}$, $1 \leqslant j'$, $h' \leqslant m(1)$, $1 \leqslant j$, $h \leqslant m(2)$, is then the conditional probability that in the interval $(0,x)$ no renewals occur in either of the two PH-renewal processes and that the product process is in the state (h', h) at time x, given that it was in the state (j', j) at time 0. This interpretation is an immediate consequence of formula (7)

Next, we consider the bivariate sequence of random variables $\{(J_n, X_n),\ n \geqslant 0\}$ with $X_0 = 0$. The random variables J_n, $n \geqslant 0$, take values in the set $\{1, \ldots, m(2), 1', \ldots, m'(1)\} = E$. The nonnegative random variable X_1 is the time until the first renewal in either of the two PH-renewal processes. The nonnegative random variables X_n, $n \geqslant 2$, are the times between the n-th and the (n + 1)st renewals occuring in either process. The label J_n, associated with the epoch $S_n = X_1 + \ldots + X_n$, $n \geqslant 1$, takes the value j, $1 \leqslant j \leqslant m(2)$, if the renewal at time S_n occurs in the first process and the Markov process $Q^{\star}(2)$ is in the phase j at that time. Correspondingly, J_n takes the value j', $1' \leqslant j' \leqslant m'(1)$, if the renewal at time S_n occurs in the second process and the Markov process $Q^{\star}(1)$ is in the phase j' at that time.

The precise definition of the probabilities $P\{J_0 = h_0\}$ and $P\{J_1 = h_1,\ X_1 \leqslant x \mid J_0 = h_0\}$, $x \geqslant 0$, $h_0 \in E$, $h_1 \in E$, is given below, but it is already clear that the sequence $\{(J_n, X_n),\ n \geqslant 0\}$ will be a Markov renewal sequence with the state space $E \times [0, \infty)$.

Its transition probability matrix $C(x)$ of order $m(2) + m(1)$ with elements $P\{J_n = h_n, X_n \leq x \mid J_{n-1} = h_{n-1}\}$, $x \geq 0$, $h_{n-1} \in E$, $h_n \in E$, is given for $n \geq 2$ by

$$C(x) = \begin{bmatrix} \underline{\alpha}(1) \otimes I \\ \\ I \otimes \underline{\alpha}(2) \end{bmatrix} \int_o^x \exp[\,T(1,2)u]\,du\;[\,\underline{T}^0(1) \otimes I \quad I \otimes \underline{T}^0(2)\,]\,,$$

$$(8)$$

The following theorem summarizes the most important properties of the matrix $C(x)$.

<u>Theorem 1.</u>

The matrix $C(x)$ is an irreducible, stochastic semi-Markov matrix of order $m(2) + m(1)$. Its Laplace-Stieltjes transform is given by

$$C^\star(s) = \begin{bmatrix} \underline{\alpha}(1) \otimes I \\ \\ I \otimes \underline{\alpha}(2) \end{bmatrix} [\,sI \otimes I - T(1,2)]^{-1}[\,\underline{T}^0(1) \otimes I \quad I \otimes \underline{T}^0(2)\,]\,.$$

$$(9)$$

The invariant probability vector of the stochastic matrix $C = C(\infty)$ is given by the partitioned vector

$$\underline{u} = [\,\underline{u}(1),\underline{u}(2)\,] = [\,\lambda_1'(1) + \lambda_1'(2)]^{-1}[\,\lambda_1'(2)\;\underline{\pi}(2),\lambda_1'(1)\underline{\pi}(1)\,]\,.$$

$$(10)$$

The column vector of row sum means of $C(x)$ is given by

$$\widetilde{\underline{u}} = \begin{bmatrix} \widetilde{\underline{u}}(1) \\ \\ \widetilde{\underline{u}}(2) \end{bmatrix} = - \begin{bmatrix} \underline{\alpha}(1) \otimes I \\ \\ I \otimes \underline{\alpha}(2) \end{bmatrix} T^{-1}(1,2)\,(\underline{e} \otimes \underline{e})\,, \qquad (11)$$

and the fundamental mean E of $C(x)$ is given by

$$E = \underline{u}(1)\ \tilde{\underline{u}}(1) + \underline{u}(2)\ \tilde{\underline{u}}(2) = \left[\frac{1}{\lambda_1'(1)} + \frac{1}{\lambda_1'(2)}\right]^{-1} \tag{12}$$

Proof

By virtue of (8) and since $\exp[\,T(1,2)u\,]$, $u \geqslant 0$, is a nonnegative matrix, the elements of $C(x)$ are nonnegative, nondecreasing and continuous. To verify that $C = C(\infty)$ is stochastic, we observe that

$$[\,\underline{T}^0(1)\ \& \ I \quad I\ \& \ \underline{T}^0(2)\,]\begin{bmatrix} \underline{e} \\[6pt] \underline{e} \end{bmatrix} = \underline{T}^0(1)\ \& \ \underline{e} + \underline{e}\ \& \ \underline{T}^0(2) \tag{13}$$

$$= -[\,T(1)\ \& \ I + I\ \& \ T(2)\,]\cdot(\underline{e}\ \& \ \underline{e}) = -T(1,2)(\underline{e}\ \& \ \underline{e})$$

It is clear from (13) that

$$C\begin{bmatrix} \underline{e} \\[6pt] \underline{e} \end{bmatrix} = \begin{bmatrix} \underline{\alpha}(1)\ \& \ I \\[10pt] I\ \& \ \underline{\alpha}(2) \end{bmatrix} (\underline{e}\ \& \ \underline{e}) = \begin{bmatrix} \underline{e} \\[6pt] \underline{e} \end{bmatrix}.$$

The irreducibility of C is established by considering the paths of the (irreducible) product process of the Markov processes with generators $Q^*(1)$ and $Q^*(2)$. Since the representations of $F_1(.)$ and $F_2(.)$ are irreducible, these Markov processes are irreducible. It is clear that in the embedded Markov chain C, every pair of states communicates.

The formulas (10), (11) and (12) are proved by standard manipulations with Kronecker products. We verify from the definition of Kronecker product that

$$[\,\underline{u}(1),\ \underline{u}(2)\,]\begin{bmatrix} \underline{\alpha}(1)\ \& \ I \\[14pt] I\ \& \ \underline{\alpha}(2) \end{bmatrix} = \underline{\alpha}(1)\ \& \ \underline{u}(1) + \underline{u}(2)\ \& \ \underline{\alpha}(2)$$

$$= [\,\lambda_1'(1) + \lambda_1'(2)\,]^{-1}[\,\underline{\alpha}(1)\ \& \ \lambda_1'(2)\ \underline{\pi}(2) + \lambda_1'(1)\ \underline{\pi}(1)\ \& \ \underline{\alpha}(2)\,]$$

$$= -[\,\lambda_1'(1) + \lambda_1'(2)\,]^{-1}[\,\underline{\alpha}(1)\ \& \ \underline{\alpha}(2)\ T^{-1}(2) + \underline{\alpha}(1)\ T^{-1}(1)\ \& \ \underline{\alpha}(2)\,].$$

Next, we note that

$$[\underline{\alpha}(1)\ T^{-1}(1)\ \otimes\ \underline{\alpha}(2)\ T^{-1}(2)][T(1)\ \otimes\ I\ +\ I\ \otimes\ T(2)]$$

$$=\ \underline{\alpha}(1)\ \otimes\ \underline{\alpha}(2)\ T^{-1}(2)\ +\ \underline{\alpha}(1)\ T^{-1}(1)\ \otimes\ \underline{\alpha}(2),$$

so that

$$[\underline{u}(1),\ \underline{u}(2)]\begin{bmatrix}\underline{\alpha}(1)\ \otimes\ I\\ \\ I\ \otimes\ \alpha(2)\end{bmatrix}[-T(1,2)]^{-1}[\underline{T}^{0}(1)\ \otimes\ I\quad I\ \otimes\ \underline{T}^{0}(2)]$$

$$=\ -[\lambda_1'(1)\ +\ \lambda_1'(2)]^{-1}$$

$$\cdot\ [\underline{\alpha}(1)\ T^{-1}(1)\ \otimes\ \underline{\alpha}(2)\ T^{-1}(2)][\underline{T}^{0}(1)\ \otimes\ I\quad I\ \otimes\ \underline{T}^{0}(2)]$$

$$=\ [\lambda_1'(1)\ +\ \lambda_1'(2)]^{-1}[\lambda_1'(2)\ \underline{\pi}(2),\lambda_1'(1)\ \underline{\pi}(1)]\ =\ [\underline{u}(1,),\underline{u}(2)].$$

The vector $[\underline{u}(1),\ \underline{u}(2)]$ is obviously a probability vector.

The vector of row sum means of $C(.)$ is found by differentiating with respect to s in (9) and is given by

$$\begin{bmatrix}\underline{\alpha}(1)\ \otimes\ I\\ \\ I\ \otimes\ \underline{\alpha}(2)\end{bmatrix}T^{-2}(1,2)\ [\underline{T}^{0}(1)\ \otimes\ I\quad I\ \otimes\ \underline{T}^{0}(2)]\begin{bmatrix}\underline{e}\\ \\ \underline{e}\end{bmatrix}.$$

By virtue of formula (13), this may be simplified to (11). In order to obtain (12), we form the inner product of the vectors $[\underline{u}(1),\underline{u}(2)]$ and $[\underline{u}'(1),\ \underline{u}'(2)]'$. This leads to

$$E\ =\ -[\lambda_1'(1)\ +\ \lambda_1'(2)]^{-1}[\lambda_1'(2)\ \underline{\alpha}(1)\ \otimes\ \underline{\pi}(2)\ +\ \lambda_1'(1)\ \underline{\pi}(1)\ \otimes\ \underline{\alpha}(2)]$$

$$\cdot\ T^{-1}(1,2)(\underline{e}\ \otimes\ \underline{e})$$

$$=\ [\lambda_1'(1)\ +\ \lambda_1'(2)]^{-1}[\underline{\alpha}(1)\ T^{-1}(1)\ \otimes\ \underline{\alpha}(2)\ T^{-1}(2)]\ (\underline{e}\ \otimes\ \underline{e})$$

$$=\ [\lambda_1'(1)\ +\ \lambda_1'(2)]^{-1}\ \lambda_1'(1)\ \lambda_1'(2).$$

Remark

It is well-known that E^{-1} may be interpreted as the rate at which transitions occur in the stationary version of the Markov renewal process. As is to be anticipated, E^{-1} is here the sum of the stationary arrival rates $1/\lambda_1'(1)$ and $1/\lambda_1'(2)$ of the component renewal processes.

We now have the ingredients to obtain the stationary version of the Markov renewal process according to a standard construction discussed in Pyke, 1961. Let h and h' denote generic elements of the set E. The probabilities $P\{J_0 = h\}$ and $P\{J_1 = h', X_1 \leqslant x \mid J_0 = h\}$ are defined by

$$P\{J_0 = h\} = E^{-1} u_h \tilde{u}_h \, ,$$

$$P\{J_1 = h', X_1 \leqslant x \mid J_0 = h\} = \tilde{C}_{hh'}(x) \tag{14}$$

$$= \frac{1}{\tilde{u}_h} \int_0^x [C_{hh'} - C_{hh'}(u)] \, du.$$

Theorem 2

The sequence of intervals between the successive epochs in the superposition of two stationary independent PH-renewal processes is stochastically equivalent to the sequence $\{X_n, n \geqslant 1\}$ of the successive sojourn intervals in the stationary Markov renewal sequence $\{(J_n, X_n)\}$.

The joint distribution of $X_1, \ldots, X_n, n \geqslant 1$, is given by

$$\hat{F}(x_1, \ldots, x_n)$$

$$= E^{-1} \underline{u} \int_0^{x_1} [C - C(u)] \, du \, C(x_2) \, \ldots \, C(x_n) \underline{e} \tag{15}$$

$$= [\underline{\pi}(1) \otimes \underline{\pi}(2)] \int_o^{x_1} \exp[\,T(1,2)u_1\,]\,du_1[\underline{T}^0(1)\underline{\alpha}(1) \oplus \underline{T}^0(2)\alpha(2)]$$

$$\cdot \int_o^{x_2} \exp[\,T(1,2)u_2\,]\,du_2[\underline{T}^0(1)\underline{\alpha}(1) \oplus \underline{T}^0(2)\underline{\alpha}(2)] \cdots$$

$$\cdot \int_o^{x_n} \exp[\,T(1,2)u_n\,]\,du_n[\underline{T}^0(1) \otimes \underline{e} + \underline{e} \otimes \underline{T}^0(2)],$$

for $x_1 \geqslant 0$, ..., $x_n \geqslant 0$, and has the joint Laplace Stieltjes transform

$$\hat{\varphi}\,(s_1, \ldots, s_n)$$

$$= [\underline{\pi}(1) \otimes \underline{\pi}(2)][\,s_1 \quad I \otimes I - T(1,2)]^{-1}[\underline{T}^0(1)\,\alpha(1) \oplus \underline{T}^0(2)\,\alpha(2)]$$

$$\cdot [\,s_2\; I \otimes I - T(1,2)]^{-1}[\underline{T}^0(1)\,\underline{\alpha}(1) \oplus \underline{T}^0(2)\,\underline{\alpha}(2)] \cdots \qquad (16)$$

$$\cdot [\,s_n\; I \otimes I - T(1,2)]^{-1}[\underline{T}^0(1) \otimes \underline{e} + \underline{e} \otimes \underline{T}^0(2)].$$

<u>Proof</u>

The first statement is evident from the construction of the
Markov renewal process. The first expression for $\hat{F}(x_1, \ldots, x_n)$ is
obvious by standard Markov renewal calculations. In order to obtain
the second expression, we note that (8) implies that

$$C - C(x) = \begin{bmatrix} \underline{\alpha}(1) \otimes I \\ \\ I \otimes \underline{\alpha}(2) \end{bmatrix}[-T(1,2)]^{-1}\exp[\,T(1,2)x\,][\underline{T}^0(1) \otimes I \quad I \otimes \underline{T}^0(2)],$$

and that by the same calculations as in the proof of Theorem 1, we
have that

$$E^{-1}\underline{u}\begin{bmatrix} \underline{\alpha}(1) \otimes I \\ \\ I \otimes \underline{\alpha}(2) \end{bmatrix}[-T(1,2)]^{-1} = \underline{\pi}(1) \otimes \underline{\pi}(2).$$

It is also clear that

$$[\underline{T}^0(1) \otimes I \quad I \otimes \underline{T}^0(2)] \begin{bmatrix} \underline{\alpha}(1) \otimes I \\ \\ I \otimes \underline{\alpha}(2) \end{bmatrix}$$

$$= \underline{T}^0(1)\, \underline{\alpha}(1) \otimes I + I \otimes \underline{T}^0(2)\, \underline{\alpha}(2)$$

$$= \underline{T}^0(1)\, \underline{\alpha}(1) \oplus \underline{T}^0(2)\, \underline{\alpha}(2),$$

so that the second form of $\hat{F}(x_1, \ldots, x_n)$ is obtained by simple substitutions. The expression $\varphi(s_1, \ldots, s_n)$ is obtained routinely.

From (16), we see that the marginal distribution of the interval X_n, $n \geqslant 1$, is the PH-distribution with the transform

$$[\underline{\pi}(1) \otimes \underline{\pi}(2)]\, D^{n-1}[\, sI \otimes I - T(1,2)]^{-1}[\underline{T}^0(1) \otimes \underline{e} + \underline{e} \otimes \underline{T}^0(2)]$$

$$(17)$$

where D is the stochastic matrix

$$D = [-T(1,2)]^{-1}[\underline{T}^0(1)\, \underline{\alpha}(1) \oplus \underline{T}^0(2)\, \underline{\alpha}(2)].$$

We notice that the representations of these distributions for successive values of n are of the form $\{\underline{\beta}(n),\, T(1,2)\}$ where

$$\underline{\beta}(n) = [\underline{\pi}(1) \otimes \underline{\pi}(2)]\, D^{n-1}, \qquad n \geqslant 1.$$

These representations involve $m(1)m(2)$ phases, but they readily yield computationally useful formulas for moments. We have

$$E[X_n^k] = (-1)^k\, k!\, \underline{\beta}(n)\, T^{-k}(1,2)(\underline{e} \otimes \underline{e}),$$

for $k \geqslant 0$, $n \geqslant 1$.

From the first form of $\hat{F}(x_1, \ldots, x_n)$ in (15), we may also derive the form

$$\tilde{F}_n(x) = E^{-1}\,\underline{u}\,\tilde{C}(x)\underline{e}, \qquad n = 1,$$

$$= E^{-1}\,\underline{u}\,\tilde{C}\,C^{n-2}\,C(x)\,\underline{e}, \qquad n \geqslant 2, \tag{18}$$

for the marginal distribution of X_n. The matrices $\tilde{C}(x)$ and $\tilde{C} = \tilde{C}(\infty)$ are then given by

$$\tilde{C}(x) = \int_o^x [C - C(u)]\,du$$

$$= \begin{bmatrix} \underline{\alpha}(1) \otimes I \\[12pt] I \otimes \underline{\alpha}(2) \end{bmatrix} T^{-2}(1,2)\{I \otimes I - \exp[\,T(1,2)x]\,\}[\,\underline{T}^0(1) \otimes I \quad I \otimes \underline{T}^0(2)]$$

and

$$\tilde{C} = \begin{bmatrix} \underline{\alpha}(1) \otimes I \\[12pt] I \otimes \underline{\alpha}(2) \end{bmatrix} T^{-2}(1,2)[\,\underline{T}^0(1) \otimes I \quad I \otimes \underline{T}^0(2)]\,.$$

In the examination of the correlation between the interval X_j, $j \geqslant 1$, and X_{j+r}, $r \geqslant 1$, of the superposition, one makes use of the following expression for the cross-moment $E[\,X_j X_{j+r}]$.

$$E[\,X_j X_{j+r}] = [\,\underline{\pi}(1) \otimes \underline{\pi}(2)]\,D^{j-1}\,T^{-1}(1,2)\,D^r\,T^{-1}(1,2)\,(\underline{e} \otimes \underline{e})\,. \tag{19}$$

This formula is obtained from (16) by routine differentiations.

3. The Perron-Frobenius eigenvalue of $C^*(s)$

For $s \geqslant 0$, the matrix $C^*(s)$ is an irreducible, nonnegative matrix. Its Perron-Frobenius eigenvalue $\chi(s)$ is therefore well-defined and we may define corresponding left and right eigenvectors $\underline{u}^*(s)$ and

$\underline{v}^*(s)$ which are <u>positive</u> vectors having analytic components and satisfying the normalizations

$$\underline{u}^*(s)\ \underline{e} = 1, \qquad \underline{u}^*(s)\ \underline{v}^*(s) = 1 ,$$

for all $s \geqslant 0$. We then have that $\lim_{s\to 0+} \underline{u}^*(s) = \underline{u}$, and $\lim_{s\to 0+} \underline{v}^*(s) = \underline{e}$.

The eigenvalue $\chi(s)$ plays an important role in the analysis of models involving the Markov renewal process with transition probability matrix $C(x)$, as discussed e.g. in M.F. Neuts (1977) and M.F.Neuts (november 1982).By applying general properties of the Perron eigenvalue of the transform of a semi-Markov matrix, we obtain that $\log \chi(s)$ is a <u>concave</u> function of s for s $\geqslant 0$, (J.F.C. Kingman , 1961) and that $\chi'(0+) = -E$, where E is the fundamental mean of $C(x)$. As shown in Neuts (1977), the triples $\{\underline{u}^{*(n)}(0+),\ \underline{v}^{*(n)}(0+),\ \chi^{(n)}(0+\}$, $n \geqslant 1$, of the n-th derivatives at 0+ of $\underline{u}^*(s)$ and $\underline{v}^*(s)$ and $\chi(s)$ may be evaluated by means of matrix recursion formulas.

It is rarely possible to give explicit analytic expressions for $\chi(s)$, $\underline{u}^*(s)$ and $\underline{v}^*(s)$. The following two theorems give particular cases where this is indeed possible, but the derivations suggest that it may be difficult to obtain the corresponding expressions for the general form of $C^*(s)$, given in formula (9).

<u>Theorem 3</u>

The Perron-Frobenius eigenvalue $\chi(s)$ for the matrix $C^*(s)$, $s \geqslant 0$, for the superposition of **two** PH-distributions with the <u>same</u> distribution $F(.)$ with representation $(\underline{\alpha},T)$ is given by

$$\chi(s) = f^*\left(\frac{s}{2}\right) , \tag{20}$$

where $f^*(s) = \underline{\alpha}(sI-T)^{-1}\underline{T}^0$, is the Laplace-Stieltjes transform of $F(.)$.

The corresponding normalized eigenvectors $\underline{u}^{\star}(s)$ and $\underline{v}^{\star}(s)$ are of the forms

$$\underline{u}^{\star}(s) = [\underline{u}(s),\underline{u}(s)], \qquad \underline{v}^{\star}(s) = \begin{bmatrix} \underline{v}(s) \\ \\ \underline{v}(s) \end{bmatrix} ,$$

where

$$\underline{u}(s) = \frac{1}{4} s [1 - \chi(s)]^{-1} \underline{\alpha}(\frac{s}{2} I - T)^{-1} , \tag{21}$$

and

$$\underline{v}(s) = 2s^{-1} [1 - \chi(s)] \underline{\alpha} (\frac{s}{2} I - T)^{-2} \underline{T}^{0}(\frac{s}{2} I - T)^{-1}\underline{T}^{0} \tag{22}$$

Proof

By symmetry, we anticipate that $\underline{u}^{\star}(s)$ is of the form $[\underline{u}(s),\underline{u}(s)]$ and we note that

$$[\underline{u}(s),\underline{u}(s)] \begin{bmatrix} \underline{\alpha} \otimes I \\ \\ I \otimes \underline{\alpha} \end{bmatrix} = \underline{\alpha} \otimes \underline{u}(s) + \underline{u}(s) \otimes \underline{\alpha}.$$

Next, we attempt to find a vector $\underline{\varphi}(s)$, such that

$$(\underline{\alpha} \otimes \underline{u}(s) + \underline{u}(s) \otimes \underline{\alpha})[sI \otimes I - T \otimes I - I \otimes T]^{-1} = \underline{\varphi}(s) \otimes \underline{\varphi}(s),$$

or equivalently

$$\underline{\alpha} \otimes \underline{u}(s) + \underline{u}(s) \otimes \underline{\alpha} = s\,\underline{\varphi}(s) \otimes \underline{\varphi}(s) - \underline{\varphi}(s) T \otimes \underline{\varphi}(s)$$
$$- \underline{\varphi}(s) \otimes \underline{\varphi}(s) T$$
$$= \underline{\varphi}(s) [\frac{s}{2} I - T] \otimes \underline{\varphi}(s) + \underline{\varphi}(s) \otimes \underline{\varphi}(s)[\frac{s}{2} I-T] .$$

This equality is clearly satisfied for

$$\underline{\varphi}(s) = c\,\underline{\alpha}(\frac{s}{2} I-T)^{-1} ,$$

$$\underline{u}(s) = c^2 \, \underline{\alpha}(\tfrac{s}{2} I - T)^{-1} = c \, \underline{\varphi}(s),$$

where c is a constant still to be determined.

In order to satisfy the equation $\underline{u}^{\star}(s)C^{\star}(s) = \chi(s)\underline{u}^{\star}(s)$, it suffices that

$$(\underline{\varphi}(s) \otimes \varphi(s)) \begin{bmatrix} \underline{T}^0 \otimes I \\[2ex] I \otimes \underline{T}^0 \end{bmatrix} = c \, \chi(s) \begin{bmatrix} \underline{\varphi}(s) \\[2ex] \varphi(s) \end{bmatrix},$$

but the left hand side simplifies to

$$[\underline{\varphi}(s) \, \underline{T}^0] \begin{bmatrix} \underline{\varphi}(s) \\[2ex] \varphi(s) \end{bmatrix} ,$$

so that

$$\underline{\varphi}(s) \, \underline{T}^0 = c \, f^{\star}(\tfrac{s}{2}) = c \, \chi(s) ,$$

so that $\chi(s) = f^{\star}(\tfrac{s}{2})$, provided $c > 0$.

The normalizing equation $\underline{u}^{\star}(s) \, \underline{e} = 1$, leads to

$$\frac{1}{2} = \underline{u}(s)\underline{e} = c^2 \, \underline{\alpha}(\tfrac{s}{2} I - T)^{-1} \, \underline{e} = c^2 \, \frac{1 - f^{\star}(\tfrac{s}{2})}{\tfrac{s}{2}} ,$$

so that $c^2 = \tfrac{1}{4} s \, [1 - \chi(s)]^{-1} $.

As shown in Corollary 2.2.2. of M.F. Neuts (1981) the vector $\underline{u}(s)$ is positive and $\underline{u}^{\star}(s)$ has been constructed to be a left eigenvector of $C^{\star}(s)$ corresponding to $\chi(s)$. This implies from F.R. Grantmacher (1959) that $\chi(s)$ is the Perron eigenvalue of $C^{\star}(s)$.

The same construction for the right eigenvector $\underline{v}^{\star}(s)$ leads to

$$\underline{v}(s) = d^2 \left(\frac{s}{2} I - T\right)^{-1} \underline{T}^0 ,$$

and the constant d^2 is determined by the requirement $2 \underline{u}(s) \underline{v}(s) = 1$. This readily leads to (22).

<u>Remark</u>

Theorem 3 may be extended to the superposition of r independent PH-renewal processes with the same underlying distribution of representation (α, T). Even though the analogue of $C^{\star}(s)$ is a matrix of order $r\, m^{r-1}$, where m is the number of phases in the representation (α, T), the symmetry of the Kronecker sums leads, by the same calculations, to the result that

$$\chi(s) = f^{\star}\left(\frac{s}{r}\right). \qquad r \geq 2.$$

The left eigenvector $\underline{u}^{\star}(s)$ consists of r copies of the $(r-1)$-fold Kronecker product

$$\underline{u}(s) \otimes \underline{u}(s) \otimes \ldots \otimes \underline{u}(s),$$

where

$$\underline{u}(s) = \left\{ \frac{s}{r^2}[1-\chi(s)]^{-1} \right\}^{\frac{1}{r-1}} \underline{\alpha}\left(\frac{s}{r} I - T\right)^{-1}.$$

Similarly, the right eigenvectors $\underline{v}^{\star}(s)$ consists of r copies of the $(r-1)$-fold Kronecker product

$$\underline{v}(s) \otimes \underline{v}(s) \otimes \ldots \otimes \underline{v}(s),$$

where

$$\underline{v}(s) = \left[\frac{r}{s}[1-\chi(s)]\right]^{\frac{1}{r-1}} \left[\underline{\alpha}\left(\frac{s}{r} I - T\right)^{-2} \underline{T}^0\right]^{-1} \left(\frac{s}{r} I - T\right)^{-1} \underline{T}^0 .$$

An application of this result is discussed in M.F. Neuts, 1982.

The next theorem deals with the case where one of the two PH-renewal processes is a Poisson process. We obtain this case by setting $\underline{\alpha}(2) = 1$, $T(2) = -\lambda$, $\underline{T}^0(2) = \lambda$. The probability distribution underlying the first renewal process has the irreducible representation $(\underline{\alpha}, T)$, and its Laplace-Stieltjes transform is denoted by $f^*(s)$.

The matrix $C^*(s)$ is now given by

$$C^*(s) = \begin{bmatrix} \underline{\alpha} \\ I \end{bmatrix} [(s + \lambda) I - T]^{-1} [\underline{T}^0 \ I]$$

$$= \begin{bmatrix} \underline{\alpha} \, M(s) \, \underline{T}^0 & \lambda \, \underline{\alpha} \, M(s) \\ \\ M(s) \, \underline{T}^0 & \lambda \, M(s) \end{bmatrix} , \tag{23}$$

where $M(s) = (sI + \lambda I - T)^{-1}$. We shall partition the vectors $\underline{u}^*(s)$ and $\underline{v}^*(s)$ as

$$\underline{u}^*(s) = [u_0(s), \underline{u}(s)] , \qquad \underline{v}^*(s) = \begin{bmatrix} v_0(s) \\ \\ \underline{v}(s) \end{bmatrix} .$$

Theorem 4

The Perron-Frobenius eigenvalue $\chi(s)$ of the matrix $C^*(s)$, $s \geqslant 0$, in (23) is the unique positive root of the equation

$$Z = f^*(s + \lambda - \lambda Z^{-1}), \tag{24}$$

and the corresponding eigenvectors $\underline{u}^*(s)$ and $\underline{v}^*(s)$ are given by

$$u_0(s) = 1 - \frac{\lambda}{s} \frac{1 - \chi(s)}{\chi(s)} ,$$

$$\underline{u}(s) = \frac{\lambda}{\chi(s)} u_0(s) \, \underline{\alpha} \, [(s + \lambda - \frac{\lambda}{\chi(s)})I - T]^{-1} , \tag{25}$$

$$v_0(s) = u_0^{-1}(s) \ \{1 + \chi^{-2}(s) \ \underline{\alpha} [(s + \lambda - \tfrac{\lambda}{\chi(s)}) I - T]^{-2} \ \underline{T}^0\}^{-1},$$

$$\underline{v}(s) = v_0(s) \ \chi^{-1}(s) \ [(s + \lambda - \tfrac{\lambda}{\chi(s)}) I - T]^{-1} \ \underline{T}^0.$$

<u>Proof</u>

It follows from (23) and the definitions that

$$u_0(s) \ \underline{\alpha} \ M(s) \ \underline{T}^0 + \underline{u}(s) \ M(s) \ \underline{T}^0 = \chi(s) \ u_0(s),$$

$$\lambda \ u_0(s) \ \underline{\alpha} \ M(s) + \lambda \ \underline{u}(s) \ M(s) = \chi(s) \ \underline{u}(s),$$

and these equations readily imply that

$$\lambda \ \underline{u}(s) \ \underline{T}^0 = u_0(s).$$

The second equation leads to

$$\underline{u}(s) = \lambda \ u_0(s) \ \chi^{-1}(s) \ \underline{\alpha} \ [(s + \lambda - \tfrac{\lambda}{\chi(s)}) I - T]^{-1}, \qquad (26)$$

provided the inverse exists. Postmultiplying by $\underline{T}^0$, we obtain

$$u_0(s) = u_0(s) \ \chi^{-1}(s) \ \underline{\alpha} \ [(s + \lambda - \tfrac{\lambda}{\chi(s)}) I - T]^{-1} \ \underline{T}^0,$$

so that

$$\chi(s) = f^{\star}[s + \lambda - \tfrac{\lambda}{\chi(s)}] \ .$$

Setting $s + \lambda - \lambda z^{-1} = \xi$, in equation (24), we obtain

$$f^{\star}(\xi) = \frac{\lambda}{s+\lambda-\xi} \ . \qquad (27)$$

An examination of the graphs of $f^{\star}(\xi)$ and $\lambda(s+\lambda-\xi)^{-1}$, shows that the equation (27) has a unique positive root ξ_0 which satisfies $0 < \xi_0 < s$ for $s > 0$. Moreover $\chi(s) = \lambda[s+\lambda-\xi_0]^{-1}$, satisfies

$$1 > \chi(s) > \max\left[\frac{\lambda}{\lambda+s}, f^*(s)\right], \quad \text{for} \quad s > 0. \tag{28}$$

It is clear from (28) that the inverse in (26) exists.

Formula (26) and the normalization $u_0(s) + \underline{u}(s)\underline{e} = 1$, lead to

$$u_0(s) + \lambda\, u_0(s)\, \chi^{-1}(s)[s+\lambda - \frac{\lambda}{\chi(s)}]^{-1}[1-f^*(s+\lambda - \frac{\lambda}{\chi(s)})] = 1$$

from which we obtain the expression for $u_0(s)$.

Entirely similar calculations for the right eigenvector lead to

$$\underline{v}(s) = v_0(s)\, \chi^{-1}(s)\, [(s+\lambda - \frac{\lambda}{\chi(s)})I-T]^{-1}\, \underline{T}^0,$$

and we obtain the stated expression for $v_0(s)$ from the normalizing equation $u_0(s)v_0(s) + \underline{u}(s)\underline{v}(s) = 1$.

4. The superposition of identical hyperexponential renewal processes

The superposition, even of identical, Ph-renewal processes quickly leads to Markov renewal processes with a huge number of states. An exception is the superposition of r renewal process with the same hyperexponential distribution

$$F(x) = p(1-e^{-\theta_1 x}) + (1-p)(1-e^{-\theta_2 x}), \tag{29}$$

$0 < p < 1$, $\theta_1 \neq \theta_2$, and a homogeneous Poisson process of rate λ_0. We shall see that such a process is equivalent to a Markov-modulated Poisson process with r+1 states.

It is well-known (M.F. Neuts, 1982) that a hyperexponential renewal process is equivalent to an interrupted Poisson process, that is a Markov-modulated Poisson process defined on the two-state Markov process with generator

$$Q = \begin{bmatrix} -\sigma_1 & \sigma_1 \\ \\ \sigma_2 & -\sigma_2 \end{bmatrix} ,$$

and rates λ and 0. The parameters λ, σ_1 and σ_2 are related to the parameters of $F(.)$ by

$$\lambda = p\theta_1 + (1-p)\theta_2,$$

$$\sigma_1 = \frac{p(1-p)(\theta_1-\theta_2)^2}{p\theta_1 + (1-p)\theta_2} ,$$

$$\sigma_2 = \frac{\theta_1\theta_2}{p\theta_1 + (1-p)\theta_2} .$$

(30)

Henceforth, let the hyperexponential renewal process be parametrized by λ, σ_1 and σ_2.

Consider now the superposition of r interrupted Poisson processes with identical parameters λ, σ_1 and σ_2. It is clear that the process obtained by recording the number of two-state Markov processes that are in state 1 is again a Markov process with the states $\{0,1,\ldots,r\}$ and with the generator $Q(r)$ with

$$Q_{jj}(r) = -j\sigma_1-(r-j)\sigma_2, \qquad 0 \leq j \leq r,$$

$$Q_{j,j-1}(r) = j\sigma_1, \qquad 1 \leq j \leq r,$$

$$Q_{j,j+1}(r) = (r-j)\sigma_2, \qquad 0 \leq j \leq r-1,$$

(31)

$$Q_{jj'}(r) = 0 , \qquad |j-j'| > 1.$$

The superposition is equivalent to the Markov-modulated Poisson process defined on the Markov process with generator $Q(r)$ and having the arrival rates 0, λ, $2\lambda,\ldots,r\lambda$. The superposition of an additional (independent) Poisson process of rate λ_0 only changes the rates to

$$\lambda_j = \lambda_0 + j\lambda, \qquad 0 \leqslant j \leqslant r. \tag{32}$$

Let Λ be a diagonal matrix with diagonal elements λ_j, $0 \leqslant j \leqslant r$. The sequence $\{(J_n,X_n)\}$, $n \geqslant 0$ where $X_0 = 0$, and X_n and J_n are respectively the times between the $(n-1)$st and the n-th arrivals, $n \geqslant 1$, and the states of the Markov process $Q(r)$ is again a Markov renewal sequence. Its initial conditions may again be chosen so as to correspond to the stationary version, but we shall limit our discussion to the transition probability matrix, which is given by the semi-Markov matrix

$$D(x) = \int_0^x \exp\{[Q(r) - \Lambda]u\}du\,\Lambda, \qquad x \geqslant 0,$$

with transform

$$D^{\star}(s) = [sI + \Lambda - Q(r)]^{-1}\Lambda. \tag{33}$$

We note that if $\lambda_0 > 0$, the matrix $D^{\star}(s)$ is irreducible (and even strictly positive) for $s \geqslant 0$. When $\lambda_0 = 0$, the first column of $D^{\star}(s)$ vanishes.

Explicit expressions for the Perron-Frobenius eigenvalue $\chi(s)$ and corresponding normalized left and right eigenvectors $\underline{u}(s)$ and $\underline{v}(s)$ of the matrix $D^{\star}(s)$ may be obtained. After some preliminaries, these items may be expressed in fairly simple forms.

Lemma 1
———

Let $G^{\star}(s)$, $\mathrm{Re}\ s \geqslant 0$, be the transform matrix $(sI+\Lambda-Q)^{-1}\Lambda$, where

Q is an irreducible generator and $\Lambda = \text{diag}(\underline{\lambda})$, with $\underline{\lambda}$ a positive vector. The maximal eigenvalue $\chi(s)$ of $G^{\star}(s)$, $s \geqslant 0$, is given by

$$\chi(s) = [\eta(s)]^{-1} , \tag{34}$$

where $\eta(s)$ is the root of $\underline{smallest}$ modulus of the equation

$$\det[sI + \Lambda(1-\eta) - Q] = 0 \tag{35}$$

The root $\eta(s)$ is real, simple and satisfies $\eta(s) \geqslant 1$, with equality holding if and only if $s=0$.

For each $s \geqslant 0$, there exist vectors $\underline{w}(s)$ and $\underline{v}(s)$, each determined up to a multiplicative constant such that

$$\underline{w}(s)[sI + \Lambda - \Lambda\eta(s) - Q] = \underline{0}, \tag{36}$$

$$[sI + \Lambda - \Lambda\eta(s) - Q]\,\underline{v}(s) = \underline{0}$$

The vectors $\underline{v}(s)$ and $\underline{w}(s)$, $s \geqslant 0$, may be chosen to be positive. The vectors $\underline{u}(s) = \eta(s)\underline{w}(s)\Lambda$, and $\underline{v}(s)$ are respectively left and right eigenvectors of $G^{\star}(s)$ corresponding to $\chi(s)$.

<u>Proof</u>

The matrix $G^{\star}(s)$ is the nonsingular transform matrix of a semi-Markov matrix. Its eigenvalues therefore all lie in the unit disk. Since

$$(sI + \Lambda - Q)^{-1}\Lambda\,\underline{z} = x\,\underline{z}, \qquad \underline{z} \neq \underline{0},$$

is obviously equivalent to

$$[sI + \Lambda(1 - \frac{1}{x}) - Q]\,\underline{z} = \underline{0} ,$$

the eigenvalues of $G^\star(s)$ are the inverses of the roots of (36) and conversely.

The stated results are now direct consequences of the Perron-Frobenius theorem. If $\underline{u}(s)$ is a left eigenvector of $G^\star(s)$ corresponding to $\chi(s)$, then upon setting $\underline{w}(s) = \underline{u}(s)[\,sI + \Lambda - Q]^{-1}$, it follows that $\underline{w}(s)$ satisfies the first equation (37) and that $\underline{w}(s)\Lambda = \chi(s)\underline{u}(s) = [\,\eta(s)]^{-1}\underline{u}(s)$. This last equation shows that if $\underline{u}(s)$ is a positive vector, so is $\underline{w}(s)$.

In particular, when

$$Q = \begin{bmatrix} -\sigma_1 & \sigma_1 \\ \\ \sigma_2 & -\sigma_2 \end{bmatrix}$$

and $\underline{\lambda} = (\lambda_1, \lambda_2)$, with $\lambda_1 > 0$, $\lambda_2 > 0$, then

$$\begin{aligned}
\eta(s) &= \eta(s; \lambda_1, \lambda_2) \\[1em]
&= 1 + (2\lambda_1\lambda_2)^{-1}\{\lambda_1(s+\sigma_2) + \lambda_2(s+\sigma_1) \\[1em]
&\quad - [[\lambda_1(s+\sigma_2)-\lambda_2(s+\sigma_1)]^2 + 4\lambda_1\lambda_2\sigma_1\sigma_2]^{\frac{1}{2}}\} \,,
\end{aligned} \tag{37}$$

and the vectors $\underline{w}(s)$ and $\underline{v}(s)$ are given by

$$\underline{w}(s) = w_1(s)\left[1, \frac{s + \lambda_1(1-\eta) + \sigma_1}{\sigma_2}\right], \tag{38}$$

$$\underline{v}(s) = v_1(s)\left[1, \frac{s + \lambda_1(1-\eta) + \sigma_1}{\sigma_1}\right]'. \tag{39}$$

The η in formulas (38) and (39) is the expression given in (37). The functions $w_1(s)$ and $v_1(s)$ may be determined by normalizing the left and right eigenvectors. For the time being, it is convenient to leave them unspecified. The derivation of the expressions in (37)-(39) is elementary.

The maximal eigenvalue $\chi(s)$ and the corresponding eigenvectors $\underline{u}(s)$ and $\underline{v}(s)$ of the matrix $D^*(s)$ in (33) are expressed in terms of the corresponding quantities for the two-phase Markov-modulated Poisson process with

$$Q = \begin{bmatrix} -\sigma_1 & \sigma_1 \\ \sigma_2 & -\sigma_2 \end{bmatrix}$$

and arrival rates $\lambda_1 = \lambda + \dfrac{\lambda_0}{r}$, and $\lambda_2 = \dfrac{\lambda_0}{r}$. The Markov renewal process with transition probability matrix $D(x)$ is indeed also the superposition of r independent two-state Markov-modulated Poisson processes with these parameters.

In what follows, let

$$\eta = \eta \left[\frac{s}{r}; \lambda + \frac{\lambda_0}{r}, \frac{\lambda_0}{r} \right], \tag{40}$$

where the right hand side is the function in (37) evaluated at the stated parameter values. We further define $\varphi(s)$ and $\psi(s)$ by

$$\varphi(s) = (r\,\sigma_2)^{-1}[s + (\lambda_0 + r\lambda)(1-\eta) + r\,\sigma_1], \tag{41}$$

$$\psi(s) = \frac{\sigma_2}{\sigma_1}\,\varphi(s).$$

We note that $\varphi(s)$ and $\psi(s)$ are just the expressions at the right of (38) and (39) evaluated at the parameter values $\lambda_1 = \lambda_0 r^{-1} + \lambda$, and $\lambda_2 = \lambda_0 r^{-1}$.

Theorem 5
___

The maximal eigenvalue of the matrix $D^*(s)$ is given by

$$\chi(s) = \eta^{-1}, \tag{42}$$

and the normalized eigenvectors $\underline{u}(s)$ and $\underline{v}(s)$ are given by

$$u_j(s) = \eta w(s)(\lambda_0 + j\lambda)\binom{r}{j}\varphi^{r-j}(s), \tag{43}$$

$$v_j(s) = v(s)\psi^{r-j}(s),$$

for $0 \leq j \leq r$. The normalizing functions $w(s)$ and $v(s)$ are given by

$$w(s) = \eta^{-1}[\lambda_0 + r\lambda + \lambda_0\varphi(s)]^{-1}[1+\varphi(s)]^{1-r}, \tag{44}$$

$$v(s) = [\eta w(s)]^{-1}[\lambda_0 + r\lambda + \lambda_0\frac{\sigma_2}{\sigma_1}\varphi^2(s)]^{-1}[1+\frac{\sigma_2}{\sigma_1}\varphi^2(s)]^{1-r}.$$

For $s=0$, we have $\chi(0)=1$, $v_j(0)=1$ and

$$u_j(0) = \frac{\lambda_0 + j\lambda}{\lambda_0\sigma_1 + (\lambda_0+r\lambda)\sigma_2}\binom{r}{j}\frac{\sigma_1^{r-j}\sigma_2^{j}}{(\sigma_1+\sigma_2)^{r-1}}, \tag{45}$$

for $0 \leq j \leq r$.

Proof

The proof proceeds by direct verification using Lemma 1. We shall show that the equations (36) are satisfied by substitution of the stated expressions for η, $v(s)$ and by the vector $\underline{w}(s)$ of the form

$$w_j(s) = w(s)\binom{r}{j}\varphi^{r-j}(s), \qquad 0 \leq j \leq r.$$

We sketch the flow of the calculations for the equations involving $\underline{w}(s)$. Those for $\underline{v}(s)$ are entirely similar.

Upon substitution into the first equation (36) and using the particular form (31) of $Q(r)$, we obtain

$$\varphi^r(s)[\,s+\lambda_0(1-\eta)+r\sigma_2] - r\,\sigma_1\,\varphi^{r-1}(s) = 0,$$

$$-(r-j+1)\sigma_2\,\binom{r}{j-1}\,\varphi^{r-j+1}(s)$$

$$+[\,s+(\lambda_0+j\lambda)(1-\eta) + j\,\sigma_1 + (r-j)\sigma_2]\,\varphi^{r-j}(s)$$

$$-(j+1)\sigma_1\,\binom{r}{j+1}\,\varphi^{r-j-1}(s) = 0, \qquad \text{for } 1 \leqslant j \leqslant r-1,$$

$$-r\,\sigma_2\,\varphi(s) + [\,s+(\lambda_0+\lambda r)(1-\eta)+r\sigma_1] = 0.$$

The first and the last of these equations are equivalent to

$$[\,1,\varphi(s)] \begin{bmatrix} s+(\lambda_0+r\lambda)(1-\eta)+r\sigma_1 & -r\sigma_1 \\[2ex] -r\sigma_2 & s+\lambda_0(1-\eta)+r\sigma_2 \end{bmatrix} = \underline{0},$$

and that equation is satisfied by virtue of the results for the case of two phases with the appropriate choices $\lambda_1 = \lambda + r^{-1}\lambda_0$, $\lambda_2 = r^{-1}\lambda_0$, of the arrival rates.

Next, we simplify the equations for $1 \leqslant j \leqslant r-1$ and substract the equation for $j-1$ from that for j. This shows that all these equations are satisfied provided that

$$-\sigma_2\,\varphi^2(s) + [\,\lambda(1-\eta)+\sigma_1-\sigma_2]\,\varphi(s) + \sigma_1 = 0.$$

Substituting the expression for $\varphi(s)$, we obtain after routine calculations that the preceding equality is equivalent to

$$(1-\eta)^2\,\lambda_0(\lambda_0+r\lambda) + (1-\eta)[\,(\lambda_0+r\lambda)(s+r\sigma_2) + \lambda_0(s+r\sigma_1)]$$

$$+ (s+r\sigma_1)(s+r\sigma_2) - r^2\sigma_1\sigma_2 = 0.$$

This is, however, the quadratic equation of which η as given by (37) and (40) is the smallest (real) root. The vector $\underline{w}(s)$ therefore satisfies the first of the equations (36).

After verifying that $\underline{v}(s)$ satisfies the second equation in (36), we calculate $\underline{u}(s) = \eta \underline{w}(s) \Lambda$, and normalize by requiring that $\underline{u}(s)\underline{e}=1$, and $\underline{u}(s) \underline{v}(s)=1$. This routinely leads to the stated expressions.

We note that the reducibility of $D^{\star}(s)$ for $\lambda_0=0$, does not cause any difficulty. The normal expressions obtained for $\lambda_0 > 0$, are clearly valid for $\lambda_0=0$, by continuity. The only difference is that $u_0(s)$ now vanishes.

5. First passage time distributions for the PH/PH/1 queue

A number of first passage time distributions for the PH/PH/1 and related queueing models may be conveniently studied in terms of the formalism developed in the preceding sections. Let us refer to the renewal processes with underlying distributions $F_1(.)$ and $F_2(.)$ as the arrival and service processes respectively. We also partition the matrix $C^{\star}(s)$ as

$$C^{\star}(s) = \begin{bmatrix} C^{\star}(1,1;s) & C^{\star}(1,2;s) \\ \\ C^{\star}(2,1;s) & C^{\star}(2,2;s) \end{bmatrix} \tag{46}$$

where

$$C^{\star}(1,1;s) = [\underline{\alpha}(1) \otimes I][sI \otimes I - T(1,2)]^{-1}[\underline{T}^0(1) \otimes I],$$

$$C^{\star}(1,2;s) = [\underline{\alpha}(1) \otimes I][sI \otimes I - T(1,2)]^{-1}[I \otimes \underline{T}^0(2)],$$

$$\tag{47}$$

$$C^{\star}(2,1;s) = [I \otimes \underline{\alpha}(2)][sI \otimes I - T(1,2)]^{-1}[\underline{T}^0(1) \otimes I],$$

$$C^{\star}(2,2;s) = [I \otimes \underline{\alpha}(2)][sI \otimes I - T(1,2)]^{-1}[I \otimes \underline{T}^0(2)].$$

The sequence of matrice $\tilde{A}_\nu(s)$, $\nu \geqslant 0$, is defined by

$$\tilde{A}_0(s) = C^\star(2,2;s),$$

$$\tilde{A}_\nu(s) = C^\star(2,1;s)[C^\star(1,1;s)]^{\nu-1} C^\star(1,2;s), \quad \nu \geqslant 1. \tag{48}$$

The element $[\tilde{A}_\nu(s)]_{j'j'_1}$, $1 \leqslant j'$, $j'_1 \leqslant m(1)$, is clearly the transform of the conditional probability that a service starting with the arrival process in the phase j', ends before time x with the arrival process at the service termination in the phase j'_1 and with ν arrivals occuring during the service time.

The matrix

$$\tilde{A}(z,s) = \sum_{\nu=0}^{\infty} \tilde{A}_\nu(s) z^\nu,$$

is given by

$$\tilde{A}(z,s) = C^\star(2,2;s) + z C^\star(2,1;s)[I-z C^\star(1,1;s)]^{-1}C^\star(1,2;s). \tag{49}$$

We note in passing that by routine differentia tions in the matrix product

$$\tilde{A}(z_1,s_1) \tilde{A}(z_2,s_2)...\tilde{A}(z_n,s_n),$$

we may examine various correlations of the numbers of arrivals in the first renewal p rocess between successive renewals in the second process. Such correlations are analytically intractable for general renewal processes, but become readily available for PH-renewal processes. This matter will not be pursued here.

Let us now consider a counter which is started with a count of $r \geqslant 1$ at an arbitrary service termination. The counter is increased by one at each subsequent arrival and is decreased by one at each subsequent service completion. The first passage time until the counter reaches zero plays an important role in the theory of the PH/PH/1 queue (M.F.Neuts, 1983).

Let $\tilde{G}(z,s)$ be the transform matrix of the sequence of matrices $G(k,x) = \{G_{j'j_1}(k,x),\ k \geqslant 1,\ x \geqslant 0\}$. $G_{j'j_1}(k,x)$ is the conditional probability that the counter reaches r-1 for the first time at the end of the k-th service and prior to time x with the arrival process in the phase j_1', given that the counter was started at the beginning of a service ant with the arrival process in the phase j'.

The following properties are well-known :
a. The transform matrix corresponding to the first passage time from r to zero is given by $[\tilde{G}(z,s)]^r$.
b. The matrix $\tilde{G}(z,s)$, $0 \leqslant z \leqslant 1$, $s \geqslant 0$, is the minimal nonnegative solution of the matrix-functional equation

$$\tilde{G}(z,s) = z \sum_{\nu=0}^{\infty} \tilde{A}_{\nu}(s)\ \tilde{G}^{\nu}(z,s). \tag{50}$$

c. For the PH/PH/1 queue, the joint transform of the duration of the busy period and of the number of customers served during it, is given by

$$\gamma(z,s) = \alpha(1)\ \tilde{G}(z,s)e. \tag{51}$$

From the equation (50), we may deduce many important properties and moment formulas for the PH/PH/1 queue. As these have been treated, often in greater generality, in V. Ramaswami (1980, 1982), we shall not dwell on these properties here. We instead exploit the special form of the matrices $\tilde{A}_{\nu}(s)$ to obtain equivalent forms of the equation (50), which are particurlarly well-suited for numerical computations. In what follows, we shall consider the equation (50) for s=0, and also for z=1. In doing so, we shall simply suppress either the variable s or z and, according to the case, write $\tilde{G}(z)$ for $\tilde{G}(z,o)$ or $\tilde{G}(s)$ for $\tilde{G}(1,s)$.

For the case $s=0$, we define $\tilde{K}(z)$ and $\tilde{N}(z)$ by

$$\tilde{K}(z) = \sum_{\nu=0}^{\infty} z^{\nu}[C^{\star}(1,1)]^{\nu} C^{\star}(1,2) \tilde{G}^{\nu}(z), \qquad (52)$$

and

$$\tilde{N}(z) = \tilde{K}(z) \tilde{G}(z). \qquad (53)$$

The matrices $\tilde{K}(z)$ and $\tilde{N}(z)$ are of dimensions $m(2) \times m(1)$. The equation (50) with $s=0$, is equivalent to the system

$$\tilde{G}(z) = z C^{\star}(2,2) + z C^{\star}(2,1) \tilde{N}(z),$$

$$\tilde{K}(z) = C^{\star}(1,2) + C^{\star}(1,1) \tilde{N}(z), \qquad (54)$$

$$\tilde{N}(z) = \tilde{K}(z) \tilde{G}(z),$$

and upon series expansion, we obtain the equations

$$K(0) = C^{\star}(1,2)$$

$$K(k) = C^{\star}(1,1) N(k), \qquad k \geq 1,$$

$$N(k) = \sum_{j=1}^{k} K(k-j) G(j), \qquad k \geq 1, \qquad (55)$$

$$G(1) = C^{\star}(2,2),$$

$$G(k) = C^{\star}(2,1) N(k-1), \qquad k \geq 2.$$

The equations (55) show that, after prior computation of the stochastic matrix $C^{\star}(0)$, the matrices $K(k)$, $N(k)$ and $G(k)$ may be computed in a simple recursive manner in the order $K(0)$, $G(1)$, $N(1)$, $K(1)$, $G(2)$, $N(2) \ldots$

For $z=1$, we introduce the matrices $\tilde{K}(s)$ and $\tilde{N}(s)$ which are similarly defined by

$$\tilde{K}(s) = \sum_{\nu=0}^{\infty} [C^{\star}(1,1;s)]^{\nu} C^{\star}(1,2;s) \tilde{G}^{\nu}(s),$$

$$\tilde{N}(s) = \tilde{K}(s) \tilde{G}(s). \tag{56}$$

The equation (50) for $z=1$, is now equivalent to the system

$$\tilde{G}(s) = C^{\star}(2,2;s) + C^{\star}(2,1;s) \tilde{N}(s),$$

$$\tilde{K}(s) = C^{\star}(1,2;s) + C^{\star}(1,1;s) \tilde{N}(s), \tag{57}$$

$$\tilde{N}(s) = \tilde{K}(s) \tilde{G}(s).$$

The numerical computation of the semi–Markov matrix $G(x)$, $x \geq 0$, with transform $\tilde{G}(s)$ is much more belabored than that of the sequence $\{G(k), k \geq 1\}$. We may, however, exploit the special form of the matrices $C^{\star}(1,1;s)$, $C^{\star}(1,2;s)$, $C^{\star}(2,1;s)$ and $C^{\star}(2,2;s)$ as follows.

We may write the first two equations in (57) as

$$\tilde{G}(s) = [I \otimes \underline{\alpha}(2)][sI \otimes I - T(1,2)]^{-1}\{I \otimes \underline{T}^{0}(2) + [\underline{T}^{0}(1) \otimes I]\tilde{N}(s)\},$$

$$\tilde{K}(s) = [\underline{\alpha}(1) \otimes I][sI \otimes I - T(1,2)]^{-1}\{I \otimes \underline{T}^{0}(2) + [\underline{T}^{0}(1) \otimes I]\tilde{N}(s)\}.$$

If we introduce the matrice $\tilde{V}(s)$ of dimensions $m(1)m(2) \times m(1)$ by setting

$$\tilde{V}(s) = [sI \otimes I - T(1,2)]^{-1}\{I \otimes \underline{T}^{0}(2) + [\underline{T}^{0}(1) \otimes I]\tilde{N}(s)\},$$

then clearly

$$s \tilde{V}(s) = T(1,2) \tilde{V}(s) + I \otimes \underline{T}^{0}(2) + [\underline{T}^{0}(1) \otimes I]\tilde{N}(s) \tag{58}$$

The equation (58) is the transform version of the equation

$$V'(x) = T(1,2) \, V(x) + [\underline{T}^0(1) \otimes I] \, N(x), \qquad x \geq 0,$$

$$V(0) = I \otimes \underline{T}^0(2),$$

and, upon inversion of the other equations, we obtain

$$G(x) = [I \otimes \underline{\alpha}(2)] \, V(x), \tag{59}$$

$$K(x) = [\underline{\alpha}(1) \otimes I] \, V(x), \tag{60}$$

$$N(x) = \int_0^x K(u) \, G'(x-u) \, du, \qquad \text{for } x \geq 0 \tag{61}$$

After routine substitutions, we see that the crux of the algorithm is the solution of the nonlinear integro-differential equation

$$V'(x) = T(1,2) \, V(x) + [\underline{T}^0(1) \, \underline{\alpha}(1) \otimes I] \int_0^x V(x-u)[I \otimes \underline{\alpha}(2)] V'(u) \, du \tag{62}$$

with the initial condition $V(0) = I \otimes \underline{T}^0(2)$.

For purposes of numerical solution, it is advantageous to partition the matrix $V(x)$ as a column of $m(1)$ blocks $V_\nu(x)$, $1 \leq \nu \leq m(1)$, of dimensions $m(2) \times m(1)$. The vectors $\underline{\alpha}(2) \, V_\nu(x)$, $1 \leq \nu \leq m(1)$, are then the rows of the matrix $G(x)$ and the equation (62) may be written as

$$V'_\nu(x) = T(2) \, V_\nu(x) + \sum_{j=1}^{m(1)} T_{\nu j}(1) \, V_j(x) + T_\nu^0(1) \int_0^x K(x-u) \, G'(u) \, du, \tag{63}$$

for $1 \leq \nu \leq m(1)$. We see that

$$K(x) = \sum_{\nu=1}^{m(1)} \alpha_\nu(1) \, V_\nu(x). \tag{64}$$

The equations (63) may be solved by the successive solution of a sequence of ordinary differential equations, obtained by starting with $G(x) = 0$, $x \geqslant 0$, and by computing the matrix $\int_o^x K(x-u) \; G'(u)du$ of dimensions $m(2)\times m(1)$ each time from the preceding iterate. One may easily show that this yields a monotone increasing sequence (for every $x \geqslant 0$) of approximations which converges to $V(x)$. The numerical convolution to evaluate the matrix $\int_o^x K(x-u) \; G'(u)du$ remains a major task, which needs to be performed only once for each iteration.

We note that even for the M/M/1 queue, the scalar function $G(x)$ may be explicitly expressed only in terms of a Bessel function (L. Takacs, 1962). For the M/G/1 and other elementary queueing models, explicit expressions for the busy period distribution involve series whose terms require delicate numerical integrations (L. Takacs, 1962). The algorithmic complexity of the evaluation of the matrix $G(x)$ for the PH/PH/1 queue is therefore not astonishing. We further note that the preceding discussion does not depend on the stability of the queue. When the queue is stable, we may draw on a number of explicitly computable moment formulas (M.F.Neuts, 1976, M.F.Neuts in preparation) to provide accuracy checks or truncation criteria. The numerical implentation of the procedures we have described in mostly very fast, except when the traffic intensity of the PH/PH/1 queue is close to one. In that case, the matrices $G(k)$ decrease very slowly with k and the computation of $G(x)$ may require many iterations.

In certain queueing problems, the minimal nonnegative solution to the matrix functional equation

$$\tilde{R}(z,s) = z \sum_{\nu=0}^{\infty} \tilde{R}^{\nu}(z,s) \; A_{\nu}(s), \qquad s \geqslant 0, \; 0 \leqslant z \leqslant 1, \qquad (65)$$

plays an important role. See Ramaswami (1982). It is clear that for the PH/PH/1 queue, the manipulations used for the equation (50), also lead to simplified forms and corresponding algorithms for that equation.

6. A cyclic queue with two stations

We now consider a cyclic queue with two stations, at each of
which services are in order of arrival. The service time distribution
at station i, i=1,2, is the PH-distribution $F_i(.)$. Upon completion of
a service at one station, a customer instantaneously joins the queue
at the other station. The system contains a constant, finite number
N of customers.

This system may be represented as a Markov process on the state
space $\{(n_1,q_1,q_2),\ 1 \leqslant n_1 \leqslant N-1,\ 1 \leqslant q_1 \leqslant m(1), 1 \leqslant q_2 \leqslant m(2)\} \cup \{(0,q_2),$
$1 \leqslant q_2 \leqslant m(2)\} \cup \{(N,q_1), 1 \leqslant q_1 \leqslant m(1)\}$, where n_1 denotes the number
of customers at Station 1 ($n_2 = N-n_1$ being the number at Station 2),
and q_1 and q_2 respectively denote the service phase at Stations 1 and
2 (if the station is not idle).

We define the <u>cycle time</u> θ as the time necessary for a given
customer, denoted as the <u>tagged</u> customer, to execute a complete loop
through the system. Different precise definitions may be used; we
choose the following as it leads to tractable expressions. Suppose
that at time 0, the tagged customer has just completed a service at
Station 2. At time $\theta > 0$, that tagged customer has just completed
another service at Station 2; in $(0,\theta)$, that customer has entered
Station 2 only once. In order to characterize the distribution of
θ completely, we need to specify the values of n_1 and q_1 at the
beginning of the interval $(0,\theta]$: either $n_1(0+) = N$, in which case q_2
is not defined, or $n_1(0+) < N$, and a new service begins at Station 2.
q_2 is then a random variable with probability density $\underline{\alpha}(2)$. If
$n_1(0+) = 1$, the tagged customer immediately begins service at Station
1; for notational convenience, we assume that his initial phase is
already specified.

6.1. Complete characterization of the cycle time

In order to analyse <u>successive</u> cycle times, we need to keep track of values for n_1 and q_1 at the end of the interval $(0,\theta]$. For that reason, we first examine the distributions

$$H_N(x;n,n';i',j') = P[\,\theta \leq x, n_1(\theta+0) = n', q_1(\theta+0) = j' \,|\, n_1(0) = n,$$
$$q_1(0+) = i'] \qquad (66)$$

$$\text{for } N \geq 0, \; x \geq 0, \; 1 \leq n, \; n' \leq N, \; 1 \leq i', \; j' \leq m(1).$$

We define their Laplace-Stieltjes transforms $H_N^{\star}(s;n,n';i',j')$ and the square matrices $H_N^{\star}(s;n,n')$ of order $m(1)$, with components $H_N^{\star}(s;n,n';i',j')$.

We shall use similar notational conventions for other matrices in the sequel and we shall not re-state them, but merely indicate the dimensions of the matrices to avoid ambiguity.

It is useful to consider θ as the sum of two random variables τ_1 and τ_2, where τ_1 is the time needed by the tagged customer to complete service at Station 1, and τ_2 is the time needed by the tagged customer to complete service at Station 2 thereafter. The distribution of τ_1 depends on $n_1(0+)$ and $q_1(0+)$, while that of τ_2 depends on $n_1(\tau_1+0)$ and $q_2(\tau_1+0)$, the two random variables τ_1 and τ_2 being <u>conditionally independent</u> given $n_1(\tau_1+0)$ and $q_2(\tau_1+0)$.

If we define

$$M_N(x;n,n';i',j) = P[\,\tau_1 \leq x, \; n_1(\tau_1+0) = n', q_2(\tau_1+0) = j \,|$$
$$n_1(0+) = n, q_1(0+) = i'],$$
$$(67)$$

$$\text{for } N \geq 1, \; x \geq 0, \; 1 \leq n \leq N, \; 0 \leq n' \leq N-1, \; 1 \leq i' \leq m(1),$$
$$1 \leq j \leq m(2)$$

and

$$\tilde{M}_N(x;k,k';i,j') = P[\,\tau_2 \leq x, \; n_2(\tau_2+0) = k', \; q_1(\tau_2+0) = j' \,|$$
$$n_2(0+) = k, q_2(0+) = i],$$
$$(68)$$

for $N \geq 1$, $x \geq 0$, $1 \leq k \leq N$, $0 \leq k' \leq N-1$, $1 \leq i \leq m(2)$,

$\quad 1 \leq j' \leq m(1)$

then it is readily seen that $H_N(x;n,n';i',j') =$

$\displaystyle\sum_{\nu'=0}^{N-1}\sum_{j=1}^{m(2)} \int_o^x P[\tau_1 \leq x - y,\ n_1(\tau_1+0) = \nu',q_2(\tau_1+0) = j | n_1(0+) = n,$

$q_1(0+) = i'] \cdot [dP\ \tau_2 \leq y,\ n_2(\tau_1+\tau_2+0) = N-n',q_1(\tau_1+\tau_2+0) = j' |$

$n_2(\tau_1+0) = N-\nu',q_2(\tau_1+0) = j]\,,$

$\displaystyle= \sum_{\nu=1}^{N}\sum_{j=1}^{m(2)} \int_o^x P[\tau_1 \leq x-y,n_1(\tau_1+0) = N-\nu,q_2(\tau_1+0) = j | n_1(0+) = n,$

$q_1(0+) = i'] \cdot dP[\tau_2 \leq y,\ n_2(\tau_2+0) = N-n',\ q_1(\tau_2+0) = j' |$

$n_2(0+) = \nu,q_2(0+) = j]\,,$

since τ_2 is conditionally independent of τ_1, given $q_2(\tau_1+0)$ and $n_1(\tau_1+0)$, and since $n_1(\tau_1+0) = N-n_2(\tau_1+0)$.

Therefore,

$$H_N^{\star}(s;n,n') = \sum_{\nu=1}^{N} M_N^{\star}(s;n,N-\nu)\widetilde{M}_N^{\star}(s;\nu,N-n')\,, \qquad (69)$$

for $N \geq 1$, $1 \leq n,\ n' \leq N$,

where the matrices $M_N^{\star}(s;n,n')$ and $\widetilde{M}_N^{\star}(s;k,k')$ are of dimensions $m(1) \times m(2)$ and $m(2) \times m(1)$ respectively.

<u>Lemma 2</u>

The matrices $M_N^{\star}(s;n,n')$, $N \geq 1$, $1 \leq n \leq N$, $0 \leq n' \leq N-1$, are recursively determined by

$$M_1^{\star}(s;1,0) = (sI - T(1))^{-1} \underline{T}^0(1) \cdot \underline{\alpha}(2), \tag{70}$$

$$M_N^{\star}(s;N,n') = (sI - T(1))^{-1}\underline{T}^0(1) \cdot \underline{\alpha}(1)M_N(s;N-1,n'), \tag{71}$$

for $N \geqslant 2$, $0 \leqslant n' \leqslant N-1$,

$$M_N^{\star}(s;n,0) = C^{\star}(2,1;s)[\,C^{\star}(1,1;s)]^{n-1}, \tag{72}$$

for $N \geqslant 2$, $1 \leqslant n \leqslant N-1$,

and $\quad M_N^{\star}(s;n,n') = \sum_{\nu=0}^{n-1} \tilde{A}_\nu(s)M_{N-1}^{\star}(s;n-\nu,n'-1), \tag{73}$

for $N \geqslant 2$, $1 \leqslant n \leqslant N-1$, $1 \leqslant n' \leqslant N-1$,

where the matrices $\tilde{A}_\nu(s)$, $\nu \geqslant 0$, of order $m(1)$ are given in (48).

Proof

The equation (70) is obvious : if N=1, then the tagged customer
is alone in the system, τ_1 merely represents his service time, for a
given initial phase of $F_1(.)$ and at the end of service, an initial
phase q is chosen according to the density $\underline{\alpha}(2)$.

If $N \geqslant 2$ and n=N, then the Station 2 is initially idle, Station
1 is continuing a service. At the end of that service, a new customer selects an initial phase of $F_1(.)$ according to the probability
vector $\underline{\alpha}(1)$; the tagged customer is the last of N-1 customers at
Station 1 and one customer then is at Station 2, ready to start service. This proves (71).

If $N \geqslant 2$ and $n \leqslant N-1$, we distinguish two cases : either Station
1 becomes idle before any service is completed at Station 2, in which
case n'=0, or a service completion occurs at Station 2 before Station
1 becomes idle. It is readily apparent that in the former case one
obtains (72).

Assume now that $N \geq 2$, $n \leq N-1$ and $n' \geq 1$; a service completion occurs at Station 2, at a time denoted by τ', after ν, $0 \leq \nu \leq n-1$, service completions at Station 1. Clearly, $[\tilde{A}_\nu(s)]_{i',j'}$, $1 \leq i'$, $j' \leq m(1)$ is the transform of the conditional probability that $\tau' \leq x$ and $q_1(\tau'+0) = j'$, given that $q_1(0+) = i'$. At time τ', a customer joins the queue at Station 1, <u>behind</u> the tagged customer; the total number of customers at that station is $n-\nu+1$, of which $n-\nu$ (including the tagged one) remain to be served at Station 1 during the interval (τ',τ_1). We have that

$$P[\tau_1-\tau' \leq x, n_1(\tau_1+0)=n', q_2(\tau_1+0)=j \mid n_1(\tau'+0)=n-\nu+1, q_1(\tau'+0)=i']$$
$$=M_{N-1}(x;n-\nu,n'-1;i',j).$$

To prove this, we observe that during the interval (τ',τ_1) $n-\nu$ customers will be served at Station 1, $n'-1$ customers will be served at Station 2, at time τ' there are $N-n+\nu-1 = (N-1)-(n-\nu)$ customers at Station 2. Equation (73) now follows by the law of total probability.

Since, in defining the functions $\tilde{M}(x;\ldots)$ after the functions $M(x;\ldots)$, we have only interchanged the roles of Stations 1 and 2, we may immediately conclude that analogously,

$$\tilde{M}_1^\star(s;1,0) = (sI-T(2))^{-1}\underline{T}^0(2).\underline{\alpha}(1) \tag{74}$$

$$\tilde{M}_N^\star(s;N,k') = (sI-T(2))^{-1}\underline{T}^0(2).\underline{\alpha}(2)\,\tilde{M}_N^\star(s;N-1,k'), \tag{75}$$
for $N \geq 2$, $0 \leq k' \leq N-1$,

$$\tilde{M}_N^\star(s;k,0) = C^\star(1,2;s)[C^\star(2,2;s)]^{k-1}, \tag{76}$$
for $N \geq 2$, $1 \leq k \leq N-1$,

$$\text{and } \tilde{M}_N^\star(s;k,k') = \sum_{\nu=0}^{k-1} \tilde{A}_\nu'(s)\tilde{M}_{N-1}^\star(s;k-\nu,k'-1), \tag{77}$$

for $N \geq 2$, $1 \leq k$, $k' \leq N-1$,

where the matrices $\tilde{A}'_\nu(s)$ of order $m(2)$ are given by

$$\tilde{A}'_0(s) = C^\star(1,1;s) \, , \tag{78}$$

$$\tilde{A}'_\nu(s) = C^\star(1,2;s)[\, C^\star(2,2;s)]^{\nu-1} C^\star(2,1;s), \quad \nu \geqslant 1.$$

6.2. Marginal distribution of the cycle time

It is possible in principle to determine the matrices $M_N^\star$, $\tilde{M}_N^\star$ and $H_N^\star$ recursively for increasing values of N. This is very cumbersome, however, and we now examine the distribution of θ for a given initial value of n_1 and q_2, regardless of the value of n_1 and q_2 at time $\theta+0$. Let

$$h_N(x;n,i') = P[\, \theta \leqslant x | n_1(0+) = n, \, q_2(0+) = i'] \, , \tag{79}$$
$$\text{for } n \geqslant 1, \, 1 \leqslant n \leqslant N, \, 1 \leqslant i' \leqslant m(1),$$

let $\underline{h}_N^\star(s;n)$, $N \geqslant 1$, $1 \leqslant n \leqslant N$, be $m(1)$-vectors with the Laplace-Stieltjes transform $h_N^\star(s;n,i')$ of $h_N(x;n,i')$ as components. It is readily apparent that

$$\underline{h}_N^\star(s;n) = \sum_{n'=1}^{N} H_N^\star(s;n,n')\underline{e}, \quad N \geqslant 1, \, 1 \leqslant n \leqslant N, \tag{80}$$

where $\underline{e}$ is an $m(1)$-vector with components identically equal to one.

Lemma 3

The $m(2)$-vectors $\underline{\tilde{m}}_N^\star(s;k)$, $N \geqslant 1$, $1 \leqslant k \leqslant N$, defined by

$$\underline{\tilde{m}}_N^\star(s;k) = \sum_{k'=0}^{N-1} \tilde{M}_N^\star(s;k,k')\underline{e}, \tag{81}$$

are given by

$$\underline{\tilde{m}}_N^\star(s;k) = [\, F_2^\star(s)]^{k-1} (sI-T(2))^{-1} \underline{T}^0(2), \quad 1 \leqslant k \leqslant N, \tag{82}$$

where $F_2^\star(s) = \underline{\alpha}(2)(sI-T(2))^{-1}\underline{T}^0(2)$ is the Laplace-Stieltjes transform of the distribution $F_2(.)$.

<u>Proof</u>

The proof is purely technical. From (5) we readily show that

$$[sI \otimes I - T(1,2)][\underline{e} \otimes (sI - T(2))^{-1}\underline{T}^0(2)]$$

$$= -(T(1) \otimes I)[\underline{e} \otimes (sI - T(2)^{-1}\underline{T}^0(2)]$$

$$+ [I \otimes (sI - T(2))][\underline{e} \otimes (sI - T(2))^{-1}\underline{T}^0(2)]$$

$$= (\underline{T}^0(1) \otimes I)(sI - T(2))^{-1}\underline{T}^0(2) + (I \otimes \underline{T}^0(2))\underline{e}.$$

When we premultiply both sides by $[\underline{\alpha}(1) \otimes I][sI \otimes I - T(1,2)]^{-1}$, we obtain (see (7))

$$(sI - T(2))^{-1}\underline{T}^0(2) = C^\star(1,2;s)\underline{e} + C^\star(1,1;s)(sI - T(2))^{-1}\underline{T}^0(2),$$
$$(83)$$

and when we premultiply both sides by $[I \otimes \underline{\alpha}(2)][sI \otimes I - T(1,2)]^{-1}$, we obtain

$$F_2^\star(s)\underline{e} = C^\star(2,2;s)\underline{e} + C^\star(2,1;s)(sI - T(2))^{-1}\underline{T}^0(2). \qquad (84)$$

The equations (82) are proved by induction on k, using the equations (81, 74 to 78, 83 and 84).

The lemma is intuitively obvious : $(\underline{\widetilde{m}}_N^\star(s;k))_i$, $1 \le i \le m(2)$, is the transform of the distribution of the time needed to serve k customers at Station 2, starting with an initial service phase i. The matrix factor in the right-hand side of (82) accounts for the first service, the scalar factor accounts for the remaining (k-1) services.

Theorem 6

The Laplace-Stieltjes transforms for the marginal distributions of the cycle time are recursively determined by

$$\underline{h}_1^{\star}(s;1) = F_2^{\star}(s)(sI-T(1))^{-1}\underline{T}^0(1), \tag{85}$$

$$\underline{h}_N^{\star}(s;n) = [F_2^{\star}(s)]^{N-1} C^{\star}(2,1;s)[C^{\star}(1,1;s)]^{n-1}(sI-T(2))^{-1}\underline{T}^0(2) \tag{86}$$

$$+ \sum_{\nu=1}^{n-1} \tilde{A}_\nu(s)\, \underline{h}_{N-1}^{\star}(s;n-\nu), \quad \text{for } N \geqslant 2,\ 1 \leqslant n \leqslant N-1,$$

$$\underline{h}_N^{\star}(s;N) = (\underline{\alpha}(1)\, \underline{h}_N^{\star}(s;N-1))(sI-T(1))^{-1}\underline{T}^0(1), \quad \text{for } N \geqslant 2 \tag{87}$$

The proof is elementary and need not be given in detail here. We merely apply Definition (80), and use Equation (69) and Lemmas 2 and 3. We observe that the result is intuitive. Consider Equation (86). At time 0, the tagged customer is the last of n < N customers at Station 1. If they all get served before any service completion at Station 2, then the tagged customer becomes the last of N customers at Station 2; the remainder of the cycle time is merely the sum of N services of type 2, the initial phase of the first service being specified. The first term in (86) corresponds to that event. If, on the other hand, a customer is served at Station 2 after $\nu \leqslant$ n-1 customers have left Station 1, that customer joins the queue at Station 1, behind the tagged customer; his presence no longer influences the duration of the cycle time and one may consider a system with one less customer, as appears in the second term of (86). The interpretation of Equations (85) and (87) is similar.

The equations (85-87) do not yield a simple form for the distribution of the cycle time, but they are well suited for recursive computations, for instance of moments of the distribution.

In practice, one would probably not be interested in the distribution of θ for every initial value of n_1 and q_1, but rather for some specific initial conditions. For instance, if one assumes that at time 0, all customers are in Station 1 and that the first service begins, then the distribution of the time until all customers are served in both stations has the Laplace-Stieltjes transform $\underline{\alpha}(1)\underline{h}_N^{\star}(s;N)$.

6.3. Stationary distribution of the cycle time

It is of particular interest to determine the distribution of θ under the assumption that at time 0, the values of n_1 and q_1 are chosen according to the stationary distribution of the system at the end of a service at Section 2. The Laplace-Stieltjes transform of the distribution of such an "arbitrary" cycle time is given by

$$g_N^{\star}(s) = \underline{x}^{(N)} \, \underline{h}_N^{\star}(s) , \tag{88}$$

where $\underline{x}^{(N)}$ and $\underline{h}_N^{\star}(s)$ are $m(1)N$-vectors,

$$\underline{h}_N^{\star}(s) = \begin{bmatrix} \underline{h}_N^{\star}(s;1) \\ \underline{h}_N^{\star}(s;2) \\ \cdot \\ \cdot \\ \cdot \\ \underline{h}_N^{\star}(s;N) \end{bmatrix}$$

and $\underline{x}^{(N)} = [\,\underline{x}_1^{(N)} \, , \, \underline{x}_2^{(N)} \, , \, \dots, \, \underline{x}_N^{(N)}\,]$,

where $(\underline{x}_n^{(N)})_{j'}$ is the stationary probability that at the end of a service in Station 2, $n_1 = n$ and $q_1 = j'$. (In order to keep the notations homogeneous, we assume as we did before that if $n \leqslant N-1$, then q_2 is a random variable with probability density $\underline{\alpha}(2)$, and that if $n=1$, then q_1 is already specified).

We develop in this section a recursive algorithm to determine $\underline{x}^{(N)}$ for increasing values of N. This, in conjunction with Theorem 6, will recursively determine the stationary distribution of the cycle time for increasing N. We denote by $P^{(N)}$ the transition probability matrix of the Markov chain embedded at epochs of service completion at Station 2. The matrix $P^{(N)}$ is decomposed in submatrices $P_{i,j}^{(N)}$, $1 \leq i$, $j \leq N$ of order m(1) and has a block-lower Hessenberg from : $P_{i,j}^{(N)} = 0$ for $j > i+1$. Writing $\tilde{A}_\nu$, $C^\star(1,1)$ and $C^\star(2,1)$ for $\tilde{A}_\nu(0)$, $C^\star(1,1;0)$ and $C^\star(2,1;0)$ respectively, we have that

$$P^{(1)} = \underline{e}.\underline{\alpha}(1), \tag{89}$$

$$P^{(N)} = \begin{bmatrix} \tilde{B}_1^N & \tilde{A}_0 & 0 & \cdots & 0 & 0 \\[2ex] \tilde{B}_2^N & \tilde{A}_1 & \tilde{A}_0 & \cdots & 0 & 0 \\[2ex] \tilde{B}_3^N & \tilde{A}_2 & \tilde{A}_1 & \cdots & 0 & 0 \\[1ex] \vdots & & & & & \\[1ex] \tilde{B}_{N-1}^N & \tilde{A}_{N-2} & \tilde{A}_{n-3} & \cdots & \tilde{A}_1 & \tilde{A}_0 \\[2ex] \tilde{B}_N^N & \tilde{D}_{N-2} & \tilde{D}_{N-3} & \cdots & \tilde{D}_1 & \tilde{D}_0 \end{bmatrix}, \tag{90}$$

where $\tilde{D}_i = \underline{e}.\underline{\alpha}(1)\,\tilde{A}_i$, $\qquad i \geq 0$, $\tag{91}$

$$\tilde{B}_i^N = C^\star(2,1)\,[C^\star(1,1)]^{i-1}\,\underline{e}.\underline{\alpha}(1), \qquad 1 \leq i \leq N-1, \tag{92}$$

$$\tilde{B}_N^N = \underline{e}.\underline{\alpha}(1)\,\tilde{B}_{N-1}^N. \tag{93}$$

The unique solution $\underline{x}^{(N)}$ of the system

$$\underline{x}^{(N)}\,P^{(N)} = \underline{x}^{(N)}, \qquad \underline{x}^{(N)}\,\underline{e} = \underline{e}, \tag{94}$$

may be obtained by various algorithms (see G. Latouche, P.A. Jacobs and D.P. Gaver, 1984)   but, as mentioned before, we are interested in one that is recursive on N.

To that end, we define matrices $U^{(N)}$ of order $m(1)N$, $V^{(N)}$ of dimensions $m(1) \times m(1)N$ and $W^{(N)}$ of dimensions $m(1)N \times m(1)$, for $N \geqslant 1$, as follows :

$$V^{(N)} = [\tilde{A}_0, 0, \ldots, 0], \qquad \text{for } N \geqslant 1 ; \tag{95}$$

$$W^{(1)} = \tilde{D}_1 , \tag{96}$$

$$W^{(N)} = \begin{bmatrix} \tilde{A}_2 \\ \vdots \\ \tilde{A}_N \\ \\ \tilde{D}_N \end{bmatrix} , \qquad \text{for } N \geqslant 2 ; \tag{97}$$

$$U^{(1)} = \tilde{D}_0 - I , \tag{98}$$

$$U^{(N)} = \begin{bmatrix} \tilde{A}_1 - I & V^{(N-1)} \\ \\ W^{(N-1)} & U^{(N-1)} \end{bmatrix} , \qquad \text{for } N \geqslant 2. \tag{99}$$

Observe that the matrices $P^{(N)}$, for $N \geqslant 2$, may then be written as

$$P^{(N)} = \left[\begin{array}{c|c} \tilde{B}_1^N & V^{(N-1)} \\ \hline \tilde{B}_2^N & \\ \vdots & U^{(N-1)} + I \\ \tilde{B}_N^N & \end{array} \right] . \tag{100}$$

Theorem 7
____

The stationary probability vector $x^{(N)}$ at end of services at Station 2 is given by

$$\underline{x}^{(1)} = \underline{\alpha}(1);$$
(101)

$$\underline{x}_1^{(N)} = \gamma_N \, \underline{\alpha}(1),$$
(102)

$$[\,\underline{x}_2^{(N)}, \ \ldots, \ \underline{x}_N^{(N)}\,] = \gamma_N \, \underline{\alpha}(1) \, Y^{(N-1)}, \quad \text{for } N \geqslant 2,$$
(103)

where $\gamma_N = (1 + \underline{\alpha}(1) \, Y^{(N-1)} \underline{e})^{-1}$ is a normalization constant, and the matrices $Y^{(N)}$ of dimensions $m(1) \times m(1)N$, $N \geqslant 1$, are recursively determined by

$$Y^{(1)} = \tilde{A}_0 (I - \tilde{D}_0)^{-1},$$
(104)

$$Y^{(N)} = [\, \tilde{A}_0 X^{(N-1)}, \ \tilde{A}_0 X^{(N-1)} Y^{(N-1)}\,], \quad N \geqslant 2$$
(105)

with

$$X^{(N)} = [\, I - \tilde{A}_1 - Y^{(N)} W^{(N)}\,]^{-1}, \quad N \geqslant 1.$$
(106)

<u>Proof</u>

The equation (101) trivially results from (89,94). For $N \geqslant 2$, we conclude from (90,92,93,94) that

$$\underline{x}_1^{(N)} = \sum_{i=1}^{N} \underline{x}_i^{(N)} \, \tilde{B}_i^N = \gamma_N \, \underline{\alpha}(1)$$

for some constant γ_N, which proves (102). Then we use (100) to obtain that

$$[\,\underline{x}_2^{(N)}, \ \ldots, \ \underline{x}_N^{(N)}\,] = \underline{x}_1^{(N)} V^{(N-1)} + [\,\underline{x}_2^{(N)}, \ldots, \underline{x}_N^{(N)}\,][\, U^{(N-1)} + I\,],$$

or $[\,\underline{x}_2^{(N)}, \ \ldots, \ \underline{x}_N^{(N)}\,][\, -U^{(N-1)}\,] = \underline{x}_1^{(N)} \, V^{(N-1)}.$
(107)

For all $N \geqslant 1$, the matrix $I + U^{(N)}$ is substochastic and irreducible, therefore the matrix $U^{(N)}$ is non-singular and its inverse is strictly negative. We may thus define the matrices $Y^{(N)}$ as follows :

$$Y^{(N)} = V^{(N)}[-U^{(N)}]^{-1}, \quad N \geq 1. \tag{108}$$

A simple probabilistic argument proves that the matrices $\tilde{A}_1 + Y^{(N)} W^{(N)}$ $N \geq 1$, are substochastic and that the inverses $X^{(N)}$, $N \geq 1$, defined by (106) exist and are positive. It is then a simple matter to verify that $[-U^{(N)}]^{-1} =$

$$\begin{bmatrix} X^{(N-1)} & X^{(N-1)}Y^{(N-1)} \\[2em] [-U^{(N-1)}]^{-1} W^{(N-1)}X^{(N-1)} & [-U^{(N-1)}]^{-1}[I+W^{(N-1)}X^{(N-1)}Y^{(N-1)}] \end{bmatrix} \tag{109}$$

The equation (105) now results from (95,108 and 109), the equation (103) results from (102,107 and 108).

It is worth noting that the computational algorithms are stable as we are only dealing with positive matrices $X^{(N)}$ and $Y^{(N)}$.

Acknowledgement

The research of M.F. Neuts was supported by the National Science Foundation of the United States (Grant Nr ECS-8205404) and by a Senior U.S. Scientist Award from the Alexander von Humboldt Foundation of the German Federal Republic. The present paper was written during stays at the Institute for Telecommunications and Data Processing at the University of Stuttgart and at the Faculty of Industrial Engineering and Management at Technion.

<u>References</u>

R. Bellman, Introduction to Matrix Analysis, McGraw Hill Book Co,
     New York, 1960.

F.R. Gantmacher, The Theory of Matrices, Chelsea, New York, 1959.

J.F.C. Kingman, A convexity property of positive matrices, Quart.J.
     Math., 12, (1961), 283-284.

G. Latouche, A phase-type semi-Markov point process. SIAM Journal on
     Algebraic and Discrete Methods, 3, (1982), 77-90.

G. Latouche, An exponential semi-Markov process, with applications to
     queueing theory, Stochastic Models (1985), 137 - 170.

G. Latouche, P.A. Jacobs and D.P. Gaver, Finite Markov chain models
     skip-free in one direction. Naval Research Logistics Quarterly,
     31, (1984), 571 - 588.

M.F. Neuts, Moment formulas for the Markov renewal branching process.
     Advances in Applied Probability, 8, (1976), 690-711.

M.F. Neuts, Some explicit formulas for the steady-state behavior of
     the queue with semi-Markovian services times. Advances in
     Applied Probability, 9, (1977), 141-157.

M.F. Neuts, Matric-Geometric Solutions in Stochastic Models. An
     Algorithm Approach. The John Hopkins University Press, Baltimore,
     1981.

M.F. Neuts, The caudal characteristic curve of queues. Technical
     report n° 82, B, Appl. Mathematics Institute, University of
     Delaware, November 1982.

M.F. Neuts, A new informative embedded Markov renewal process for the
     PH/G/1 queue. Technical report n° 94, B, Appl. Mathematics
     Institute, University of Delaware, November 1983.

M.F. Neuts, Partitioned Stochastic Matrices of M/G/1 Type and their
     Applications. In preparation.

M.F. Neuts and K.S. Meier, On the use of phase type distributions in
     reliability modelling of systems with a small number of compo-
     nents. OR Spektrum, 2, (1981), 227-234.

R. Pyke, Markov renewal processes with finitely many states. Ann.
   Math. Statist., 32, (1961), 1243-1259.

V. Ramaswami, The N/G/1 queue and its detailed analysis. Advances
   in Applied Probability, 12, (1980), 222-261.

V. Ramaswami, The busy period of queues which have a matrix-geometric
   steady state vector. Opsearch, 19, (1982), 238-261.

L. Takacs, Introduction to the Theory of Queues. Oxford University
   Press, New York, (1962).

SECTION IV. SEMI MARKOV MODELS IN ECONOMY AND INSURANCE

PENSION ACCUMULATION AS A SEMI-MARKOV REWARD PROCESS,
WITH APPLICATIONS TO PENSION REFORM

Yves Balcer[1] and Izzet Sahin[2]

[1]Dept of Economics, University of Wisconsin-Madison
[2]School of Business Administration, University of
Wisconsin-Milwaukee

1. Introduction

A fundamental point of departure of the economic analysis of
savings by individuals is Samuelson's consumption loan model.
Although prevalent in the analysis of social security and saving
behavior, this "representative consumer life cycle model" is inappropriate for the study of private (employment) pensions due to
a number of important characteristics of these latter arrangements
that distinguish them from social security. In addition to wage
and benefit levels, accrual of employment pension benefits are
based on the applicable vesting provisions (see below) and therefore on mobility. In addition, while the wage factors entered in
social security benefit calculations are indexed until retirement,
in the case of employment pensions there is either no indexing or
only partial indexing until employment termination which may precede
retirement by many years. It is well-known that these differences
between the two systems have very important consequences in the
face of high labor mobility, especially during periods of high wage
growth. It is the interaction of labor mobility, inflation, and
pension coverage and portability (see below) that results in an

unequal distribution of pension benefits to individuals with comparable working lives and lifetime wage profiles. These characteristics are not reflected by the consumption loan model.

In general, employment pension plans may be divided into two
categories : <u>defined benefit plans</u> and <u>defined contributions plans</u>.
In turn, a defined benefit plan may be either a <u>unit benefit plan</u>
or a <u>flat benefit plan</u>. In unit benefit plans, pension benefits
are determined by taking a specified percentage (called the <u>benefit
level</u>) of the remuneration of an employee for each year (<u>career
average plans</u>) or for a selected number of years of service (<u>final
earnings plans</u>). In flat benefit plans, pension benefits are expressed as a fixed amount for each year of <u>pensionable service</u> (see
below). Most of the large pension plans in the United States
and Canada are defined benefit plans. In defined contribution (or
"money purchase") plans, pension benefits are determined on retirement by the accumulated amount of past contributions, made according
to a specified <u>contribution level</u>, and returns on investment. These
plans are favored by small employees, partnerships, and the non-
profit sector.

In almost all employment pension plans, a terminating employee
is entitled to pension benefits at retirement upon completion of a
prescribed minimum number of years of service with the organization.
In addition to this "service requirement" some plans also feature
an "age requirement" in terms of a prescribed minimum age on separation. If these provisions are met, then the pension is said to
be <u>vested</u> in the employee in that the employee collects benefits
from it upon retirement even if he or she never again works for
that organization. Otherwise, the employee is generally entitled
only to a return of his or her own contributions with little or no
interest. These provisions are called <u>vesting rules</u> and a length
of service that meets them is referred to as <u>pensionable</u> (or qua-
lifying) <u>service</u>. Vesting rules are subject to government regulation. In the United States, under the Federal Employee Retirement

Income Security Act (ERISA) of 1974, the service requirement may not
be more stringent than full vesting after 10 years of service or
graded vesting from 5 to 15 years of service. If an age requirement
is also used, then a graded (partial) vesting should start when the
age plus length of service equals 45 with at least 5 years of service. In Canada, regulation of the employment pension system falls
under provincial jurisdiction. Under the Ontario Pension Benefits
Act of 1965, which was subsequently adopted by most other Canadian
provinces and the Federal Government, the age and service requirements may not be more stringent than 10 years of service and 45
years of age, respectively. Although permitted by law to the above
extent, the age requirement has been gradually losing its prominence
in both countries. At the present time, a majority of employment
pension plans in North America feature the vesting rule of 10 years
of service.

The service requirement for vesting may be viewed as an instrument for the firm in reducing labor mobility. On the other hand,
by prolonging an employee's obligation to the firm and by increasing
the risk of benefit forfeitures, it results in a misallocation of
human resources and adversely affects pension accumulation. This
latter effect is compounded by limited pension <u>coverage</u> and <u>porta-
bility</u>.

Pension coverage varies from one segment of the labor force to
another, being more prevalent in large unionized firms. Generally
left out are workers in small non-union enterprises, marginal firms
and industries, and those in self-employement and agriculture.
Coverage rates in Canada and the United States have stabilized
around 40 and 50 percent respectively. Portability refers to the
transferability of pension rights and the underlying assets from
one employment to the next. It is typical of multi-employer plans
which allow for intra-system mobility without loss of pension rights,
irrespective of the vesting status. These plans have experienced a
rapid growth in the United States over the last decade, now involving

about 30 percent of all covered workers. Inter-system portability and portability between multi-employer and single-employer plans are much less common. While vesting provisions are subject to government regulation, coverage and portability are not. There is nothing in the current North American pension legislation that would oblige an employer to establish or maintain a pension plan.

Given the institutional and structural characteristics of employment pension plans outlined above, important issues related to private pensions, such as capital accumulation, mobility and efficient allocation of labor, cross subsidies, income distribution at retirement, and the impact of pension legislation, have been discussed in the literature from different perspectives. An objective examination of some of the problems involved requires a formalization of the process by which employment pension benefits accumulate over time. Such a formalization, in terms of a stochastic theory of pension accumulation, was recently presented by the authors (1983) in which the accumulation of private pension benefits throughout the working lives of an age cohort of workers was modelled as a semi-Markov reward process. This model allows for partial coverage, portability, and coverage-dependent completed length of service distributions. It also allows a unified treatment of different plan types through alternative specfications of the reward function.

In this paper, we present some extensions of the model to pension reform, designed to predict the evolution of pension benefits following a liberalization in the statutory vesting requirements, or an increase in coverage rates.

2. Previous research

Assume that completed lengths of service (CLS) in covered and non-covered employments are mutually independent random variables with distribution functions $F(x)$ and $G(x)$, respectively. Upon termination of an employment, an individual moves into a covered employment with probability c, or into a non-covered employment with probability $1-c$. Likewise, upon termination of a covered employment, if the next job is also covered, the pension rights are transferred, irrespective of the vesting status, with probability π, or not transferred with probability $1-\pi$.

Under these assumptions (see Balcer and Sahin, 1983 for a discussion of their implications) the employment termination process can be represented by a 2-state semi-Markov process with the individual regarded as being in state 1 or 2 at time t, depending on the coverage or non-coverage at that time of his employment by a pension plan. The semi-Markov matrix is :

$$(1) \quad A(t) = \begin{pmatrix} c(1-\pi) \sum_1^\infty (c\pi)^{n-1} F_n(t) & (1-c) \sum_1^\infty (c\pi)^{n-1} F_n(t) \\[2em] c \sum_1^\infty (1-c)^{n-1} G_n(t) & 0 \end{pmatrix}$$

where $F_n(t)$ and $G_n(t)$ are n-fold convolutions. For $i, j = 1, 2$, let

$$K_i(t) \equiv A_{i1}(t) + A_{i2}(t)$$

be the distribution function of the sojourn times in state i, and $M_{ij}(t)$ the conditional renewal functions. (Unless otherwise noted the lower case symbols will denote the corresponding density functions; thus $a_{ij}(t) \equiv A'_{ij}(t)$, etc.). The process is relevant over a finite interval $(0,t]$ where 0 is the beginning of the working life and t the time of retirement. However, a formal extension beyond t is required for simplicity, so that the last employment that terminates with retirement is an incomplete interval (i.a., a backward recurrence time).

This semi-Markov process was first constructed and studied in (Sahin, 1978) in relation to the accumulation of pensionable service. If $c = 1$ and $\pi = 0$ (full coverage, no portability) it reduces

to a renewal process with interval distribution F(x). This case
was investigated in (Sahin and Balcer, 1979), again in relation to
pensionable service.

Accrual of pension benefits is associated with pensionable
lengths of stay in state 1. Let s represent the service requirement for vesting. Consider a sojourn time of $\geqslant s$ in state 1. Suppose that such an interval commences y time units <u>before</u> and terminates y − u time units <u>before</u> retirement, resulting in the
<u>reward</u> V(y,u). Depending on the context, the reward function
V(y,u) may describe pensionable service, pension benefits or value
of wages received. Clearly, if we take V(y,u) = u, u $\geqslant$ s, then
the reward function represents pensionable service. Appropriate
specifications for pension benefits in flat-benefit (FB), career-
average (CA), final-earnings (FE), and money-purchase (MP) plans
would be, for $y \geqslant u \geqslant s \geqslant 0$:

$$
(2) \quad V(y,u) = \begin{cases} fu & \text{(FB)} \\[2mm] b \int_{y-u}^{y} W(x)\,dx & \text{(CA)} \\[2mm] bu\,W(y - u) & \text{(FE)} \\[2mm] \dfrac{d}{a} \int_{y-u}^{y} e^{rx} W(x)\,dx & \text{(MP)} \end{cases}
$$

where W(x) is the wage at x time units <u>before</u> retirement, f is
the fixed benefit per unit time (year) of pensionable service, b
is the benefit level in defined benefit plans (which may differ
from CA to FE plans), d is the contribution level, a is the cost
of unit annuity purchases at retirement, and r is the investment
rate of return. We take the wage function W(x) as continuous and
normalized such that W(0) = 1. (Note that W(0) is the wage at
retirement). This is for the representation of pension benefits as
a fraction of the wage at retirement (i.e., as a <u>replacement rate</u>).
Specification of the reward function V(y,u) in terms of the
value of wages received leads to characterizations of pension costs
(see Balcer and Sahin, 1983).

Accumulation in time of rewards (pensionable service, pension benefits, or pension costs) can now be modelled by superimposing the reward structure on the semi-Markov process introduced. Let

$$R_i(s,t) \ , \ t \geqslant s \geqslant 0, \ i = 1,2,$$

denote the sum of rewards, accumulated during $(0,t]$ under the vesting rule s, conditional on the status i of the initial employment $(i = 1$: covered, $i = 2$: non-covered). Let

$$\mu_i^{(n)}(s,t) = E[R_i(s,t)^n], \ n = 1,2,\ldots, \ ; \ \mu_i^{(0)}(s,t) \equiv 1.$$

It is proved in Balcer and Sahin (1983) that :

$$(3) \quad \mu_i^{(n)}(s,t) = \delta_{i1}Q_n(t) + \int_0^{t-s} Q_n(t - u)m_{i1}(u)du$$

where

$$(4) \quad Q_n(t) = V(t,t)^n[1 - K_1(t)]$$

$$+ \ \sum_{j=1}^{2} \sum_{m=1}^{n} \binom{n}{m} \int_s^t V(t,y)^m \mu_j^{(n-m)}(s,t - y)a_{ij}(y)dy$$

This establishes recursively all the moments of the ultimate reward resulting from a career membership in pension plans. For the expected reward, we have :

$$(5) \quad \mu_i(s,t) = \delta_{i1}[V(t,t)[1 - K_1(t)] + \int_s^t k_1(y)V(t,y)dy]$$

$$+ \int_{u=s}^t m_{i1}(t - u)[V(u,u)[1 - K_1(u)] + \int_{y=s}^u k_1(y)V(u,y)dy]du$$

In applications to pension benefits, $\mu_i(s,t)$ was called <u>termination benefits function</u> in Balcer and Sahin (1983). It would underestimate expected retirement benefits if the last employment that terminates with retirement must be pensionable irrespective of its length, if it is covered. In the United States, for example, ERISA requires that an employee must be fully vested in his or her accrued benefit when he or she attains the normal or

stated retirement age, regardless of the vesting rules in effect.
In the above, the last employment, as all others, is regarded as
pensionable only if it lasts s time units. If we take the last
employment pensionable (provided that it is covered) and denote by
$\bar{\mu}_i(s,t)$ the expected value of total pension benefits at retirement,
we find (see Balcer and Sahin, 1983) for $t \geqslant s \geqslant 0$, $i = 1,2$, that :

$$(6) \quad \bar{\mu}_i(s,t) = \mu_i(s,t) + \int_0^s m_{i1}(t - u) [1 - K_1(u)] V(u,u) du$$

This function was called the <u>retirement benefits function</u>.

To isalte the loss in expected pension benefits due to delayed
vesting after s years of service, relative to full and immediate
vesting (s = 0), a third function, called the <u>loss function</u>, was
also introduced in Balcer and Sahin (1983) as :

$$(7) \quad L_i(s,t) = \bar{\mu}_i(0,t) - \bar{\mu}_i(s,t)$$

$$= \int_{y=0}^s k_1(y) [\delta_{i1} V(t,y) + \int_{u=y}^t V(u,y) m_{i1}(t - u) du] dy$$

These functions were investigated in Balcer and Sahin (1983),
both analytically and numerically, in terms of the impact on lifetime pension benefits of plan types, vesting rules, coverage, and
portability. They were also linked to pension costs under non-
contributory (only the employer contributes into the pension fund)
and contributory (employer and employee both contribute) pension
plans. Some of the results obtained show that the impact of policy
parameters reflecting vesting, coverage, and portability cannot be
isolated from the plan type; different plans would have substantially different responses to changes in these parameters. It was
proved, for example, that while the marginal incremental change in
expected termination benefits would be larger in FE plans than in
CA plans, the marginal relative change would be smaller in response
to a change in the vesting rule. It was also demonstrated that
while being more sensitive to changes in the vesting rules, MP
plans result in smaller relative variabilities in termination

benefits than FE plans. These observations raise some questions
about the effectiveness of statutory vesting provisions as a regu-
latory instrument in the face of potential market response by way
of changes in plan types and other plan characteristics. However,
attempts to further regulate the private pension system in North
America in terms of plan types and benefit levels have not been
successful. Instead, further government regulation appears to be
heading in the direction of more liberal vesting provisions, a
degree of indexation to maintain the real value of pensions during
an inflationary environment, and, possibly, higher rates of coverage.

In the context of pension reform through legislation, an
important consideration on the part of all economic agents is
related to the evolution of pension benefits and costs as the
"reform package" takes effect. Thus, suppose, for example, that
the statutory vesting rule is changed from 10 to 5 years of service.
What would the replacement rates at retirement be for cohorts who
were of various ages at the time the change is legislated ? What
are the additional liabilities generated by the change over the
remaining work life of each age cohort ? In the next section we
provide a framework for the investigation of these issues, through
some extensions of the semi-Markov reward model.

3. Dynamics of pension reform

Although the North America pension legislation in place has
been working well in terms of its original purpose, the private
sector has been encountering renewed criticism for its inability
to provide adequate pensions. Pension coverage has stabilized
under 50 percent and portability is generally lacking. In addition,
it is becoming increasingly difficult for employment pension plans
to maintain the real value of pensions during an inflationary
environment. The statutory vesting rule of 10 years of service
(or its equivalents) in the United States, and 10 years of service

at age 45 in Canada, once regarded as substantial improvements, are now considered too stringent.

Studies in the North American context suggest that, at low to moderate income levels, a preretirement income equivalency would be attained by a postretirement income of about 75 percent of the preretirement total. The social security primary benefit can account for about 40 percent of the final earnings in these income groups, leaving 35 percent substantially to private pension benefits. But if we use, as a reasonable expectation from the employment pension system, a replacement ratio of 35 percent, results in Balcer and Sahin (1983) show that far more "generous" plan designs than those in place today are needed, and that under the existing vesting provisions, the "pension promise" could be fulfilled only for relatively immobile workers covered by final earnings and money purchase plans.

While it could be argued that the primary responsibility to provide for the retirement years should rest with the individual, it is generally recognized in North America that the provision of a reasonable level of retirement income cannot be expected from the current system of employer plans under the legislation in place. Additional government compulsion is, therefore, needed. Public debate on the nature and form of such "reform" legislation is now entering its final phase in Ontario, after a long period of hearings and studies since 1978 (see Report of the Royal Commission on the Status of Pensions in Ontario, 1980). Some of the possible policy initiatives discussed in Ontario during this period have been : inflation indexation of benefits, more liberal vesting rules, and increased rates of pension coverage, possibly through a mandatory employer pension plan. Pension benefit and pension cost implications of these and other proposed changes in the employment pension system of Ontario were investigated by the authors for the Ontario

Government in Balcer and Sahin (1980, 1982 and forthcoming). These studies were based on discrete models constructed under the assumptions of full coverage and no portability. Two different modes of analysis were used : (1) a "steady state" analysis of comparing pension benefits and pension costs before the introduction of a "reform package" with those benefits and costs after the reform package has fully impacted, and (2) a "dynamic analysis" of describing the evolution of pension benefits and costs as the reform package takes effect. In the sequel, we extend the semi-Markov reward model to the second mode of analysis in relation to reform initiatives involving vesting rules and coverage rates.

3.1. Vesting

Suppose that as part of a new pension legislation, the statutory vesting rule will be changed from s_1 to s years of service $(s_1 > s)$ as of a given effective date. Individuals entering the system after the effective date will be fully affected by the change, while those retiring before this date will not be affected at all. Of interest are cohorts that will be affected by the change during a part of their working lives. Consider such an age cohort of workers who entered their working lives t_1 time units before the effective date and who plan to retire at working age $t \geqslant t_1$. All pension benefits related to terminations during $(0,t_1]$ are understood to have taken place under the vesting rule s_1, and all terminations during $(t_1,t]$ will be creditable under the vesting rule s. Let $\mu_i(s_1,t_1;s,t)$ denote the expected total termination benefits during $(0,t]$. (We assume, unless noted otherwise, that the probabilities c and π, related to coverage and portability, and the CLS distributions in covered and non-covered employments remain stationary over time).

Theorem 1

$$(8) \quad \mu_i(s_1,t_1;s,t) = \mu_i(s,t) - \int_{y=s}^{\mathrm{Min}(s_1,t_1)} k_1(y)\, [\delta_{i1} V(t,y)$$

$$+ \int_{u=y}^{t_1} m_{i1}(t_1 - u) V(t - t_1 + u,y)\, du]\, dy$$

where $\mu_i(s,t)$ is given by (5).

Proof

Let $R_i(s_1,t_1;s,t)$ be the cumulative reward (benefit) under the conditions stated above. Let $R_i(s,t)$ be the cumulative reward during $(0,t]$ as defined in Section 2, and $R_i^{(t)}(w,t_1)$ the cumulative reward during $(0,t_1]$, accumulated under the vesting rule w, and expressed as a fraction of the wage at t. In addition, let $X(y)$ denote the state of the semi-Markov process at time y and $U(y)$ the time since the last transition at time y. For $t_1 \geqslant s_1$ ($> s$) we can write :

$$(9) \quad R_i(s_1,t_1;s,t) = \begin{cases} R_i(s,t) - R_i^{(t)}(s,t_1) + R_i^{(t)}(s_1,t_1) + V(t-t_1+u,u) \\ \qquad \text{if } X(t_1) = 1 \text{ and } s < U(t_1) \equiv u < s_1 \\[6pt] R_i(s,t) - R_i^{(t)}(s,t_1) + R_i^{(t)}(s_1,t_1), \text{ otherwise} \end{cases}$$

We note that the same relationship also holds for $t_1 < s_1$ with the interpretation that $R_i^{(t)}(s_1,t_1) = 0$. We have :

$$(10) \quad P\,[X(t_1) = 1,\ u - du < U(t_1) \leqslant u] = m_{i1}(t_1 - u)\,[1 - K_1(u)]\,du,$$

$$u < t_1$$

$$(11) \quad P\,[X(t_1) = 1,\ U(t_1) = t_1] = \delta_{i1}\,[1 - K_1(t_1)].$$

On passing to expectations in (9) we obtain :

$$(12) \quad \mu_i(s_1,t_1;s,t) = \mu_i(s,t) - \mu_i^{(t)}(s,t_1) + \mu_i^{(t)}(s_1,t_1)$$

$$+ \int_s^{s_1} V(t - t_1 + u,u)\, m_{i1}(t_1 - u)\,[1 - K_1(u)]\,du,\ t_1 \geqslant s_1$$

$$(13) \quad \mu_i(s_1,t_1;s,t) = \mu_i(s,t) - \mu_i^{(t)}(s,t_1)$$

$$+ \int_s^{t_1} V(t - t_1 + u,u)m_{i1}(t_1 - u)[1 - K_1(u)]\,du$$

$$+ \delta_{i1} V(t,t_1)[1 - K_1(t_1)], \quad t_1 < s_1.$$

where $\mu_i^{(t)}(w,t_1)$ can be written by a slight modification of (5) as :

$$(14) \quad \mu_i^{(t)}(w,t_1) = \delta_{i1} V(t,t_1)[1 - K_1(t_1)] + \int_w^{t_1} k_1(y)V(t,y)\,dy$$

$$+ \int_{u=w}^{t_1} m_{i1}(t_1 - u)[V(t - t_1 + u,u)[1 - K_1(u)]$$

$$+ \int_{y=w}^{u} k_1(y)V(t - t_1 + u,y)\,dy]\,du.$$

If the CLS distributions and the estimates for probabilities c and π are different for periods $(0,t_1]$ and $(t_1,t]$, $\mu_i(s,t)$ and $\mu_i^{(t)}(w,t_1)$ could be computed separately with applicable data, with a further decomposition of $\mu_i(s,t)$. Under appropriate assumptions of stationarity, substitution of (14) in (13) and (12) results in (8). This completes the proof. $\square$

The first term in the right-hand side of (8) is the expected termination benefits if the cohort in question were subjected to the more liberal vesting requirement during its entire working life. The second term, then, accounts for the reduction related to having been subjected to the more stringent rule s_1 during $(0,t_1]$. An alternative expression can be written down in terms of termination benefits during $(0,t]$ under s_1, $\mu_i(s_1,t)$, plus an increase induced by liberalization during $(t_1,t]$. This expression turns out to be, for $t_1 \leq s_1$:

$$(15) \quad \mu_i(s_1,t_1;s,t) = \mu_i(s_1,t) + \int_s^{s_1} m_{i1}(t - u)V(u,u)[1 - K_1(u)]\,du$$

$$+ \int_{y=s}^{Min(s_1,t_1)} k_1(y) \int_{u=y}^{t-t_1+y} m_{i1}(t - u)V(u,y)\,du\,dy$$

$$+ \int_{y=t_1}^{s_1} k_1(y) [\delta_{i1} V(t,y) + \int_{u=y}^{t} m_{i1}(t-u)V(u,y)du]dy$$

The same expression holds for $t_1 > s_1$, with the last term missing.

As t_1 increases, $\mu_i(s_1,t_1;s,t)$ provides a basis for the evolution of expected pension benefits and related costs as the liberalization takes effect. Since t_1 represents the working age at inception, as t_1 increases, we expect a decrease in expected replacement rate at retirement. Conversely, as $t - t_1$ increases, this measure will increase as the replacement rate at retirement of a younger cohort. This observation leads to a measure representing the rate of evolution of pension benefits from inception to "full impact". If we denote this rate by $r(u)$, $0 \leq u \leq t - s$, we have from (15) that :

$$(16) \quad r(u) = \frac{d}{du} \mu(s_1, t - u; s, t)$$

$$= \delta_{i1} k_1(t - u)V(t, t - u)$$

$$+ \int_{s}^{Min(t-u,s_1)} k_1(y)m_{i1}(t-u-y)V(u+y,y)dy,$$

$$u \geq t - s_1$$

For $0 \leq u < t - s_1$, the same expression holds without the first term.

3.2. Coverage

In the semi-Markov reward model, partial coverage is accounted for in terms of the probability c of moving into a covered job upon termination of the previous employment. This representation implies that the probability of moving into a covered job is typical of the particular jobs an individual is likely to encounter, or that an individual chooses his jobs at random. The second interpretation requires that the parameter c be formally related to termination and coverage rates. Based on the stationary behavior of the underlying semi-Markov process, an approximate relationship was derived in (Balcer and Sahin, 1983) to be :

$$(17) \quad c = \frac{p\gamma_1}{p\gamma_1 + (1 - p)\phi_1}$$

where p is the proportion of covered jobs and γ_1 and ϕ_1 are the means of the CLS distributions $G(\cdot)$ and $F(\cdot)$, respectively. Note that the parameter p is readily observable.

Suppose now that the coverage rate is increased by legislation to 1 as of an effective date. Suppose further that all pension plans are to be portable after the effective date. Such a legislative proposal has been formulated and submitted to the Ontario Government by the Royal Commission on the Status of Pensions in Ontario. Called PURS (for Provincial Universal Retirement System) this proposal calls for the institution by legislation of a mandatory pension plan with full and immediate vesting and complete portability, for all workers aged 18 to 64 (see Report of the Royal Commission on the Status of Pensions in Ontario, 1980). Not surprisingly, one of the strongest criticisms directed to PURS has been that the private pension system would be unable to meet the steep increases in pension costs implied by the expected increases in pension benefits. However, such criticisms have been generally speculative in nature.

As in Section 3.1, consider an age cohort of workers who entered their working lives t_1 time units before the effective date and who plan to retire at working age $t \geqslant t_1$. Let $\mu_i(s,t_1;t)$ denote the expected total pension benefits during $(0,t]$, accumulated under the vesting rule s and coverage and portability measures c and π during $(0,t_1]$, and under full coverage $(c = 1)$ and complete portability $(\pi = 1)$ during $(t_1,t]$. Note that the vesting rule does not play any rôle during $(t_1,t]$, and that the pension reform in question represents the ultimate liberalization of the employment pension system. We now show the following :

Theorem 2

$$(18)\quad \mu_i(s,t_1;t) = \delta_{i1}V(t,t)\,[1 - K_1(t_1)]$$

$$+ \int_{y=s}^{t_1} k_1(y)\{\delta_{i1}V(t,y) + \int_{u=y}^{t_1} m_{i1}(t_1-u)V(t-t_1+u,y)\,du\}dy$$

$$+ V(t-t_1,t-t_1)\{\delta_{i2}\,[1 - K_2(t_1)]$$

$$+ \int_0^{t_1} m_{i2}(t_1-u)\,[1 - K_2(u)]\,du\}$$

$$+ \int_0^{t_1} V(t-t_1+u,t-t_1+u)m_{i1}(t_1-u)\,[1-K_1(u)]\,du,\quad t \geqslant t_1.$$

Proof

Let $R_i(s,t_1;t)$ be the cumulative reward during $(0,t]$, and $R_i^{(t)}(s,t_1)$ the reward during $(0,t_1]$ as defined in section 3.1. We can write :

$$(19)\quad R_i(s,t_1;t) = \begin{cases} R_i^{(t)}(s,t_1) + V(t-t_1,t-t_1),\ \text{if}\ \ X(t_1) = 2 \\[2ex] R_i^{(t)}(s,t_1) + V(t-t_1+u,t-t_1+u), \\[1ex] \qquad \text{if}\ \ X(t_1) = 1\ \ \text{and}\ \ U(t_1) \equiv u < s \\[2ex] R_i^{(t)}(s,t_1) + V(t-t_1+u,t-t_1+u) \\[1ex] \qquad - V(t-t_1+u,u), \\[1ex] \qquad \text{if}\ \ X(t_1) = 1\ \ \text{and}\ \ U(t_1) \equiv u \geqslant s \end{cases}$$

where $X(t_1)$ is the state at time t_1 and $U(t_1)$ is the time since the last transition at time t_1. The joint distribution of $X(t_1)$ and $U(t_1)$ is given by (10) and (11). We also have :

$$(20)\quad P\,[X(t_1) = 2] = \delta_{i2}\,[1 - K_2(t_1)] + \int_0^{t_1} m_{i2}(t_1-u)\,[1 - K_2(u)]\,du$$

The result follows on passing to expectations in (19) and using (14) for $E\,[R_i^{(t)}(s,t_1)]$. $\square$

As an example, we consider the evolution of pensionable service (PS). On taking $V(s,y) = y$ in (18),we obtain, after some simplifications, that :

$$(21)\quad \mu_i^{(PS)}(s,t_1;t) = t - t_1 + \int_0^{t_1} [\delta_{i1} + M_{i1}(t_1 - u)] [1 - K_1(u)]\,du$$

$$- \int_0^{Min(s,t_1)} uk_1(u) [\delta_{i1} + M_{i1}(t_1 - u)]\,du$$

From this expression, the expected rate of increase in pensionable service, from inception to full impact of the reform, can be obtained as :

$$(22)\quad r(u) = \delta_{i2}[1 - K_2(t - u)] + \int_0^{t-u} m_{i2}(t - u - y) [1 - K_2(y)]\,dy$$

$$+ \delta_{i1}(t - u)k_1(t - u) + \int_0^{Min(s,t-u)} yk_1(y)m_{i1}(t - u - y)\,dy,$$

$$u \geqslant t - s$$

For $u < t - s$, the same result holds without the third term.

4. Concluding remarks

We have considered two simple extensions of the semi-Markov model of pension accumulation to the dynamics of pension reform. These have been in terms of expected benefits. The results obtained can be converted easily to those involving expected pension costs, using the methodology outlined in Balcer and Sahin (1983). However, although the basic model has been solved for all moments of the total pension benefits over time (cf. (3)), it is not clear how higher moments of total benefits defined in Section 3 could be characterized in useful forms. The problem of determining the distribution of total pension benefits also remains to be unsolved even for the basic model.

Distribution of pension benefits and a number of distributional measures of public policy interest were computed and discussed in

Balcer and Sahin (1979 and 1982) by way of a set of discrete
models, under the assumptions of full coverage and no portability.
Within the framework of the semi-Markov model, probability of no
vested pension over a given period of time was characterized explicitly and computed in Sahin (1984) as a first passage time
problem. It is not difficult to provide some recursive expressions
for the distribution of benefits, but useful results are difficult
to obtain.

Finally, as we pointed out elsewhere, we assumed in the above
developments that labor mobility is independent of the wage-pension
mix and plan characteristics. This avoids the estimation problems
related to the extent to which varying plan parameters would affect
labor mobility or cause compensating adjustments in wages. In
principle, prospects for higher pension benefits induced by a given
liberalization in vesting rules, for example, may result in higher
employee mobility and higher costs to employers due in part to
additional hiring and training. This in turn may lead to lower
rates of coverage as some plans may be terminated. To incorporate
these and other relationships assumed away in this paper, a coherent
model of behavior maximization is needed. In spite of these limitations, however, the methodology presented in this and related
papers constitute an improvement over the traditional literature.
Most of this literature is based on even more idealized scenarios
in which there is full coverage, no portability and no job mobility
(i.e., a single employer).

References

Balcer, Y. and I.Sahin (1979). Probabilistic Models for Pension
     Benefits, The Journal of Risk and Insurance, 46, 99-123.
Balcer, Y. and I.Sahin (1980). A Study of Private Pensions in
     Ontario. Report of the Royal Commission on the Status of

Pensions in Ontario. Volume 9 : Background Studies and Papers, Government of Ontario, 151-207.

Balcer, Y. and I.Sahin (1982). Modelling the Impact of Pension Reform - A Case Study. The Journal of Risk and Insurance, 49, 158-191.

Balcer, Y. and I.Sahin (1983). A Stochastic Theory of Pension Dynamics. Insurance : Mathematics and Economics, 2, 179-197.

Balcer Y. and I. Sahin (1984). Dynamics of Pension Reform: The Case of Ontario. The Journal of Risk and Insurance, 51, 652-686.

Report of the Royal Commission on the Status of Pensions in Ontario. (1980), (9 volumes), Government of Ontario.

Sahin, I. (1978). Cumulative Constrained Sojourn Times in Semi-Markov Processes with an Application to Pensionable Service. Journal of Applied Probability, 15, 531-542.

Sahin, I. (1984). Bruce's Spider and the Employee's Risk under a Pension System. The Journal of Risk and Insurance, 51, 143-149.

Sahin, I. and Y.Balcer (1979). Stochastic Models for Pensionable Service. Operations Research, 27, 888-903.

THE STRUCTURE OF A FIRM'S OPTIMAL NON-DECREASING WAGE POLICY WHEN
RECRUITMENT IS A WAGE DEPENDENT POISSON PROCESS

Nils Henrik Schager

Industrial Institue for Economic and Social Research

Grevgatan 34, S-114 53 Stockholm, Sweden

1. Introduction

During the last fifteen years there has been a growing number
of studies, in which the firm is supposed to face a labor market
where job applicants are imperfectly informed about the exact
location of job and wage offers. Job matching is thus characterized
by search under uncertainty. Nonetheless the models used assume
almost universally that the firm acts as if the flows of labor were
known with certainty at different levels of its wage offer. Consequently the optimal wage policy is derived by using deterministic
control methods (see e.g. Leban, 1982a; Mortensen, 1971; Pissarides,
1976; Salop, 1973; Virén, 1979).

To my knowledge there has as yet been no attempt to apply the
established theory of stochastic processes and their control to
derive analytical properties of a firm's wage and employment
policy. However, it is really a straightforward procedure to formulate the labor flows to and from the firm in terms of such birth-
and-death-processes that are used e.g. in queueing models. One
reason, why this approach has been unpopular among economists, is

admittedly the difficulty in obtaining any strong analytical results
as soon as the model reaches even a modest level of complexity.

In this paper a simple model of a firm's optimal wage and
recruitment policy will be presented, from which analytical results
can be derived. The structure of the optimal wage policy will be
established as well as the closed form of the value function of the
firm's optimization problem. For a further discussion of the
economic interpretation of the properties of solution, especially
in the form of "comparative statics" analysis, see (Schager, forthcoming).

Before we embark on the formal treatment, let me point to the
assumptions that are crucial for arriving at analytically tractable
results. First of all, there exist only hires - "arrivals" - and
no quits - "departures" - in the process to be controlled by the
firm's wage level. This is really the most disturbing simplification in the present model. Further research should give priority
to amending the model in this respect. Secondly, the assumption
that only such wage control policies are feasible that are non-
decreasing in time might at first blush seem to complicate matters.
While this is true in the first stage of the analysis, when the
general structure of the optimal policy is to be derived, we have
the off-setting gain that we obtain directly a closed form solution
of the value functional equation. It should be stressed that this
assumption is introduced not because of analytical convenience but
because it reflects accurately observed wage behavior (at least
according to European writers, cf. Leban, 1982b; Virén, 1979). The
remaining assumptions of the model have indeed a simplifying character, but not to such an extent that would endanger its usefullness
as a first approximation of a real-life firm's recruitment situation
- at least not as long as the search theory approach has any
relevance at all for depicting labor market phenomena.

It might also be pointed out that as far as the analysis of this paper goes the assumption of downward wage rigidity is the outstanding feature that distinguishes the present model from the formally related models in (Deshmukh and Chitke, 1976; Deshmukh and Winston, 1979; Lippman, 1980).

2. Basic assumptions

The firm is in the model regarded as a simple one-product production unit, using homogeneous labor as the only variable factor of production. The production volume is linear in the number of employees, i, up to a fixed capacity limit (in terms of i), N. The constant productivity is denoted a. All production is sold at a fixed price p.

At any instant of time, t, all employees are paid the same wage rate, $v(t)$. We assume that $v(t)$ is non-decreasing in time.

When the wage v is offered by the firm (and vacancies are announced), job-applicants are recruited according to a Poisson process with the intensity $\lambda(v) = \gamma \cdot F(v)$. γ is the intensity of a Poisson process, describing the realization of contacts between job-applicants and the firm. It reflects labor market conditions and is here regarded as fixed. $F(\cdot)$ is the distribution function of the job-applicants' reservation wages; according to results from labor market search literature, each job-applicant should under fairly wide conditions apply a search strategy imply-ing the calculation of a reservation wage, such that a job offer should be accepted if and only if the corresponding wage offer exceeds the reservation wage. If contacts between the firm and job-applicants occur independently of the applicants' reservation wages, the recruitment process is of the postulated character.

Little is known about the job-applicants' reservation wages in practice. For the purpose of this paper it is enough, however, to make the weak assumption that the reservation wages are distributed over job-applicants according to a probability density function $f(v)$ which is continuous over an interval of wage levels with a lower bound $\underline{v} \geqslant 0$.

3. The decision problem

We suppose that the firm pursues the traditional objective of maximizing the net worth of its production opportunities, i.e., it maximizes the expected total discounted profits over an horizon, that we assume extends to infinity. Generally speaking, the firm faces the problem of choosing a higher (or an unchanged) wage level such that the larger costs incurred by the higher wage is balanced against the increased probability that higher employment states and hence higher profits rates will be reached more rapidly in the future.

As the interarrival times of successive hires are exponentially distributed in the Poisson recruitment process, we may without loss of generality restrict the class of possible wage increase policies to those where increases take place only at those points of time when a transition to a higher employment state occurs (cf. Stidham, 1974). Thus we can write the profit rate function as follows :

$$r(i,w_i,x_i) = i \cdot [pa - (w_i + x_i)], \quad i = 1,2,\ldots,N; \quad x_i \geqslant 0$$

where w_i is the wage level before the transition into employment state i and x_i is the wage increase chosen as this transition has occurred. Obviously it holds that $w_{i+1} = w_i + x_i$.

We have implicitly assumed that the firm can stop recruiting whenever it wants to (by not announcing vacancies). There is really a vacancy creation control under the surface in our problem, but its rôle is trivial. It is easy to convince oneself that it is always optimal for the firm to recruit as long as $i < N$ (and $w_i < pa$), but likewise optimal to stop recruiting as soon as i reaches N. Hence we need only consider employment states $i \leqslant N$. (It is even more easy to realize, that $w_i > pa$ will lead to a stop of recruitment process).

It is well known (see e.g. Stidham, 1974) that the functional equation corresponding to our firm's decision problem reads (α being the discount intensity) :

$$(1) \begin{cases} H^*(i,w_i) = \max_{x_i \geqslant 0} \{H(i,w_i,x_i)\}, \\[2ex] H(i,w_i,x_i) = \dfrac{r(i,w_i,x_i)}{\lambda(w_i + x_i) + \alpha} + \dfrac{\lambda(w_i + x_i)}{\lambda(w_i + x_i) + \alpha} \cdot H^*(i+1,w_{i+1}) \\[2ex] \qquad i = 0,1,\ldots,N-1. \\[2ex] H(N,w_N,x_N) = \dfrac{r(N,w_N,x_N)}{\alpha} \end{cases}$$

The last equation reflects the terminal condition that the process is stopped when $i = N$ is reached and that the firm will earn a profit rate $r(N,w_N,x_N)$ forever after. We immediately observe that it must hold

$$H^*(N,w_n) = \frac{r(N,W_N,0)}{\alpha} = \frac{N(pa - w_N)}{\alpha}$$

4. The structure of the optimal wage policy : some preparatory results

In order to derive the optimal wage – or more strictly expressed wage increase – policy we will make use of the following

Lemma 1

Let $V(n,w)$ be the expected total discounted profits of establish-

ing the wage level w when employment state n is reached and
keeping that wage level fixed, regardless of all future employment
states.   Then $V(n,w) = \dfrac{(pa - w)}{\alpha} \{n + \dfrac{\lambda}{\alpha} [1 - (\dfrac{\lambda}{\lambda + \alpha})^{N-n}]\}$, $\lambda = \lambda(w)$,
$n = 0,1,\ldots,N$.

<u>Proof</u>

More illuminating proofs, based on first principles, may be given.
However, the shortest one is obtained by writing down directly the
function equation corresponding to any choice of w

$$V(n,w) = \frac{(pa - w) \cdot n}{\lambda + \alpha} + \frac{\lambda}{\lambda + \alpha} \cdot V(n + 1,w), \quad n = 0,\ldots,N-1$$

$$V(N,w) = \frac{(pa - w) \cdot N}{\alpha}$$

and observing that this is equivalent to the following first order
linear difference equation with constant coefficients

$$y_j = \frac{\lambda}{\lambda + \alpha} \cdot y_{j-1} + \frac{pa - w}{\lambda + \alpha} \cdot N - \frac{pa - w}{\lambda + \alpha} \cdot j$$

$$y_0 = \frac{pa - w}{\alpha} \cdot N$$

where we have substituted $j = N - n$ for n in order to transform
our terminal condition into an initial one.

    Standard theory of linear difference equations offers the
general solution

$$y_j = \frac{pa - w}{\alpha} \cdot [N - j + \frac{\lambda}{\alpha}] + A \cdot (\frac{\lambda}{\lambda + \alpha})^j$$

and the initial condition determines

$$A = - \frac{\lambda}{\alpha^2} (pa - w)$$

Substituting back $n = N - j$ and noting that to be j employees
from the capacity limit N is to have n employees, so that
$y_j = V(n,w)$, the desired result obtains and the proof is completed. $\square$

We also need the following

Lemma 2

The derivative of $V(n,w)$ with respect to w, $V'(n,w)$, is strictly decreasing in n, $0 \leq n \leq N - 1$, $0 \leq w < pa$.

Proof

From the result of the previous lemma we can easily express $V(n,w)$ in a recursive fashion.

$$V(n,w) = V(N,w) - \sum_{j=n}^{N-1} [V(j + 1,w) - V(j,w)]$$

$$= \frac{(pa - w) \cdot N}{\alpha} - \sum_{j=n}^{N-1} \frac{pa - w}{\alpha} [1 - (\frac{\lambda}{\lambda + \alpha})^{N-j}].$$

Denoting by V' the derivative of V with respect to w we then have

$$\alpha V'(n,w) + N = \sum_{j=n}^{N-1} [1 - (\frac{\lambda}{\lambda + \alpha})^{N-j}]$$

$$+ \frac{(pa - w) \cdot \alpha \cdot \lambda'}{[\lambda + \alpha]^2} \cdot \sum_{j=n}^{N-1} (N - j)(\frac{\lambda}{\lambda + \alpha})^{N-j-1}.$$

As $\lambda' = \gamma \cdot f(w) \geq 0$, every term in the summations on the R.H.S. of the equation is non-negative and in the first summation they are strictly positive for every w, $0 \leq w < pa$. An increase in n thus means a deletion of positive and possibly zero terms in $V'(n,w)$. Consequently, $V'(n,w)$ must be strictly decreasing in n for every w, $0 \leq w < pa$, and $n = 0,\ldots,N-1$, and the lemma is proved. $\square$

5. The structure of the optimal wage policy : basic results

We now return to the basic functional equation (1) and repeat with minor modifications in notation

$$
\begin{cases}
H^*(i,w_i) = \max_{x_i \geqslant 0} \{H(i,w_i,x_i)\} \\[2em]
H(i,w_i,x_i) = \dfrac{[pa - (w_i + x_i)] \cdot i}{\lambda(w_i + x_i) + \alpha} + \lambda_\alpha(w_i + x_i) \cdot H^*(i+i,w_{i+1}), \\[1em]
\qquad i = 0,1,\ldots,N-1. \\[1em]
H^*(N,w_N) = \dfrac{(pa - w_N) \cdot N}{\alpha} \\[2em]
\lambda_\alpha = \dfrac{\alpha}{\lambda + \alpha}; \quad w_{i+1} = w_i + x_i.
\end{cases}
$$

Let us look at the situation when state $(N - 1, w_{N-1})$ is reached.
The firm will make a wage increase $x_{N-1} \geqslant 0$ and establish the
wage level $w_{N-1} + x_{N-1} = w_N$ to hold for all future. Consequently,
recalling the definition of $V(n,w)$ from Lemma 1, we have

$$
H(N - 1, w_{N-1}, x_{N-1}) = V(N - 1, w_{N-1} + x_{N-1})
$$

and

$$
H^*(N - 1, w_{N-1}) = \max_{x \geqslant 0} \{V(N - 1, w_{N-1} + x)\} = V(N - 1, w_{N-1} + x^*_{N-1}).
$$

We are now in the position to state the following

Proposition

When the firm is in the initial state (n, w_n), $0 \leqslant n < N$, $0 \leqslant w_n$
$< pa$, the optimal wage policy is to make one instantaneous increase
$x^*_n \geqslant 0$ and keep the resulting wage level $w_n + x^*_n$ fixed for all
future.

Proof

Let the firm have entered state (i,w_i), $n \leqslant i \leqslant N-2$, and assume
that it has been established that the optimal policy when state
$(i + 1, w_{i+1})$ is reached is to make one wage increase $x^*_{i+1} \geqslant 0$
and keep the resulting wage level $w_{i+1} + x^*_{i+1} = w_{i+2}$ fixed for all
future. This corresponds to the following formulation of (1) :

$$H^*(i,w) = \max_{x_i \geqslant 0} \; H(i,w_i,x_i)$$

$$H(i,w_i,x_i) = \frac{[pa - (w_i + x_i)] \cdot i}{\lambda(w_i + x_i) + \alpha} + \lambda_\alpha(w_i + x_i)$$

$$\cdot \; V(i + 1, w_{i+1} + x^*_{i+1})$$

where $\;V(i + 1, w_{i+1} + x^*_{i+1}) = \max_{x \geqslant 0} \{V(i + 1, w_{i+1} + x)\} = H^*(i + 1, w_{i+1})$

Some rearranging yields

$$H(i,w_i,x_i) = V(i,w_i + x_i) + \lambda_\alpha(w_i + x_i) \cdot \int_0^{x^*_{i+1}} V'(i + 1, w_{i+1} + x)dx.$$

We intend to show that an optimal choice of $\;x_i\;$ such that $\;H^*(i,w_i)$ is obtained, implies $\;x^*_{i+1} = 0$.

Let $\;\hat{x}_i \geqslant 0\;$ be any choice of $\;x_i\;$ that establishes a $\hat{w}_{i+1} = w_i + \hat{x}_i\;$ such that the corresponding $\;x^*_{i+1} = \hat{x}^*_{i+1} > 0$. The value equation, when applying the policy $\;(\hat{x}_i; \hat{x}^*_{i+1})\;$ is

$$H(i,w_i,\hat{x}_i) = V(i,\hat{w}_{i+1}) + \lambda_\alpha(\hat{w}_{i+1}) \cdot \int_0^{\hat{x}^*_{i+1}} V'(i + 1, \hat{w}_{i+1} + x)dx$$

Consider now another, strictly different policy $\;x_i = \hat{x}_i + \hat{x}^*_{i+1}$; $x_{i+1} = 0$. The corresponding value of $\;w_{i+1}\;$ is then $\;\hat{w}_{i+1} + \hat{x}^*_{i+1}$. The new policy gives the value equation

$$U(i,w_i,\hat{x}_i + \hat{x}^*_{i+1}) = V(i,w_i + \hat{x}_i + \hat{x}^*_{i+1})$$

$$= V(i,\hat{w}_{i+1}) + \int_0^{\hat{x}^*_{i+1}} V'(i,\hat{w}_{i+1} + x)dx$$

So we have

$$H(i,w_i,\hat{x}_i) - U(i,w_i,\hat{x}_i + \hat{x}^*_{i+1})$$

$$= \lambda_\alpha(\hat{w}_{i+1}) \cdot \int_0^{\hat{x}^*_{i+1}} V'(i + 1, \hat{w}_{i+1} + x)dx - \int_0^{\hat{x}^*_{i+1}} V'(i,\hat{w}_{i+1} + x)dx$$

According to Lemma 2 it must hold that

$$\int_0^{\hat{\hat{x}}^*_{i+1}} V'(i,\hat{w}_{i+1} + x)dx - \int_0^{\hat{\hat{x}}^*_{i+1}} V'(i + 1,\hat{w}_{i+1} + x)dx > 0.$$

As $\hat{\hat{x}}^*_{i+1} > 0$ by assumption, we have

$$\int_0^{\hat{\hat{x}}^*_{i+1}} V'(i + 1,\hat{w}_{i+1} + x)dx \geqslant 0$$

and as we also have $0 \leqslant \lambda_\alpha(\hat{w}_{i+1}) < 1$, we conclude that

$$H(i,w_i,\hat{x}_i) - U(i,w_i,\hat{x}_i + \hat{x}^*_{i+1}) < 0$$

Thus $\hat{x}_i$ cannot be an optimal wage increase policy. But $\hat{x}_i$ is any policy establishing a wage level $\hat{w}_{i+1}$ such that $x^*_{i+1} > 0$. Hence there exists no optimal wage increase policy x^*_i such that $x^*_{i+1} > 0$ and we conclude that $x^*_{i+1} = 0$ when x_i is optimally chosen, provided that it is an optimal policy to keep the established wage level $w_{i+1} + x^*_{i+1} = w_{i+2}$ fixed for all future.

But from what we observed, in state $(N - 1,w_{N-1})$ it must thus be optimal to choose an $x^*_{N-1} = 0$, if w_{N-1} has been established by an optimal choice of x_{N-2}. This in turn implies that in state $(N - 2,w_{N-2})$ the structure of the decision problem is to make an increase x^*_{N-2} and keep the resulting wage level $w_{N-2} + x^*_{N-2} = w_{N-1}$ fixed afterwards. Induction thus gives the sequence of optimal wage increases $(x^*_N =)$ $x^*_{N-1} = x^*_{N-2}, \ldots, = x^*_{n+1} = 0$, when (n,w_n) is the initial state, in which the optimal policy is determined. As w_n is by definition given (and not determined as a part of the optimal policy), $x^*_n \geqslant 0$. By this the proposition is proved. $\square$

The proposition yields the following

<u>Corollary</u>

When the firm is in the initial state (n,w_n), $0 \leqslant n < N$, $0 \leqslant w_n < pa$, the optimal non-decreasing wage policy is to choose a wage level $w^* \geqslant w_n$ to hold for all future, irrespective of realized employment states, w^* being the solution to

$$H^*(n,w_n) = \max_{w \geqslant w_n} \{V(n,w)\}$$

$$V(n,w) = \frac{pa - w}{\alpha} \{n + \frac{\lambda}{\alpha} [1 - (\frac{\lambda}{\lambda + \alpha})^{N-n}] \}, \quad \lambda = \lambda(w)$$

<u>Proof</u>

From the proposition, we have

$$H(n,w_n,x_n) = V(n,w_n + x_n)$$

$$H^*(n,w_n) = \max_{x \geqslant 0} \{V(n,w_n + x)\} = \max_{w \geqslant w_n} \{V(n,w)\}$$

Lemma 1 gives the explicit expression for $V(n,w)$.

6. Concluding remark

The structure of the optimal policy and the closed form of the value function for the firm's recruitment decision problem under downward wage rigidity have been established. The value function has been shown to be dependent on a one-dimensional wage control variable only. (This result can be shown to hold also when the production function is <u>strictly</u> concave over some employment intervals). The properties of the optimal wage level, especially its response to changes in the values of the parameters of the problem are analysed and interpreted in (Schager, forthcoming). For such an analysis to yield conclusive results, the further restrictions laid on the reservation wage distribution are crucial.

Acknowledgments

Besides to the members of the refereeing committee of this volume my thanks go to Dr. Tomas Björk, Royal Institute of Technology, Stockholm, for giving me advice so to enable me to make my original proof of the basic proposition of my paper more stringent. Thanks are also due to Dr. Harald Lang, Department of Mathematics, University of Stockholm, for stimulating discussions on the subject of the paper. Remaining deficiencies are solely my own responsibility.

References

Deshmukh, S.D. and S.D.Chitke (1976). Dynamic Pricing with Stochastic Entry. Review of Economic Studies, 43, 91-97.

Deshmukh, S.D. and W.Winston (1979) . Stochastic Control of

Leban, R. (1982a).Employment and Wage Strategies of the Firm over a Business Cyle. Journal of Economic Dynamics and Control, 4, 371-394.

Leban, R. (1982b).Wage Rigidity and Employment Policy of the Firm in a Dynamic Environment. in G.Feichtinger (ed.), Optimal Control Theory and Economic Applications, Amsterdam : North Holland.

Lippman, S.A. (1980). Optimal Pricing to Retard Entry. Review of Economic Studies, 47, 723-731.

Moretensen, D.T. (1971). A Theory of Wage and Employment Dynamics. in E.S.Phelps, (ed.), Microeconomic Foundations of Employment and Inflation Theory. London : Macmillan.

Pissarides, C.A. (1976). Labour Market Adjustment : Microeconomic Foundations of Short-run Neoclassical and Keynesian Dynamics. Cambridge : Cambridge University Press.

Salop,S.C. (1973). Wage Differentials in a Dynamic Theory of the Firm. Journal of Economic Theory, 6, 321-344.

Schager, N.H. (forthcoming). The Wage Policy of a Firm, when
    Recruitment is a Wage Dependent Poisson Process and Wages
    are Downward Rigid.

Stidham, S. Jr. and N.U.Prabhu (1974). Optimal Control of Queueing
    Systems. in A.B.Clarke, ed. Mathematical Methods in Queueing
    Theory. Lecture Notes in Economics and Mathematical Systems,
    98, Berlin : Springer-Verlag.

Virén, M. (1979). Human Capital and Wage Differentials in a Dynamic
    Theory of the Firm. Helsingfors : Societas Scientiarum Fennica.

SECTION V. SEMI-MARKOV PROCESSES AND RELIABILITY THEORY

MARKOV RENEWAL PROCESSES IN RELIABILITY ANALYSIS

V.S. Korolyuk

Ukrainian Academy of Sciences

Repui Street 3, Kiev 4, USSR

1. Markov renewal processes

A Markov renewal processes (MRP) can be defined as a two-dimensional homogeneous Markov chain $(x_r, \theta_r; r \geqslant 0)$, where x_r takes its values in the measurable state space E with σ-algebra of measurable sets $\&$, and θ_r in $[0, +\infty)$ respectively.

The MRP transition probabilities are defined by semi-Markov kernel

$$(1) \quad Q(x,A,t) = P\{x_{r+1} \in A, \; \theta_r \leqslant t \mid x_r = x\}.$$

The definition (1) implies that the MRP transition probabilities do not depend on the second component, it is the main point which differs the MRP from any two-dimensional Markov chain with non-negative second component.

The first component $(x_r; r \geqslant 0)$ is the Markov chain called the imbedded Markov chain (IMC). It's transitional probabilities are defined by the relation

$$(2) \quad P(x,A) = P\{x_{r+1} \in A \mid x_r = x\} = Q(x,A,+\infty).$$

Non-negative random variables θ_r, $r \geqslant 1$ describe the intervals between Markov renewal moments

$$(3) \quad \tau_r = \sum_{k=1}^{r} \theta_k, \quad r \geq 1, \quad \tau_o = \theta_o = 0.$$

Distribution functions of renewal times (intervals between Markov renewal moments) θ_r depend on the states of IMC :

$$(4) \quad G_x(t) = P\{\theta_{r+1} \leq t \mid x_r = x\} = Q(x, E, t).$$

It's convenient to denote Markov renewal times with distribution functions $G_x(t)$ as $\theta_x : G_x(t) = P\{\theta_x \leq t\}$.

MRP is the convenient constructive model for a wide class of the step-wise processes representation. We introduce the counting process

$$(5) \quad \nu(t) = \max\{r : \tau_r \leq t\}, \quad t \geq 0.$$

This process is counting the numbers of renewal moments in the interval $[0, t]$.

<u>Definition.</u>

The semi-Markov process (SMP) is defined by

$$(6) \quad x(t) = x_{\nu(t)}, \quad t \geq 0.$$

The <u>renewal times</u> defined by $\theta_{r+1} = \tau_{r+1} - \tau_r$ are naturally interpreted as the sojourn times in the states x_r.

The <u>renewal moments</u> τ_r are the moments at which the SMP changes its states. The IMC $(x_r, r \geq 0)$ is indeed imbedded into the SMP because

$$(7) \quad x_r = x(\tau_r), \quad r > 0.$$

The definition of SMP implies that its trajectories are continuous from the right.

2. The semi-Markov systems phase-lumping algorithms

SMP's represent sufficiently general and constructive mathematical models of complex multicomponent systems, whose states are randomly changed by the shocks.

The only essential restriction is the semi-Markov property : changing of the system states must be described by a homogeneous Markov chain, and sojourn times only depend on a current state of the system.

In many cases it is possible to remove this restriction if we extend the physical state space of the system to the semi-Markov one. The system whose evolution can be modelled by SMP will be called in what follows semi-Markov systems (SMS).

Suppose now that the first stage construction of the MRP modelling the evolution of some real system is realized. Its realization will be illustrated below on a concrete example.

By no means any extension of the physical state space to the semi-Markov one complicates mathematical analysis of the SMS. The most radical approach to overcome the analysis of real systems gives the construction of more simple so called "lumped" model for its analysis as one hopes can be essentially simpler that of the initial system, and the main reliability characteristics of the lumped model may be taken as the approximate characteristics of the real system. We suggest the phase lumping algorithms (PLA) of SMS states based on decomposition of the initial state space

$$(8) \quad E = \bigcup_{v \in V} E_v, \quad E_v \cap E_{v'} = \emptyset, \quad v \neq v', \quad E_v \in \mathcal{E}$$

and constructing of the lumped MRP $(\hat{x}_r, \hat{\theta}_r; \; r \geqslant 0)$ with lumped MRP (factorized) state space V. Every subset $E_v \subseteq E$ corresponds to some state $v \in V$.

The function realizing this correspondence

$$(9) \quad v(x) = v, \quad \text{for} \quad x \in E_v$$

is called the lumping function.

An essential feature of the PLA is in exploiting of basic MRP $(x_r^o, \theta_r^o; \; r \geqslant 0)$ which is ergodic in every subset of states E_v, $v \in V$ and in some sense close to initial MRP $(x_r, \theta_r; \; r \geqslant 0)$; we suppose the closedness of the IMC's :

$$(10) \quad P - P^o = 0(\varepsilon),$$

where P and P^o are operators in Banach space $\mathcal{B}$ of measurable bounded functions which are generated in a natural way by the transition probabilities $P(x,A)$ and $P_o(x,A)$ – corresponding to initial IMC $(x_r;\ r \geqslant 0)$ and basic one $(x_r^o;\ r \geqslant 0)$ respectively. ε is some small positive parameter. In other words it is supposed that

$$(11) \quad P(x,A) = P_o(x,A) + P_1(x,A)$$

and stochastic kernel $P_o(x,A)$ is consistent with the decomposition of initial state space (8) in the sense that

$$(12) \quad P_o(x,E_v) = \mathbb{1}_v(x) = \begin{cases} 1, & x \in E_v, \\ 0, & x \notin E_v, \end{cases} \quad v \in V.$$

The conditions (11) and (12) mean that the initial IMC $(x_r;\ r \geqslant 0)$ passes from one class E_v to the other $E_{v'}$ with probabilities approaching to zero together with the small parameter ε.

Let $\rho_v(dx),\ v \in V$ be a family of invariant distributions of the basis IMC.

The PLA is realized in the following manner. The lumped MRP $(\hat{x}_r, \theta_r;\ r \geqslant 0)$ is defined by the SM-kernel

$$(13) \quad \hat{Q}(v,\Gamma,t) = \hat{P}(v,\Gamma)[1 - e^{-\Lambda(v)t}]$$

where the transition probabilities of the IMC are defined by the relations :

$$(14) \quad \hat{P}(v,\Gamma) = \int_{E_v} \rho_v(dx)P(x,E_v).$$

Intensities $\Lambda(v)$ of sojourn times in the states $v \in V$ are calculated by the formulas

$$(15) \quad \Lambda(v) = q(v)/m(v)$$

$$(16) \quad q(v) = 1 - \int_{E_v} \rho_v(dx)P(x,E_v)$$

$$(17) \quad m(v) = \int_{E_v} \rho_v(dx)m(x), \quad m(x) = \int_0^\infty \bar{G}_x(t)dt = E\theta x.$$

In particular, in the simplest and practically the most often met cases, i.e. when $V = \{0,1\}$, we have

$$(18) \quad \hat{\rho}_{ij} = \int_{E_i} \rho_i(dx)P(x,E_j), \quad j = 0,1, \quad i = 0,1;$$

$$(19) \quad \Lambda_i = 1/m_i, \quad m_i = \int_{E_i} \rho_i(dx)m(x), \quad i = 0,1.$$

So, the lumped process is the Markov chain with two states one of which, for example, 1 is interpreted as the function state of the system, and 0 – as the refusal state. Decomposing on initial state space $E = E_o \cup E_1$ into two subclasses : function states E_1 and refusal states E_o and then lumping this classes up to the states 1 and 0 respectively we get instead of initial SMS with complicated state space E, the simplest Markov system with two states, whose reliability analysis does not contain any difficulties. For example, the time of unrefusal work τ of the lumping system has approximately exponential distribution with the parameter $\Lambda = q/m_1$, $q = \hat{P}_{10}$:

$$(20) \quad P\{\tau > t\} = e^{-\Lambda t}: \quad q = \int_{E_1} \rho_1(dx)P(x,E_o).$$

The stationary coefficient of readiness of the lumped system can be estimated by the formula

$$(21) \quad K = m_1\hat{P}_{01}/(m_1\hat{P}_{01} + m_o\hat{P}_{10})$$

under condition of high reliability of SMS which means in mathematical sense the smallness of averaged transition probabilities of jumps between the state classes E_1 and E_o

$$(22) \quad \hat{P}_{10}, \hat{P}_{01} \ll 1.$$

The reliability characteristics of the lumped system approximate the reliability characteristics of real SMS. Here we will not discuss in detail the merits of PLS's but we would like to mention

only two of them. First, as a result of states lumping of SMS's we
get Markov systems (MS) with essentially simpler lumped state space
V. The Markov property of lumped systems arises not from initial
assumptions but as a corollary of lumping operation. Second, there
is a real opportunity for classification of SMS's via the properties
of close basic systems. Namely the lumped state space V is defined according to the basic system, more exactly, it's IMC $(x_r^o,$
$r \geqslant 0)$.

Here it is relevant to remind the words of the well-known
specialist in system theory U.Ashby at one of the first conferences
(1964) "... theory of systems must be built on simplification methods
and in essence to be represented a science of simplification", and
further : "Not be subject to doubt that the science of simplification has it's own methods and refinements. I am sure that in the
future the theoretist of systems must become a specialist in simplification". PLA as we expect is just the mean for simplification
using analytic instrument of the theory of semi-Markov processes
and theory of asymptotic inversion of operators perturbed on spectrum.

<u>3. Phase lumping limit theorems</u> (Korolyuk and Turbin, 1976 and 1978)

The proof of the PLA is realized in the following way.
Initial assumptions. Consider the SMP $x_\varepsilon(t)$ in the scheme series
i.e. depending on some small parameter ε with the semi-Markov
kernel

$$(23) \quad Q_\varepsilon(x,A,t) = P_\varepsilon(x,A)G_x(t),$$

$$(24) \quad P_\varepsilon(x,A) = P_o(x,A) + \varepsilon P_1(x,A).$$

The transition probabilities P_o of the non-perturbed Markov chain
are coordinated with the phase space splitting $E = \bigcup_{v \in V} E_v$ by the
condition

$$(25) \quad P_o(x, E_v) = \mathbf{1}_v(x) = \begin{cases} 1, & x \in E_v; \\ 0, & x \bar{\in} E_v \end{cases}$$

and satisfy to uniformly ergodic hypothesis

$$\lim_{r \to \infty} \frac{1}{r} \sum_{k=1}^{r} P_o^k = \Pi,$$

$$(26) \quad \Pi f(x) = \int_{E_v} \rho_v(dx) f(x), \quad x \quad E_v.$$

The projector Π is the projector in Banach space $\mathcal{B}$ of bounded functions with sup-norm, and is generated by the stationary distributions $\rho_v(dx)$, $v \in V$ of non-perturbed Markov chain.

The perturbation kernel $P_1(x, A)$ generated the bounded operator in $\mathcal{B}$ and the following condition is fulfilled :

$$(27) \quad q_v = - \int_{E_v} \rho_v(dx) P_1(x, E_v) > 0, \quad v \in V.$$

The mean sojourn times in SMP states are uniformly bounded :

$$(28) \quad \sup_{x \in E} m(x) = \sup_{x \in E} \int_0^\infty \bar{G}_x(t) dt < + \infty.$$

The following conditions are also fulfilled :

$$(29) \quad m_v = \int_{E_v} \rho_v(dx) m(x) > 0, \quad v \in V;$$

$$(30) \quad \lim_{T \to \infty} \sup_{x \in E} \int_T^\infty \bar{G}_x(t) dt = 0.$$

The lumped phase space V with σ-algebra $\mathcal{V}$ of measurable sets generates the sup-σ-algebra $\mathcal{E}_o \subset \mathcal{E}$ of sets $E_\Gamma = \bigcup_{v \in \Gamma} E_v \in \mathcal{E}_o$.

<u>Theorem.</u> For $x \in E_v$, $v \in V$, $\Gamma \in V$

$$(31) \quad \lim_{\varepsilon \to 0} P\{x_\varepsilon(t|\varepsilon) \in E_\Gamma \mid x_\varepsilon(0) = x\} = P\{\hat{x}(t) \in \Gamma \mid \hat{x}(0) = v\}.$$

Here the Markov process $\hat{x}(t)$ with the lumped phase space V is represented by the semi-Markov kernel

$$(32) \quad \hat{Q}(v,\Gamma,t) = \hat{P}(v,\Gamma)[1 - e^{-\Lambda(v)t}],$$

$$(33) \quad \hat{P}(v,\Gamma) = \int_{E_v} \rho_v(dx)P_1(x,E)/q_v, \quad v \notin \Gamma,$$

$$(34) \quad q_v = - \int_{E_v} \rho_v(dx)P_1(x,E_v), \quad \Lambda(v) = q_v/m_v.$$

The proof of the theorem is based on an asymptotic analysis of
Laplace transform of Markov renewal equation (MRE) for transition
probabilities of SMP $x_\varepsilon(t|\varepsilon)$. It is easily to see that

$$(35) \quad u_\varepsilon(t,x_o,\Gamma) = \mathbb{P}\{x_\varepsilon(t|\varepsilon) \in \Gamma \mid x_\varepsilon(0) = x\}$$

satisfies the following MRE

$$(36) \quad u_\varepsilon(t,x) - \int_E \int_0^{t|\varepsilon} P_\varepsilon(x,dy)G_x(d\ell)u_\varepsilon(t - \varepsilon\ell,y)$$

$$= \bar{G}_x(t|\varepsilon) \, \mathbb{1}_{E_\Gamma}(x)$$

or in Laplace transforms

$$(37) \quad \tilde{u}_\varepsilon(s,x) - \tilde{g}_x(\varepsilon s) \int_E P_\varepsilon(x,dy)\tilde{u}_\varepsilon(s,y) = \varepsilon d_\varepsilon(\varepsilon s) \, \mathbb{1}_{E_\Gamma}(x).$$

The proof makes use asymptotic representation of (37) :

$$(38) \quad [1 - P_o + \varepsilon[sM - P_1] + 0(\varepsilon)]\tilde{u}_\varepsilon = \varepsilon m \, \mathbb{1}_{E_\Gamma} + 0(\varepsilon).$$

Now the limit theorem follows from

$$(39) \quad \lim_{\varepsilon \to 0} \tilde{u}_\varepsilon = \lim_{\varepsilon \to 0} \varepsilon[I - P_o + \varepsilon[sM - P_1] + 0(\varepsilon)]^{-1}m \, \mathbb{1}_\Gamma = \hat{u}(s,v),$$

where $\hat{u}(s,v)$ satisfies the equation

$$(39') \quad (s + \Lambda(v))\hat{u}(s,v) - \int_V \hat{P}(v,dv')\hat{u}(s,v') = \mathbb{1}_\Gamma(v).$$

Using algorithms of inversion of perturbed on spectrum operators
(Korolyuk and Turbin, 1978) we obtain the statement of our theorem.

Note that the operator M is defined as an operator of multiplication in the means $m(x)$:

$$(40) \quad Mf(x) = m(x)f(x).$$

We have investigated different situations of asymptotic phase lumping (see Korolyuk and Turbin, 1976 and 1978).

4. Superposition of MRP (SMPP)

The phase lumping algorithms can be applied for simplified description of SMS. Certainly, in applications first of all we deal with the problem of how to choose the basic system. That's clearly the engineer's problem. However, for high-reliable systems this problem is easily solved : the basic system must be absolutely reliable, the unrefusal one. There exists a class of SMS's can be described by the superposition of MRP's (Korolyuk and Turbin, 1978 and 1982).

Let us have a finite collection of independent MRP's $(x_r^{(k)}, \theta_r^{(k)}; r \geqslant 0)$, $k = 1,N$ with semi-Markov kernel $Q^{(k)}(t) = \{Q_{ij}^{(k)}(t); i, j \in E_k\}$, E_k – are discrete phase spaces; $Q_{ij}^{(k)}(t) = P_{ij}^{(k)} E_i^{(k)}(t)$.

By the superposition of MRP's we mean the MRP whose renewal moments contain all renewal moments of the processes composing our superposition.

An heuristic definition of MRP's superposition can be formalized (Korolyuk and Turbin, 1982). In this connection the phase space of semi-Markov states E of superposition contains except discrete components $E_1 \times E_2 \times \ldots \times E_r$ and continuous components

$$(41) \quad E = \prod_{k=1}^{N} E_k \times R_+^N ,$$

$$(42) \quad R_N^+ = \{x_1, x_2, \ldots, x_N; \ \exists\, i : x_i = 0; \ \forall\, k \neq i \quad x_k > 0\}.$$

The remarkable property of the superposition of independent MRP with ergodic IMC as one can show is the superposition of MRP's has an invariant distribution which density is presented by the formula

$$(43) \quad \rho \underset{ik}{\longrightarrow} (\vec{x}) = \rho_o \prod_{\ell=1}^{N} \rho_{k_\ell}^{(\ell)} \prod_{\ell \neq i} \bar{G}_{k_\ell}^{(\ell)}(x_\ell).$$

Here $\rho_{k_\ell}^{(\ell)}$ – are stationary distributions of IMC's $x_r^{(\ell)}$, $\ell = 1,N$.

The normalizing constant ρ_0 is defined by the obvious formula

$$(44) \quad \rho_0 = \left[\sum_{i=1}^{N} \sum_{k_\ell \in E_\ell} \prod_{\ell=1}^{N} \rho_{k_\ell}^{(\ell)} \prod_{\ell \neq i} m_{k_\ell}^{(\ell)} \right]^{-1}.$$

Here $m_{k_\ell}^{(\ell)} = E\theta_{k_\ell}^{(\ell)}$ is the mean sojourn times in the states of the initial MRP generating the superposition.

The existence of invariant distribution of MRP-superposition allows us to use such MRP's as basic ones in the PLA.

It is quite obvious that with the help of MRP-superpositions we can describe multi-components SMS whose separate components operate independently from one of another.

Of course, the condition of independence for SMS-components is restrictive, however only under this condition we succeed to use effectively simplified description of SMS by the help of PLA. Note that the independence of components operating does not indicate the absence of functional dependence between them.

5. Heuristic principles of phase lumping

An unexpected discovery for us was the possibility to realize the reliability analysis on the basic of some heuristic principles which essentially simplify the mathematical analysis and at the same time gives the results equivalent to obtained by PLA (Korolyuk and Turbin, 1983).

1. The lack consequence principle

The time of unrefusal work τ of high-reliable SMS has an exponential distribution :

$$(45) \quad \mathbb{P}\{\tau > t\} = e^{-\Lambda t}.$$

In this connection refusal intensity Λ does not depend on an initial state of the system.

Explanation. As a result of a high-reliability of the SMS and it's proximity to the basic ergodic system the time of unrefusal work is large with respect to sojourn times, so that a great number, of system states changes will occur before the system's refuse. As the final result the system refuses almost in stationary regime and the explaines the presence the property of the lack consequence and independence of the intensity on an initial state of the system.

2. Principle of refuses superposition

Intensity of system's refuses Λ presents the sum of system's refuses intensities on renewal period of separate elements :

(46) $\Lambda = \Lambda_1 + \Lambda_2 + \ldots + \Lambda_N.$

Here N is the number of elements of the system; Λ_k is the intensity of system's refuses on the renewal period of the k-th element of the system.

3. Principle of monotonicity

The refusal intensities Λ_k can be calculated by the formula

(47) $\Lambda_k = q_k / \mathbb{E}\,\theta_1^{(k)}$, $k = \overline{1,N}.$

Here q_k is the probability of system's refuse on the period $\theta_o^{(k)}$ of renewal of the k-element, $\theta_1^{(k)}$ is the working interval of k-element.

4. Principle of ergodicity

In stationary working regime every element of the system is defined by the invariant distribution of the basic IMC $\rho_i^{(k)}$, $i \in E_k$ and stationary random sojourn time in states $\theta_i^{(k)*}$ with the distribution density

$$(48) \quad g_i^{(k)*}(t) = [1 - G_i^{(k)}(t)]/\mathbb{E}\,\theta_i^{(k)}.$$

6. Example

As an illustration of PLA's and heuristic principles given
above we consider one of well-known example of SMS investigated in
the number of papers.

System of two lifts.

The system consists of two devices (lifts), each of them can
be in two possible state : 1 is the working state and 0 is the
renewing one.

The devices operate independently one of another. The distribution functions of working times and renewals repairs of the lifts
are $G_k^{(i)}(t) = \mathbb{P}\{\theta_k^{(i)} \leqslant t\}$, $i = 1,2$, $k = 0,1$.

The devices of the system are modelled by the alternating
renewal processes and the whole system by superposition of two alternating renewal processes in phase state $E = \{ik_1k_2x;\ i = 1,2;\ k_1,$
$k_2 = 0,1;\ x > 0\}$. Here i is the index of device changing it's
state; k_1, k_2 are devices states codes, x is continuous component
fixing the time period from the moment of changing of state by the
other device.

Phase space of the MRP which describes the evolution of the
duplicated system consists of eight semi-lines, six of which consti-
tutes the set E_1 of function states and two $E_0 = \{100x, 200x\}$ of
refusal ones.

Getting the invariant distribution of SMS (see (43)) one can
estimate the invariant reliability characteristics of the system.

For estimation of non-stationary characteristics namely the
time of unrefusant work under the conditions of fast renewal
$(\mathbb{E}\,\theta_0^{(i)} \leqslant \mathbb{E}\,\theta_1^{(i)})$, it is possible to apply the PLA. As the basic
system we put one in which devices are renewed instantly $(\theta_0^{(i)} = 0,$
$i = 1,2)$.

Thus the basic MRP is represented by the superposition of

MRP's in the phase space E_1 and for this system we can easily find it's invariant distribution.

In my opinion to apply PLA in considering problem is being demonstrated quite enough.

According to the heuristic principles formulated above intensity of refuses is estimated by the formula

(49) $\Lambda = \Lambda_1 + \Lambda_2,\ \Lambda_k = q_k / \mathbb{E}\, \theta_1^{(k)},\ k = 1,2.$

For estimation of refusal probability of our system on the renewal period of the 1-st element, note that according to the principle of ergodicity from the beginning of renewal of the 1-st element the residual sojourn time of the 2-nd element in working state is $\theta_1^{(2)*}$ Therefore the next formula is obvious

(50) $q_1 = P\{\theta_1^{(2)*} < \theta_o^{(1)}\}.$

The formulas (49) and (50) solve the problem of estimating the probability of unrefusal work of the system of two lifts in suppositing of highreliability of the system which is described by the relation

(51) $q_k \ll 1,\ k = 1,2.$

7. Conclusion

Note that heuristical principles of reliability estimating of high-reliable systems are the fruitfull interpretation of PLA's, based on the limit theorems. Thus the results obtained by applying of heuristic principles can be always controlled by PLA's.

References

Ashby,W. (1964). Introductory remarks at Panel discussion. Views on general system theory. – Views on general system theory. – Proceedings of the 2^d oystems theory symposium at Case Institute of Technology.

Korolyuk, V.S. and A.F.Turbin (1976). Semi-Markov processes and
     their applications. Kiev, "Naukova Dumka" (in Russian).
Korolyuk, V.S. and A.F.Turbin (1978). Mathematical foundations of
     the complex systems phase lumping. Kiev, 22"Naukova Dumka"
     (in Russian).
Korolyuk, V.S. and A.F.Turbin (1982). Markov renewal processes in
     the problems of system's reliability. Kiev, " aukova Dumka"
     (in Russian).
Korolyuk, V.S. and A.F.Turbin (1983). Analytic methods to estimate
     reliability properties of renewal systems. Kiev.Soc."Znaniye".
     Ukr.SSR (in Russian).

DETERIORATION PROCESSES

Mohamed Abdel-Hameed

Department of Mathematics, University of North Carolina

and Kuwait University

1. Introduction and summary

In reliability studies, the question of assessing the behavior
of the failure rate of a given device often arises. In practice,
it is assumed that the life length of the device has a certain
distribution, such as exponential, Weibull, gamma, or that its life
length belongs to a given family of distributions, such as increasing failure rate, increasing failure rate average. Based on field
data collected about the failure times of identical devices optimal
estimates of the failure rate are obtained and hypothesis testing
for the parameters of the assumed distribution function are carried
out. In many cases collecting enough data for sound statistical
conclusions to be drawn is not possible either because of prohibit-
ive cost or insufficient time available to observe all failure
times of items on test. Even if enough data can be collected, the
validity of the inference procedures are questionable due to the
sometimes unfounded but necessary assumptions that must be imposed
concerning the distribution function of the failure time. One way
to avoid the above difficulties is to examine the failure mechanism
of the given device and thus determine in a proper fashion the form

of its distribution function. In this paper we concern ourselves
with deterioration models, and we discuss three such models. In
the first model a device is subject to damage and wear. The damage
is assumed to be an increasing strong Markov pure jump process and
the wear occurs at a constant rate. This amounts to saying that
damage occurs because of shocks and that the times and magnitudes
of shocks form a Poisson random measure on $R_+ \times (0,\infty)$ and the
rate of wear is a constant $a > 0$. We call this model <u>Pure</u> <u>Jump</u>
<u>Damage</u> <u>Process</u> <u>With</u> <u>Drift</u>. The second model differs from the first
in one regard : between shocks the device wears at a rate which is
equal to the damage accumulated right before the occurrence of the
shock. For example, between the first and second shock the device
fails at a constant rate which is equal to the left hand limit of
the damage level at the time of occurrence of the first shock. We
call this model <u>Deterioration</u> <u>Processes</u> <u>With</u> <u>Wear</u> <u>Depending</u> <u>On</u>
<u>Damage</u> <u>Level</u>. The third model differs drastically from the first
two. In this model, the deterioration process is assumed to be a
Markov additive process (X,Z). The process $X = (X_t)$ describes
the state of the environment, and the increasing process $Z = (Z_t)$
describes the accumulated deterioration the device suffers. The
process Z has conditionally independent increments given the
paths of the environment process. In all three models, the device
is assumed to have a threshold and it fails once the accumulated
deterioration exceeds or is equal to the threshold. We study life
distribution properties of such devices and the effect of the parameters of the deterioration process on the failure rate.

2. Pure jump damage process with drift

Suppose that a device is subject to damage and wear. The
amount of damage that the device suffers over time is assumed to be
an increasing pure jump Markov process. We denote such a process
by $X = (X_t, t \geq 0)$. For each $t \geq 0$ we have

$$X_t = X_0 + \sum_{s \leqslant t} (X_s - X_{s-}).$$

The following figure illustrates a typical sample path of such a process. Çinlar and Jacod (1981) show that there exists a Poisson random measure N on $R_+ \times R_+$ whose mean measure at the point (s,z) is $ds\, dz/z^2$, and a deterministic function c defined on $R_+ \times (0,\infty)$ that is increasing in the first argument such that

$$\sum_{s \leqslant t} f(X_{s-},X_s) = \int_{[0,t] \times R_+} N(ds,dz) f(X_{s-},X_{s-} + c(X_{s-},z))$$

almost everywhere for each function f defined on $R_+ \times R_+$ with $f(x,x) = 0$ for all x in R_+. In particular, it follows that

$$X_t = X_0 + \int_{[0,t] \times R_+} N(ds,dz) c(X_{s-},z).$$

The above formula has the following interpretation :

$t \to X_t(\omega)$ jumps at s if the Poisson random measure $N(\omega,.)$ has an atom (s,z) and then the jump is from the left-hand limit $X_{s-}(\omega)$ to the right-hand limit : $X_s = X_{s-} + c(X_{s-},z)$.

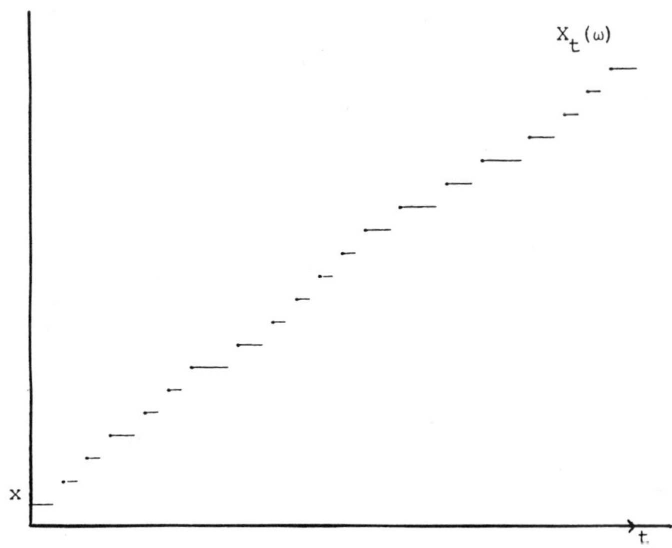

Fig. 1. A realization of the damage process X.

The function $c(x,z)$ represents the damage due to a shock of magnitude z occuring at a time when the previous cumulative damage is equal to x, and the function $z \rightarrow c(x,z)$ is an increasing function. Figures (2) and (3) illustrate the above result graphically for a given ω. Between jumps the device is subject to wear which occurs at a constant rate a, $a \geqslant 0$. Then the total amount of damage and wear the device suffers in $[0,t]$ is given by

$$Z_t(\omega) = at + X_t(\omega)$$

We refer to the process $Z = (Z_t, \ t \geqslant 0)$ by the damage and wear process. Figure (4) below illustrates a typical realization of the process Z starting at x.

For technical reasons we will only deal with those pure jump processes whose damage function $c(x,z)$ satisfies the condition : there exists a constant $k_c > 0$ such that

$$\int_0^1 |c(x,z)| \ dz/z^2 \leqslant k_c \quad \text{for each} \quad x \geqslant 0.$$

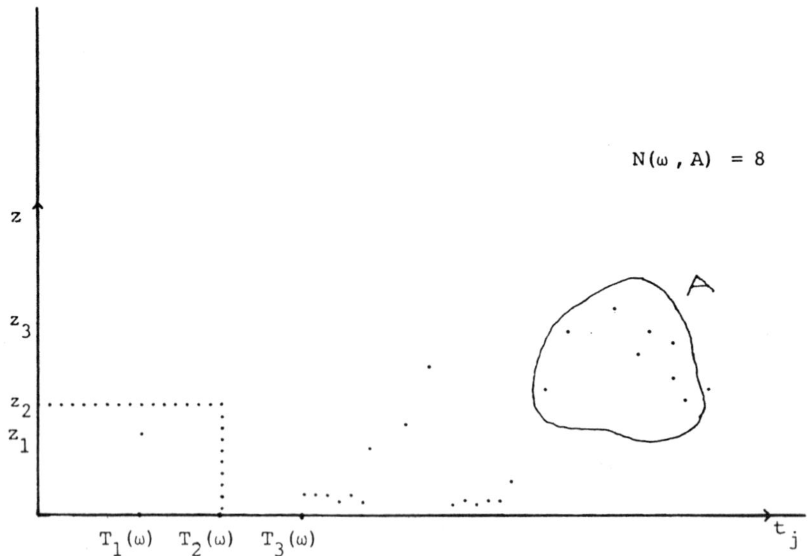

Fig.2. A realization of the Poisson random measure.

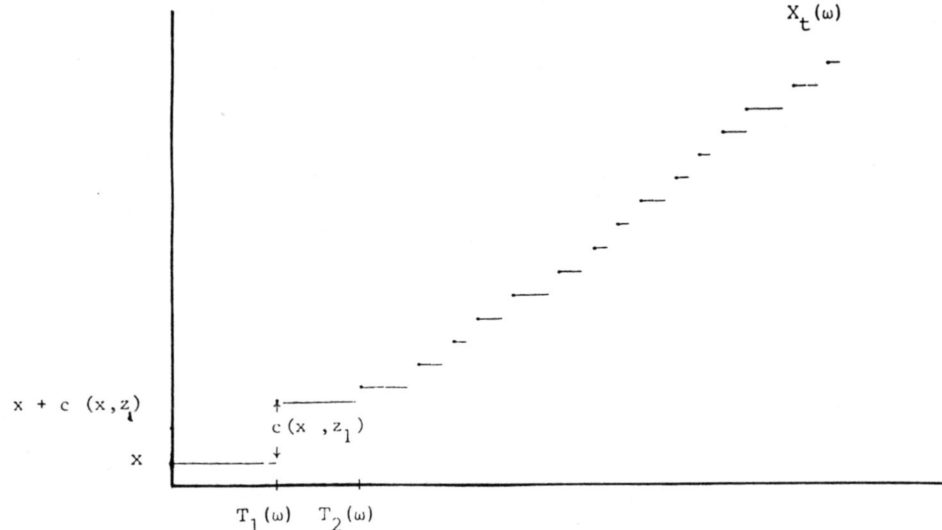

Fig. 3. At the time of occurence of the first shock the damage increases by an amount $c(s,z_1)$ if the damage level just before beta e the shock is x and the magnitude of the shock is z_1.

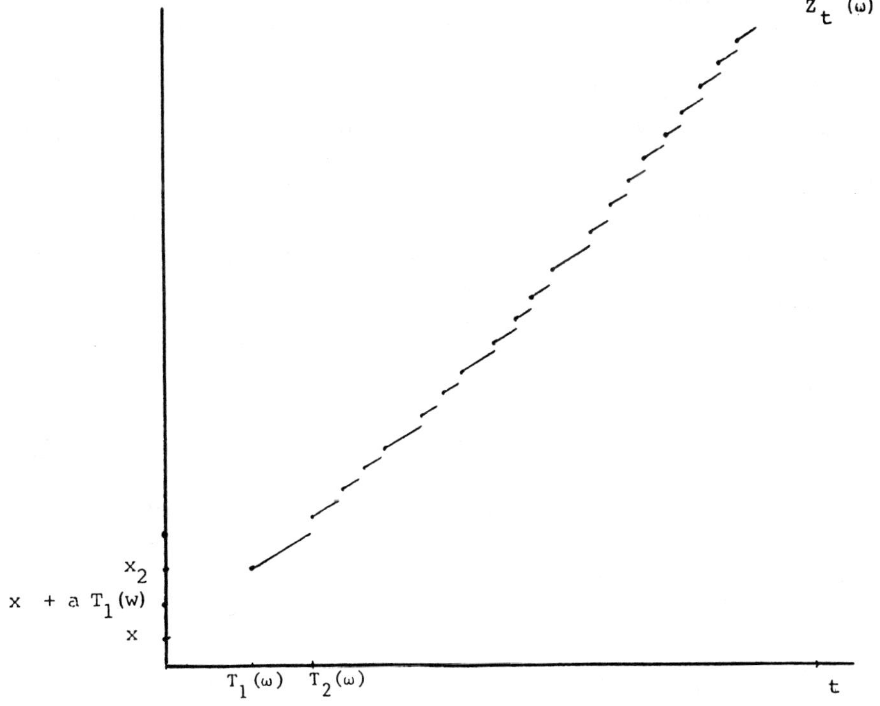

Fig.4. A realization of the deterioration process Z.

The device has a threshold Y which is independent of the damage and wear process Z and it fails once the damage and wear exceeds the threshold. Therefore the time at which the device fails is given by

$$\zeta = \inf\{t : Z_t \geqslant Y\}$$

For $y \geqslant 0$, let

$$\bar{G}(y) = P\{Y > y\}.$$

It follows that the survival probability of ζ is given by

$$(1) \quad \bar{F}(t) = P(\zeta > t)$$
$$= P(Y > Z_t)$$
$$= E\bar{G}(Z_t).$$

We are interested in life distribution properties of $\bar{F}$ and how they are influenced by the corresponding life distribution properties of survival probability $\bar{G}$. Throughout the interval $[0,\infty)$ will be denoted by R_+, $R_+^n = [0,\infty) \times \ldots \times [0,\infty)$ is the nth product of R_+ with itself, and $N = \{0,1,\ldots\}$. The word "increasing" will be used to mean "non-decreasing" and the word "decreasing" will mean "non-increasing". Moreover, we will use positive to mean "$\geqslant 0$" and negative to mean "$\leqslant 0$".

Let $H = -\ln \bar{F}$, throughout we will refer to H be the hazard function of $\bar{F}$.

Definition

The survival probability $\bar{F}$ is said to be or to have :

(i) increasing failure rate (IFR) if the hazard function of $\bar{F}$ is a convex function on $[0,\infty)$. If the failure rate exists, then $\bar{F}$ is IFR if and only if the failure rate is an increasing function on $[0,\infty)$.

(ii) decreasing failure rate (DFR) if the hazard function of $\bar{F}$ is concave function on $[0,\infty)$. If the failure rate exists, then $\bar{F}$ is DFR if and only if the failure rate is a decreasing function on $[0,\infty)$.

(iii) increasing failure rate average (IFRA) if $t^{-1}H(t)$ is an increasing function on $[0,\infty)$. When the failure rate r(r) exists, then $\bar{F}$ is IFRA if and only if $t^{-1}\int_0^t r(u)du$ is an increasing function on $[0,\infty)$.

(iv) decreasing failure rate average (DFRA) if $t^{-1}H(t)$ is a decreasing function on $[0,\infty)$. When the failure rate exists then $\bar{F}$ is DFRA if and only if $t^{-1}\int_0^t r(u)du$ is a decrasing function on $[0,\infty)$.

(v) new better than used (NBU) if $\bar{F}(t + s) \leq \bar{F}(t)\bar{F}(s)$ for each s, $t \geq 0$. To say that F is NBU is equivalent to saying that the failure time of a new device is stochastically larger than the remaining failure time of an old device that is still alive at time t, for each $t \geq 0$.

(vi) new worse than used (NWU) if $\bar{F}(t + s) \geq \bar{F}(t)\bar{F}(s)$ for each s, $t \geq 0$. To say that $\bar{F}$ is NWU is equivalent to saying that the failure time of a new device is stochastically smaller than the remaining failure time of an old device that is still alive at time t, for each $t \geq 0$.

Assuming that $\bar{F}(0) = 1$, the following implications are readily verified :

$$F \text{ IFR} \Rightarrow F \text{ IFRA} \Rightarrow F \text{ NBU}$$

and

$$F \text{ DFR} \Rightarrow F \text{ DFRA} \Rightarrow F \text{ NWU}$$

Knowing the behavior of the failure rate and/or the survival probability of a given device enables us to determine reliable estimates of the failure rate and the form of the optimal replacement and maintenance policies for such devices. We refer the reader to (Hollander and Proschan, 1972) and (Marshall and Proschan, 1965) in regard to the problem of optimaly estimating the failure rate, to (Feldman, 1977),and (Marshall and Proschan, 1965) with respect to determining optimal maintenance and replacement policies.

Definition

Let $Y = (\Omega, F, F_t, Y_t, P_y)$ be a Markov process. An operator A with domain $\mathcal{D}_A$ is said to be an extended generator of the process Y provided that for any function f in $\mathcal{D}_A$

$$L_t = f(Y_t) - f(Y_0) - \int_0^t Af(Y_s)ds$$

is an F_t-martingale for each P_y.

Theorem

Let X be an increasing pure jump process. Define

$$A = \{f : R_+ \to R \text{ that are Lipschitz of order 1 and bounded}\}.$$

Then $A \subset \mathcal{D}_A$ and for each f in $\mathcal{D}_A$

$$Af(x) = \int_0^\infty (f(x + c(s,z)) - f(x))dz/z^2.$$

Definition

A function $f : R^2 \to R_+$ is said to be totally positive of order r (TP_r) provided that $\det(f(x_i, y_j)) \geqslant 0$ for each choice of $x_1 \leqslant x_2 \leqslant \ldots \leqslant x_r$ and $y_1 \leqslant y_2 \leqslant \ldots \leqslant y_r$.

Definition

A function $f : R \to R_+$ is said to be a Pólya frequency function of order r if the function $g : R^2 \to R_+$ defined by $g(x,y) = f(x - y)$ is a totally positive function of order r.

Theorem

Let $f : R^2 \to R_+$ be TP_r and $g : R \to R$ changes sign at most $j \leqslant r - 1$ times, then the function $h : R \to R$ defined by

$$h(x) = \int_R f(x,y)g(y)dy$$

changes sign at most j times; moreover, if h actually changes sign j times, then it must change sign in the same order as g.

The above theorem is known as the "variation diminishing property theorem". For more details on total positivity and its application the reader is referred to Karlin (1969).

Suppose that a device has a device has a survival probability of the form given in (1). We prove that life distribution properties of the threshold righ tail probability are inherited as corresponding properties of the survival probability $\bar{F}$, under suitable assumptions on the damage function $c(x,z)$.

Theorem

Suppose that the damage function $c(x,z)$ is increasing in x,z. Then :

(i) $\bar{F}$ is IFR provided that $\bar{G}$ is IFR, $a \equiv 0$ and the process X has a totally positive density function of order two.

(ii) $\bar{F}$ is IFRA, provided that $\bar{G}$ is IFRA, $a \equiv 0$, and the process X has a total posivite density function of order two.

(iii) $\bar{F}$ is NBU if $\bar{G}$ is NBU.

Proof of (i)

The proof of (i) is given in Theorem 2.1 of (Abdel-Hameed, 1984b).

Proof of (ii)

The proof of (ii) is given in Theorem 2.1 of (Abdel-Hameed, 1984b).

Proof of (iii)

For $x,\ t \geqslant 0$, let

$$\bar{G}(x,t) = \bar{G}(x + at)$$

Define $V(x,t) = - \ln \bar{G}(x,t)$ and observe that V is a superadditive function on R_+^2 whenever $\bar{G}(x)$ is NBU. Thus, the hypothesis of part (ii) of Theorem (2.3)of (Abdel-Hameed, 1984b) are satisfied and it follows that $\bar{F}$ is NBU.

Theorem

Suppose that the function $x \rightarrow c(x,z)$ is a decreasing function for each $z \geqslant 0$. Then :

(i) $\bar{F}$ is DFR provided that $\bar{G}$ is DFR, $a \equiv 0$, and the process X has a totally positive density function of order two.

(ii) $\bar{F}$ is DFRA provided that $\bar{G}$ is DFRA, $a \equiv 0$, and the process X has a totally positive density function of order two.

(iii) $\bar{F}$ is NWU if $\bar{G}$ is NWU and the function $x \to x + c(x,z)$ is an increasing function.

<u>Proof of (i)</u>

The proof of (i) is given in Theorem 2.2 of (Abdel-Hameed, 1984b).

<u>Proof of (ii)</u>

The proof of (ii) is given in Theorem 2.2 of (Abdel-Hameed, 1984b).

<u>Proof of (iii)</u>

For $x, t \geqslant 0$, let

$$\bar{G}(x,t) = \bar{G}(x + at)$$

Define $V(x,t) = - \ell n \, \bar{G}(x,t)$, for positive x_1, x_2, t_1, t_2

$$\bar{G}(x_1 + x_2, t_1 + t_2) = \bar{G}(x_1 + x_2 + at_1 + at_2)$$

$$\geqslant \bar{G}(x_1 + at_1).\bar{G}(x_2 + at_2)$$

where the inequality follows since $\bar{G}(x)$ is NWU. Hence, $V(x,t)$ is a sub-additive function R_+^2. Thus the hypothesis of part (iv) of Theorem 2.3 of (Abdel-Hameed, 1984b) are satisfied and F is NWU.

Suppose that a device is subject to the damage and wear process Z described above and have a failure time ζ with survival probability of the form (1). For $t \geqslant 0$, let

$$Z_t^* = \begin{cases} Z_t & \text{if } t < \zeta, \\ + \infty & \text{if } t \geqslant \zeta. \end{cases}$$

Devices subject to the above deterioration process can be replaced before or at failure. The cost of a replacement before failure depends on the deterioration level at the time of replacement and is denoted by $c(x)$. That is to say, $c(x)$ is the cost of a replacement when the deterioration level before failure is equal to x. The cost of a replacement at failure is equal to a constant c, $c(x) \leqslant c$ for each $x \geqslant 0$. The deterioration process resets at zero every time a replacement is made. Every stopping time with respect to Z^* that does not exceed the failure time ζ almost everywhere is de-fined as a <u>replacement time</u>. Let F be the class of such replacement times. Observe that there is no loss of generality in dealing only with stopping times that are less than or equal to ζ almost every-where, if τ is any stopping time with respect to Z^*, then $\tau \wedge \zeta$ is again a stopping time with respect to Z^* that does not exceed ζ. For any τ in F, the long-run average cost per unit time is equal to

$$\psi_\tau(x) = [E^x(c(Z^*_\tau), \tau < \zeta) + cP^x(\tau = \zeta)]/E^x(\tau)$$
$$= [c - E^x c_1(Z^*_\tau)]/E^x(\tau)$$

where $c_1(x) = c - c(x)$, and $c_1(+\infty) = 0$.

A replacement time τ^*_x is said to be optimal if for each $x \geqslant 0$

$$\psi_{\tau^*_x}(x) = \inf_{\tau \in F} \psi_\tau(x)$$

Throughout the following discussions we assume that $G \equiv 1 - \bar{G}$ is absolutely continuous with respect to the Lebesgue measure and has a bounded failure rate. We extend the damage function $c(x,z)$ to the function $c^*(x,z) : R \times R_+ \to R_+$ by defining

$$c^*(x,z) = \begin{cases} c(x,z) & x,z > 0, \\ 0 & x \leqslant 0. \end{cases}$$

For any bounded function f defined on R_+ satisfying a Lipschitz condition of order 1, we define for $x,\ t \geqslant 0$

$$Tf(t,x) = \int_{R_+} [f(x + c^*(x - at,z)) - f(x)]dz/z^2$$

and

$$T_1 f(t,x) = \bar{G}^{-1}(x) \cdot T(\bar{G}f)(t,x).$$

Let A be the space of bounded functions on R_+ vanishing at $+\infty$ and

$$\beta = \{f \in A : f \text{ is absolutely continuous and is Lipschitz of order } 1\},$$

and let f' be the functional derivative of the function f.

The proof of the following theorem follows in a manner similar to the proof of Theorem 3.6 of (Abdel-Hameed, 1984a).

Theorem

Let f belong to β, τ is a stopping time with $E^x(\tau) < \infty$ for each $x \geq 0$. Then

$$E^x f(Z_\tau^*) - f(x) = E^x \int_0^\tau T_1 f(t,Z_t^*)dt + aE^x \int_0^\tau f'(Z_t^*)dt$$

Let

$$b_x = \inf_{\tau \in F} \psi_\tau(x).$$

The proof of the following theorem follows from the Theorem above.

Theorem

Assume that c_1 belongs to β, $\bar{G} \cdot c_1$ is a concave function on R_+, $c(\cdot,z)$ is an increasing function for each $z \geq 0$, and $E^x(\zeta) < \infty$ for each $x \geq 0$. Then

$$\tau_x^* = \inf\{t : b_x + T_1 c_1(t,Z_t^*) + ac_1'(Z_t^*) \leq 0\}$$

is optimal.

3. Deterioration processes with wear rate depending on damage level

This model assumes that the damage process is a pure jump

process. Thus, the damage is caused by shocks; the number of shocks
and the magnitude of shocks form a Poisson random measure on
$R_+ \times (0,\infty)$, if a shock of magnitude z occurs at a time when the
previous damage level is equal to x the device suffers an additional amount of damage equal to $c(x,z)$ and the damage accumulates
additively. Between shocks, the device suffers wear with a rate
equal to the accumulated damage. Let (T_n) be the sequence describing the successive times at which shocks occur, $\lim_{n\to\infty} T_n = +\infty$. For
t in $[T_n, T_{n+1})$, the wear is equal to tX_{T_n}, $n = 0,1,\ldots$, $T_0 \equiv 0$.
In this case the accumulated damage and wear process is given by,
for each $t \geqslant 0$,

$$Z_t^* = X_t + tX_t$$
$$\quad = X_t(t + 1) .$$

Motivated by the above discussion we now consider the case
where the damage and wear process is of the form

$$Z_t = g(t)X_t$$

where $g : R_+ \to R_+$ and X is the pure jump increasing Markov
process discussed in section 1. The device has a threshold Y that
is independent of X and it fails once the damage and wear process
exceeds or equal to the threshold. Let $\bar{G}$ be the right-tail probability of Y and ζ be the failure time of the device. Then,
as in the first model, the survival probability of the device is
given by

$$\bar{F}(t) = E\bar{G}(Z_t) .$$

Theorem

Suppose that the damage function $c(x,z)$ is increasing in both
arguments. Then :

(i) $\bar{F}$ is IFR provided that $\bar{G}$ is IFR, $g \equiv 1$, and X has a
totally positive density function of order two.

(ii) $\bar{F}$ is IFRA provided that $\bar{G}$ is IFRA and X has a totally positive density function of order two.

(iii) $\bar{F}$ is NBU provided that $\bar{G}$ is NBU.

Proof of (i)

The proof of (i) follows from Theorem 2.1 of (Abdel-Hameed, 1984b).

Proof of (ii)

For $t, x \geq 0$, define

$$\bar{G}(x,t) = \bar{G}(xg(t)).$$

Let $0 \leq t_1 \leq t_2$, then $g(t_1) \leq g(t_2)$ and hence $xg(t_1) \leq xg(t_2)$ for each $x \geq 0$; since $\bar{G}$ is a decreasing function, $\bar{G}(x,t_1) \geq \bar{G}(x,t_2)$. Therefore, the map $\bar{G}(x, \ ) : R_+ \to [0,1]$ is a decreasing function. Moreover, for each $t \geq 0$ and α in $(0,1)$, $- \ln G(\alpha x,t)$ $= - \ln \bar{G}(\alpha xg(t)) \leq - \alpha \ln \bar{G}(xg(t))$, where the last inequality follows since $\bar{G}$ is assumed to be IFRA and $\bar{G}$ is IFRA if and only if $- \ln \bar{G}(\alpha t) \leq - \alpha \ln \bar{G}(t)$ for each α in $(0,1)$. Hence, for each $t \geq 0$, the survival probability $\bar{G}(\cdot,t) : R_+ \to [0,1]$ is IFRA. From part (ii) of Theorem 2.3 of (Abdel-Hameed, 1984b) it follows that $\bar{F}$ is IFRA as desired.

Proof of (iii)

Let $0 \leq x_1 \leq x_2$ and $0 \leq t_1 \leq t_2$ be given. Then

$$\begin{aligned}
\bar{G}(x_1 + x_2, t_1 + t_2) &= \bar{G}((x_1 + x_2)g(t_1 + t_2)) \\
&\leq \bar{G}(x_1 g(t_1 + t_2))\bar{G}(x_2 g(t_1 + t_2)) \\
&\leq \bar{G}(x_1 g(t_1))\bar{G}(x_2 g(t_2)) \\
&= \bar{G}(x_1, t_1)\bar{G}(x_2, t_2)
\end{aligned}$$

where the first inequality follows because $\bar{G}$ is NBU, the second inequality follows because the functions $g(t)$ and $\bar{G}(x)$ are

increasing and decreasing respectively, and the last equality follows
from the definition of $\bar{G}(x,t)$. From part (iii) of Theorem 2.3 of
(Abdel-Hameed, 1984b) it follows that $\bar{F}$ is NBU.

Theorem
<u>Theorem</u>
Suppose that $x \to c(x,z)$ is a decreasing function for each $z \geqslant 0$.
Then :

(i) $\bar{F}$ is DFR provided that $\bar{G}$ is DFR, $g \equiv 1$, and X has a
totally positive density function of order two.

(ii) $\bar{F}$ is DFRA provided that $\bar{G}$ is DFRA, g is a decreasing
function and X has a density function that is totally positive
of order two.

(iii) $\bar{F}$ is NWU provided that $\bar{G}$ is NWU, g is a decreasing
function, and the map $x \to x + c(x,z)$ is increasing for each $z \geqslant 0$.

<u>Proof of (i)</u>
The proof of (i) follows from Theorem 2.2 of (Abdel-Hameed, 1984b).

<u>Proof of (ii)</u>
Let $0 \leqslant t_1 \leqslant t_2$, since g is a decreasing function, $g(t_1) \geqslant g(t_2)$
and, for $x \geqslant 0$, $xg(t_1) \geqslant xg(t_2)$. Since $\bar{G}$ is a decreasing function,
$\bar{G}(xg(t_1)) \leqslant \bar{G}(xg(t_2))$.

Let $\bar{G}(x,t) : R_+^2 \to [0,1]$ be as defined in the proof of part (ii)
of the Theorem above. It follows that the map $\bar{G}(x,\cdot) : R_+ \to [0,1]$
is increasing for each $x \geqslant 0$. Moreover, for each $t \geqslant 0$ and α
in $(0,1)$, $-\ln \bar{G}(\alpha x,t) = -\ln \bar{G}(\alpha xg(t)) \geqslant -\alpha \ln \bar{G}(xg(t))$, where
the last inequality follows since $\bar{G}$ is DFR. Hence, for each
$t \geqslant 0$, the survival probability $\bar{G}(\cdot,t) : R_+ \to [0,1]$ is DFRA. From
part (ii) of Theorem 2.3 of (Abdel-Hameed, 1984b) it follows that
$\bar{F}$ is DFRA.

<u>Proof of (iii)</u>
Let $0 \leqslant x_1 \leqslant x_2$ and $0 \leqslant t_1 \leqslant t_2$ be given. Then

$$\bar{G}(x_1 + x_2, t_1 + t_2) = \bar{G}((x_1 + x_2)g(t_1 + t_2))$$

$$\geqslant \bar{G}(x_1 g(t_1 + t_2))\bar{G}(x_2 g(t_1 + t_2))$$

$$\geqslant \bar{G}(x_1 g(t_1))\bar{G}(x_2 g(t_2))$$

$$= \bar{G}(x_1, t_1)\bar{G}(x_2, t_2)$$

where the first inequality follows because $\bar{G}$ is NWU and the second inequality follows since $\bar{G}$ is a decreasing function in its argument and g is assumed to be a decreasing function. From (iv) of Theorem 2.3 of (Abdel-Hameed, 1984b) it follows that $\bar{F}$ is NWU.

4. Markov additive deterioration processes

The model treated here is the one discussed in Çinlar (1976). Our terminology and notation coincides with his. Let $X = (X_t)$ and $Z = (Z_t)$ be two stochastic processes defined on the same sample space Ω and $t \to Z_t(\omega)$ is assumed to be increasing for almost all ω in Ω. The state space for X is denoted by E while the state space for Z is taken to be $R_+ = [0, \infty)$. For each x in E

$$P^x\{X_0 = x, Z_0 = 0\} = 1.$$

Define,

$$Q_t(x, A, B) = P^x(X_t \in A, Z_t \in B)$$

for t in R_+, x in E and Borel subsets $A \subset E$ and $B \subset R_+$.

Let H_s be the history generated by the process (X, Z) until time s, $s \geqslant 0$, that is H_s is the sigma algebra generated by $\{X_u, Z_u, u \leqslant s\}$. The process (X, Z) is called a Markov additive process if

$$(2) \quad P^x(X_{t+s} \in A, Z_{t+s} \in B \mid H_s) = Q_t(X_s, A, B)$$

for each $t, s \geqslant 0$, x in E, and Borel subsets $A \subset E$ and $B \subset R_+$.

As a result, the transition probability of (X,Z) satisfies

$$Q_{t+s}(x,A,B) = \int_{E \times R_+} Q_s(x,dy,dz)Q_t(y,A,B-z)$$

where $B - z = \{b - z \geq 0 : b \in B\}$. Such (Q_t) are called semi-Markov transition functions. It follows from the above that given X, the probability law of Z during the interval $(t, t + dt)$ is that of an increasing Lévy process whose parameter depends on X_t. The Markov additive process describes the deterioration process a device is subject to a random environment. The process X describes the states of the environment and the process Z describes the total deterioration. Cinlar (1976) gives an excellent summary of such processes in an informal setting. A more rigorous treatment of Markov additive processes is given in his papers (Çinlar, 1972 and 1975). When the state space E of X is finite, when X is in state i, Z increases according to a Lévy process with drift rate $a(i)$ and Lévy measure $\nu(i,dz)$. Increases in deterioration over different environments are added up. In addition, every change of the environment from state i to state j is accompanied by a shock which causes an additional amount of damage with distribution $F(i,j,\cdot)$.

Definition

For a device with failure time ζ the conditional failure rate is given by

$$r(t \mid X,Z) = \lim_{u \downarrow 0} \frac{1}{u} P^x(\zeta \leq t + u \mid X,Z; \zeta > t).$$

Çinlar assumes that a device subject to deterioration and (X,Z) is the Markov additive process describing the environment and the deterioration processes respectively; the conditional survival probability given (X,Z) is assumed to be in the form

$$P^x(\zeta > t \mid X,Z) = \exp(-\lambda Z_t).$$

Definition

Let ζ be the failure time of device and $r(t \mid X,Z)$ be its conditional failure rate. The unconditional failure rate is defined by

$$r_u(t) = E(r(t \mid X,Z))$$

Define for j,k in E,

$$F_\lambda(j,k) = \int_0^\infty F(j,k,dz)(1 - e^{-\lambda z}), \quad j \neq k,$$

$$F_\lambda(k,k) = 1.$$

Denote the generator of X by G and define for j in E

$$r(j) = \lambda a(j) + \int_0^\infty \nu(j,dz)(1 - e^{-\lambda z}) + GF_\lambda(j)$$

Theorem

For $j \geqslant 0$ let $r(j)$ be given above. Then

$$r_u(t) = Er(X_t)$$

Proof

$$r_u(t) = Er(t \mid X,Z)$$

$$= E \lim_{u \downarrow 0} \frac{1}{u} P(\zeta \leqslant t + u \mid X,Z; \ \zeta > t)$$

$$= E \lim_{u \downarrow 0} \frac{1}{u} [1 - P(\zeta > t + u \mid X,Z; \ \zeta > t)]$$

$$= E \lim_{u \downarrow 0} \frac{1}{u} [1 - e^{-\lambda(Z_{t+u} - Z_t)}]$$

$$= \lim_{u \downarrow 0} \frac{1}{u} E[E(1 - e^{-\lambda(Z_{t+u} - Z_t)} \mid H_t)]$$

$$= \lim_{n \downarrow 0} \frac{1}{u} E[E_{X_t}(1 - e^{-\lambda Z_u})]$$

$$= E[\lim_{u \downarrow 0} E_{X_t} \frac{1}{u}(1 - e^{-\lambda Z_u})]$$

$$= Er(X_t)$$

where the interchange of the order of expectation and limits follows

from the bounded convergence theorem and the last equality follows
from Theorem (3.7) of Cinlar (1975).

<u>Remark</u>

Çinlar (1976) uses a similar argument to the one used in proving
Theorem above to wrongly show that $r(t) = Er(X_t)$, where

$$r(t) = \lim_{u \downarrow 0} \frac{1}{u} P(\zeta \leqslant t + u \mid \zeta > t).$$

<u>Corollary</u>

Let the maps $x \to \nu(x,[0,z))$, $x \to F(x,y,[0,z))$, $x \to GF_\lambda(x)$ and
$t \to X_t(\omega)$ be increasing functions for each y, $z \geqslant 0$. Then
$t \to r_u(t)$ is increasing.

Assume that a device is subject to deterioration, $X = (X_t)$ is
the process describing the states of the environment and $Z = (Z_t)$
describes the accumulated deterioration. The process (X,Z) is
taken to be a Markov additive process with a semi-Markov transition
function $Q_t(x,A,B)$. The device has a threshold Y and it fails
once the deterioration hits or exceeds the threshold; Y, Z are
assumed to be independent. As before, the survival probability is
given by

(3) $\bar{F}(t) = E\bar{G}(Z_t)$, $t \geqslant 0$

<u>Theorem</u>

Assume that

$$Q_t(x,E,[y,\infty)) \geqslant Q_t(0,E,[y,\infty))$$

for each $t \geqslant 0$, $y \geqslant 0$, and x in E. Then $\bar{F}$ is NBU.

<u>Proof</u>

For s, $t \geqslant 0$,

$$\bar{F}(t + s) = E\bar{G}(Z_{t+s})$$

$$= E\bar{G}(Z_{t+s} - Z_t + Z_t)$$

$$\leq E[\bar{G}(Z_{t+s} - Z_t).\bar{G}(Z_t)]$$

$$= E[\bar{G}(Z_t)E(\bar{G}(Z_{t+s} - Z_t) \mid H_t)]$$

$$= E[\bar{G}(Z_t)E_{X_t}\bar{G}(Z_s)]$$

$$\leq E[\bar{G}(Z_t)E\bar{G}(Z_s)]$$

$$= E\bar{G}(Z_t)E\bar{G}(Z_s)$$

$$= \bar{F}(t)\bar{F}(s)$$

where the first inequality holds because $\bar{G}$ is NBU and the fourth equality follows from (2) and the second inequality follows from the assumption given on the semi-Markov transition function.

The proof of the following theorem is similar to the proof of the Theorem above and can be obtained by reversing the directions of the inequalities given in the proof of that theorem.

Theorem

Assume that

$$Q_t(x,E,[y,\infty)) \leq Q_t(0,E,[y,\infty))$$

for each t, $y \geq 0$ and x in E. Then $\bar{F}$ is NWU.

The special case, where E consists of only one point, the deterioration process Z is an increasing Lévy process with a Lévy measure $\nu(dz)$. In this case we have the following theorems whose proofs are similar to the proofs given in Theorems 2.3 and 2.5 of Abdel-Hameed (1984a), respectively.

Theorem

Let $\bar{F}$ be given by (3). Then :

(i) $\bar{F}$ is IFR if $\bar{G}$ is IFR and $\nu \ll$ Leb with a density that is a Pólya frequency function of order two.

(ii) $\bar{F}$ is IFRA if $\bar{G}$ is IFRA.

(iii) $\bar{F}$ is NBU if $\bar{G}$ is NBU.

Theorem

Let $\bar{F}$ by given by (3). Then :

    (i) $\bar{F}$ is DFR if $\bar{G}$ is DFR.

    (ii) $\bar{F}$ is DFRA whenever $\bar{G}$ is DFRA, and $\nu \ll$ Leb with a density that is Pólya frequency function of order two.

    (iii) $\bar{F}$ is NWU whenever $\bar{G}$ is NWU.

Optimal replacement policies for devices subject to Markov additive deterioration processes are discussed in Feldman (1977).

References

Abdel-Hameed, M.S. (1984a).Life distribution properties of devices subject to a Lévy wear process. Math.Oper.Research 9, 606-614.

Abdel-Hameed, M.S. (1984b).Life distribution properties of devices subject to a pure jump damage process. J.Appl.Prob. 21, 816-825.

Cinlar, E. (1972). Markov additive processes, II. Z.Wahrscheinlich-keitstheorie Verw.Geb. 24, 94-121.

Cinlar, E. (1975). Lévy systems of Markov additive processes. Z.Wahrscheinlichkeitstheorie Verw.Geb. 31, 175-185.

Cinlar, E. (1976). Shocks and wear models and additive processes. Proceedings of the Conference on Theory and Applications of Reliability with Emphasis on Bayesian and Nonparametric Models, Edited by C.P.Tsokos and I.N.Shimi, Academic Press, 193-214.

Cinlar, E. and J.Jacod (1981). Representation of semimartingales Markov processes in terms of Wiener processes and Poisson random measure. Seminar on Stochastic Processes, Edited by E.Cinlar, K.L.Chung and R.K.Getoor, Birkhauser, Boston, 159-242.

Feldman, R.M. (1977). Optimal replacement for a system governed by Markov additive shock processes. Ann.Prob. 5, 413-429.

Hollander, M. and F.Proschan (1972). Testing whether new is better than used. Ann.Math.Statist. 43, 1136-1146.

Karlin, S. (1969). Total Positivity. Standford University Press,
    Standford.

Marshall, A.W. and F.Proschan (1965). Maximum likelihood estimation
    for distributions with monotone failure rate. Ann.Math.Statist.
    36, 69-77.

Marshall, A.W. and F.Proschan (1972). Classes of distributions
    applicable in replacement, with renewal theory implications.
    Proceedings of the 6th Berkeley Symposium on Mathematical
    Statistics and Probability Vol.1, Edited by J.Neyman, and
    E.L.Scott, University of California Press, 395-415.

STOCHASTIC PROCESSES WITH AN EMBEDDED POINT PROCESS AND THEIR
APPLICATION TO SYSTEM RELIABILITY ANALYSIS

Peter Franken and Arnfried Streller

Humboldt-Universität zu Berlin, Sektion Mathematik

Unter den Linden, 6 - PSF1297, 1086 Berlin (G.D.R.)

1. Introduction

In applied probability it is fairly familiar to investigate
the temporal behaviour of a certain system considered by means of
some appropriately chosen (in general random) embedded epochs.
This is the idea of the famous method of an embedded Markov chain
due to A.Ya.Khinchin and D.G.Kendall. Furthermore, several well-
known classes of stochastic processes such as regenerative, semi-
Markov and semi-regenerative processes are bases on the concept
of embedded points.

The notion of a process with an embedded point process, introduced recently by several authors, provides a far-reaching generalization and a unified treatment of these ideas, cf. Franken et al.
(1981) and Rolski (1981) for detailed references. Using some
results for stationary point processes several formulas for calculating stationary state probabilities by means of embedded probabilities were derived.

The purpose of the present paper is to survey the main results
for this class of stochastic processes. The technical details are
almost completely omitted; the emphasis lies on the interpretation

of notions and results. On this basis we provide a unified approach
for deriving formulas for the stationary state probabilities, the
stationary availability and interval reliability of a complex repairable system.

The paper is organized as follows : In section 1 we introduce
the notion of a process with an embedded point process (abbr. PEP)
and discuss regenerative, semi-regenerative, semi-Markov and piecewise Markov processes as partial cases. In section 2 the definitions
of synchronous and stationary PEP are contained. We show that there
is a one-to-one correspondence between the distributions of synchronous and stationary PEP and derive formulas which express the distribution of a stationary PEP in terms of the distribution of the
associated synchronous PEP. Moreover, the distribution of the associated synchronous PEP appears as the "embedded distribution" of
the stationary PEP. Then, we formulate a general ergodic theorem
for PEP and its specification for semi-regenerative processes.
Finally the stationary distributions of semi-Markov, piecewise
Markov and semi-regenerative processes with general state spaces
are given. In section 3 we derive several interesting and important
specifications of the general formulas stated in section 2. There
are relationships between so-called time-stationary and embedded
stationary distributions and sojourn times as well. On this basis
we obtain in section 4 general formulas for stationary reliability
characteristics of repairable systems and discuss how to apply them.
As an example, a two-unit warm standy system is considered in
section 5.

2. Definition of a process with an embedded point process. Examples

Let $(\Omega, \mathcal{F}, \mathrm{Pr})$ be a basic probability space and $\mathcal{Z}$ a Polish
space endowed with the σ-field $\mathfrak{Z}$ of Borel subsets. Consider a
stochastic process $(Z(t),\ t \geq 0)$ with state space $\mathcal{Z}$ describing
the temporal behaviour of a given system. Throughout the paper

we shall assume that the sample paths of $(Z(t))$ belong to the space $D(0,\infty)$ of all right continuous $\mathcal{Z}$-valued functions on $[0,\infty)$ having left hand limits. Let be $\psi = (T_n, n \geqslant 0)$ a sequence of random variables with the property

(1) $\quad 0 = T_0 \leqslant T_1 \leqslant T_2 \leqslant \ldots$ and $\lim_{n\to\infty} T_n = \infty$ $\quad Pr - a.s.$

2.1. Definition

A couple $\chi = [(Z(t), t \geqslant 0,(T_n, n \geqslant 0)]$ consisting of a stochastic process $(Z(t))$ and a sequence (T_n) satisfying (1), both defined on $(\Omega,\mathcal{F},Pr)$ is called a process with an embedded point process (abbr. PEP). The points $T_n, n \geqslant 0$, are called embedded points of $(Z(t))$.

Often the embedded points T_n are determined by the process $(Z(t))$. For example, T_n may be some stopping times of the process $(Z(t))$; in particular, T_n may be epochs of crossing a fixed level or of entrance into a given subset of $\mathcal{Z}$. But $T_n, n \geqslant 1$, can also be moments of certain "influence from outside the system", e.g. if the system load changes. In general, the phrase "both defined on $(\Omega,\mathcal{F},Pr)$" only means that both (T_n) and $(Z(t))$ describe the same stochastic phenomenon together.

Consider the distances $D_n = T_{n+1} - T_n, n \geqslant 0$, between the embedded points. The embedded points T_n partition the process $(Z(t))$ into so-called cycles describing the behaviour of the system considered in the interval $[T_n, T_{n+1})$ (supposed that $T_n < T_{n+1}$). Moreover, it is very convenient to describe the dynamic of a PEP by the sequence of its cycles as will be shown below by examples.

2.2. Definition

Let be $(X(t), t \geqslant 0)$ a stochastic process and D a non-negative random variable on $(\Omega,\mathcal{F},Pr)$. The couple $[D,(X(t), 0 \leqslant t < D)]$, if $D > 0$, and $[0,(X(0)]$, if $D = 0$, respectively, is called a <u>cycle of length</u> D of the process $(X(t))$.

We remark that the cycle $[D, (X(t) \setminus 0 \leqslant t < D)]$ can be considered as a random element ξ taking values in the space $[0, \infty) \times D(0, \infty)$, if one does not disinguish between $(X(t), 0 \leqslant t < D)$ and $(X^D(t), t \geqslant 0)$ where

$$X^D(t) = \begin{cases} X(t) & \text{for } 0 \leqslant t < D \\ X(D-) & \text{for } t \geqslant D. \end{cases}$$

Consider a sequence $(\xi_n, n \geqslant 0)$ of cycles $\xi = [D_n, (Z_n(t), 0 \leqslant t < D_n)]$ where $(Z_n(t))$ are $\mathcal{Z}$-valued stochastic processes, with the property

$$(2) \quad \sum_{n=0}^{\infty} D_n = \infty \ \text{Pr} - \text{a.s.}$$

and define

$$(3) \quad T_0 = 0, \ T_n = \sum_{k=0}^{n-1} D_k, \ n \geqslant 1,$$

$$(4) \quad Z(t) = Z_n(t - T_n) \quad \text{for } T_n \leqslant t < T_{n+1}.$$

Obviously, $[(Z(t), t \geqslant 0), (T_n, n \geqslant 0)]$ is a PEP in the sense of Definition 2.1. We shall say that this PEP <u>is generated by the given sequence of cycles</u> (ξ_n).

The simplest example of PEP are regenerative processes where the embedded points T_n are the regeneration points. In this case, the generating cycles ξ_n, $n \geqslant 0$, are independent and for $n \geqslant 1$ identically distributed. Several important classes of stochastic processes can be considered as PEP where the generating sequence of cycles forms a time-homogeneous Markov chain, cf. Arndt and Franken (1979) and Franken and Streller (1979).

Semi-Markov processes (SMP)

Assume that the system does not change its state within a cycle, i.e. $Z_n(t) = Z_n(0) = Z_n$ for $0 \leqslant t < D_n$. Then the cycles have the form $\xi_n = [D_n, Z_n]$. Let be $p(x,E)$, $E \in \mathcal{Z}$, a given stochastic kernel from $\mathcal{Z}$ to $\mathcal{Z}$ and $q(\cdot, x, y, B)$ a stochastic kernel from $\mathcal{Z} \times \mathcal{Z}$ to $[0, \infty)$; $F_{xy}(t) = q(x, y, [0, t))$. If the transition probabilities of the Markov chain (ξ_n) are given by

$$(5) \quad \begin{aligned} P_r(Z_{n+1} \in E \mid Z_n = x) &= p(x, E) \\ P_r(D_n \leq t \mid Z_n = x, Z_{n+1} = y) &= F_{xy}(t) \end{aligned}$$

then the stochastic process $(Z(t))$ defined by (3), (4), i.e. $Z(t) = Z_n$ for $T_n \leq t < T_{n+1}$, is a SMP with state space $\mathcal{Z}$. (For a complete definition of $(Z(t))$ we need besides $p(x, E)$ and $F_{xy}(t)$ the "initial distribution" of $\xi_0 = [D_0, Z_0]$).

Semi-regenerative processes

Assume that the cycles $[D_n, (Z_n(t), 0 \leq t < D_n)]$, $n \geq 0$, constitute a Markov chain with the property

$$Pr(Z_n(0) \in \mathcal{L}, n \geq 1) = 1$$

for some subset $\mathcal{L} \subseteq \mathcal{Z}$. Further, assume that the distribution of $\xi_{n+1} = [D_{n+1}, (Z_{n+1}(t), 0 \leq t < D_{n+1})]$ depends on the past only via $Z_n(0)$, i.e.

$$(6) \quad \begin{aligned} &P_r([D_{n+1}, (Z_{n+1}(t), 0 \leq t < D_{n+1})] \in (\cdot) \mid \xi_0, \ldots, \xi_n) \\ &= Pr([D_{n+1}, (Z_{n+1}(t), 0 \leq t < D_{n+1})] \in (\cdot) \mid Z_n(0)). \end{aligned}$$

Using the notations

$$p(x, E) = Pr(Z_{n+1}(0) \in E \mid Z_n(0) = x),$$

$$Q_y(C) = Pr([D_{n+1}, (Z_{n+1}(t), 0 \leq t < D_{n+1})] \in C \mid Z_{n+1}(0) = y)$$

we obtain from (6)

$$(7) \quad \begin{aligned} &Pr(Z_{n+1}(0) \in E, [D_{n+1}, (Z_{n+1}(t), 0 \leq t < D_{n+1})] \in C \mid Z_n(0) = x) \\ &= \int_E Q_y(C) p(x, dy). \end{aligned}$$

Under these assumptions, the stochastic process $(Z(t), t \geq 0)$ defined by (3), (4) is called semi-regenerative process. For the complete description we need additionally the "initial distribution" of ξ_0 and $Z_1(0)$. Obviously, the process $(Y(t), t \geq 0)$ defined by

$$Y(t) = Z_n(0) \quad \text{for} \quad T_n \leq t < T_{n+1}$$

is a SMP (the embedded SMP). If $\mathcal{L}$ is at most countably infinite, a semi-regenerative process can be considered as a regenerative process with several types of regeneration points. Semi-regenerative processes were introduced by Pyke and Schafele (1966) where they were called SMP with auxiliary paths, cf. also Çinlar (1975), Arndt and Franken (1979), Franken and Streller (1979), Streller (1982).

Piecewise Markov processes, cf. Kuczura (1973), Jankiewicz and Rolski (1977).

The transition probabilities of the Markov chain of cycles (ξ_n) have the form

$$Pr(\xi_{n+1} \in (\cdot) \mid \xi_n) = Pr(\xi_{n+1} \in (\cdot) \mid Z_n(D_n - 0)).$$

Furthermore, assume that for given $Z_{n+1}(0) = x$, say, the conditional distribution Q_x of $[D_{n+1}, (Z_{n+1}(t), 0 \leq t < D_{n+1})]$ is given in the following way : There exists a time-homogeneous Markov process $(Z_{n+1}(t), t \geq 0)$ with transition probabilities $p(x,t,E)$ and a random variable $D_{n+1} \geq 0$ with distribution function $F_x(t)$; $(Z_{n+1}(t), t \geq 0)$ and D_{n+1} are conditionally independent under $Z_{n+1}(0) = x$. Using the notation

$$R(y,E) = Pr(Z_{n+1}(0) \in E \mid Z_n(D_n-) = y)$$

we obtain

$$(8) \quad \begin{aligned} &Pr(Z_{n+1}(0) \in E, [D_{n+1}, (Z_{n+1}(t), 0 \leq t < D_{n+1})] \in (\cdot) \mid Z_n(D_n-) = y) \\ &= \int_E Q_x(\cdot)R(y,dx), \end{aligned}$$

$$p(x,E) = Pr(Z_{n+1}(0) \in E \mid Z_n(0) = x)$$

(9)
$$= \int_0^\infty \int_{\mathcal{Z}} R(y,E)p(x,u,dy)dF_x(u).$$

For the complete description of the process we have, again, to fix the distribution of the initial cycle ξ_0.

3. Stationary PEP. Ergodic theorems

Consider a system modelled by a PEP. There are two possibilities to describe the steady behaviour of such a system which are determined by the choice of the time-origin. Consider a PEP $\chi = [(Z(t),\ t \geqslant 0),(T_n,\ n \geqslant 0)]$ and define the shift operator S_u, $u \geqslant 0$, by

$$S_u\chi = [(Z(t + u),\ t \geqslant 0),(T_n^u,\ n \geqslant 0)]$$

where

(10) $T_0^u = 0,\ T_n^u = T_{n+N(u)} - u,\ n \geqslant 1,\ N(u) = max(j : T_j \leqslant u)$

Obviously, the application of S_u to χ means that the process $(Z(t))$ is shifted in the usual way; each embedded point T_n with $T_n \geqslant u$ is shifted for the value u to the left and is renumbered according to (10); the embedded points T_n with $T_n u$ are cancelled.

3.1. Definition

a) A PEP $\bar{\chi} = [(\bar{Z}(t),\ t \geqslant 0),(\bar{T}_n,\ n \geqslant 0)]$ with distribution $\bar{P}$ is said to be <u>stationary</u> if, for any $u \geqslant 0$, it is invariant with respect to the shift operator S_u, i.e. $S_u\bar{\chi}$ has distribution $\bar{P}$, too.

b) A PEP χ generated by the sequence of cycles $(\xi_n,\ n \geqslant 0)$ according to (3) and (4) is said to be <u>synchronous</u> if $(\xi_n,\ n \geqslant 0)$ is strictly stationary.

For a stationary PEP $\bar{\chi}$ with distribution $\bar{P}$ the value

$$\lambda(\bar{P}) = E_{\bar{P}}\bar{N}(1), \quad \bar{N}(t) = \max(j : \bar{T}_j \leq t)$$

is called <u>intensity</u> of $\bar{\chi}$.

The main tool for obtaining statements about certain stationary characteristics of the system considered is the following theorem.

<u>3.2. Theorem</u>

Let be χ a synchronous PEP with distribution P satisfying $0 < \mu(P) = E_P D_0 < \infty$. Then, by formula

$$(11) \quad \bar{P}((\cdot)) = (\mu(P))^{-1} \int_0^\infty P(D_0 > t, S_t \chi \in (\cdot)) dt$$

a probability distribution $\bar{P}$ of a stationary PEP $\bar{\chi} = [(\bar{Z}(t)),$ $(\bar{T}_n)]$ is defined. In particular, we have the following equations :

$$(12) \quad \lambda(\bar{P}) = (\mu(P))^{-1};$$

$$(13) \quad \bar{P}(\bar{T}_1 \leq t) = (\mu(P))^{-1} \int_0^t P(D_0 > u) du;$$

$$(14) \quad \bar{P}(\bar{Z}(0) \in E) = (\mu(P))^{-1} \int_0^\infty P(D_0 > t, Z(t) \in E) dt.$$

If we assume that

$$(15) \quad \Pr(D_n > 0 \text{ for all } n \geq 0) = 1$$

then there exists a one-to-one mapping between the families $\{P :$ P is the distribution of a synchronous PEP with $0 < \mu(P) < \infty\}$ and $\{\bar{P} : \bar{P}$ is the distribution of a stationary PEP with $0 < \lambda(\bar{P})$ $< \infty\}$, and one direction of this mapping is given by formula (11). This correspondence between the distributions of synchronous and stationary PEP can also be forced if $\Pr(D_n = 0) > 0$; in this case we have to include some instantaneous states (in the order of their occurrence) into the definition of a PEP. These instantaneous states then correspond to the cycles of length 0 in the sequence (ξ_n), cf. Streller (1982) for details.

Theorem 3.2 is a straightforward consequence of the analogous result for stationary marked point processes, cf. König, Matthes and Nawrotzki (1971) and Port and Stone (1973). Remember that a marked point process can be considered as a sequence of couples ($[T_n, K_n]$, $n \geqslant 0$) where the points T_n, $n \geqslant 0$, satisfy condition (1), and the "marks" K_n, $n \geqslant 0$, are random elements of a certain "mark space". Then, a PEP generated by a sequence of cycles (ξ_n) can be considered as a marked point process where T_n are defined by (3) and the "mark" K_n coincides with the cycle ξ_n, $n \geqslant 0$. A direct and self-contained proof of theorem 3.2 is given in Streller (1982).

By means of standard techniques of integration theory we get the following equivalent form of equation (14) : for every measurable non-negative function f on $\mathcal{Z}$

$$(16) \quad E_{\bar{P}} f(\bar{Z}(0)) = (\mu(P))^{-1} E_P \left(\int_0^{D_0} f(Z(t)) dt \right).$$

Formula (16) is well-known for regenerative processes as the "stochastic mean value theorem", cf. Cohen (1976). A regenerative process is synchronous if the cycles are i.i.d. for $n \geqslant 0$. The structure of the corresponding (in the sense of theorem 3.2) stationary regenerative process is very simple : The cycles $\xi_n = (\bar{Z}_n(t), \, 0 \leqslant t < \bar{D}_n)$, $n \geqslant 1$, are i.i.d. and have the same distribution as the cycles of the synchronous process. The distribution of the initial cycle $\xi_0 = (\bar{Z}_0(t), \, 0 \leqslant t < \bar{D}_0)$ can be calculated by means of (11).

Theorem 3.2 can be considered as a far-reaching generalization of the stochastic mean value theorem for regenerative processes, because we do not have to require any independence assumptions. In fact, the stochastic mean value theorem and formula (11) have nothing to do with any independence of the cycles, they are rather based on the assumed stationarity of the generating sequence of cycles.

Theorem 3.2 only provides a formal relationship between the distribution P and $\bar{P}$ of a synchronous and the corresponding stationary PEP, respectively. On the other hand, for regenerative, semi-Markov (with $\mathcal{Z}$ being at most countably infinite) and semi-regenerative processes (with $\mathcal{L}$ being at most countably infinite) it is well known that, under some additional assumptions, $\bar{P}$ arises as the limiting distribution of the PEP with the same parameters but with an arbitrary initial distribution. In particular, $\bar{P}$ arises as the limiting distribution of the corresponding synchronous PEP with distribution P. The following ergodic theorem shows that a similar result holds for general PEP.

3.3. Theorem

Let P and $\bar{P}$ be the distributions of a synchronous PEP $\chi = [(Z(t),\ t \geq 0),(T_n,\ n \geq 0)]$ and the corresponding stationary PEP $\bar{\chi} = [(\bar{Z}(t),\ t \geq 0),(\bar{T}_n,\ n \geq 0)]$, respectively. If the synchronous PEP satisfies the property

$$\lim_{n \to \infty} n^{-1} \sum_{j=0}^{n-1} D_j = E_P D_0 = \mu(P) \quad P - a.s.$$

then we have the equation

$$(17) \quad \lim_{t \to \infty} t^{-1} \int_0^t P(S_u \chi \in (\cdot))du = \bar{P}(\bar{\chi} \in (\cdot)).$$

In particular,

$$\lim_{t \to \infty} t^{-1} \int_0^t P(Z(u) \in E)du = \bar{P}(\bar{Z}(0) \in E).$$

The proof of theorem 3.3 is based on the individual ergodic theorem, cf. Franken and Streller (1979) and Nawrotzki (1975).

Notice that we have a specified initial distribution here characterized by the stationarity of the sequence of cycles. In general, without any assumptions concerning the independence of the cycles, it does not make sense to speak about an "arbitrary initial distribution". However, for semi-regenerative processes with arbitrary state spaces $\mathcal{L}$ and $\mathcal{Z}$, we can use the initial distribution ν of ξ_0 and $Z_1(0)$, cf. section 1.

3.4. Theorem

Let P_ν be the distribution of a semi-regenerative process $(Z(t),\ t \geqslant 0)$ defined by (3) and (4), (6) and (7) and by the initial distribution ν. If the embedded Markov chain $(Z_n(0),\ n \geqslant 0)$ is Harris-recurrent with respect to a non-trivial σ-finite measure ρ on $\mathfrak{z}$ (i.e. $\rho(A) > 0$ implies that $P_\nu(\inf(n : n \geqslant 1, Z_n(0) \in A) < \infty \mid Z_0(0) = y) = 1$ for every $y \in \mathcal{L}$) then the equation

$$\lim_{t \to \infty} t^{-1} \int_0^t I_A(Z(u))du = \bar{P}(\bar{Z}(0) \in A)$$

holds P_ν - almost surely. More generally, for the PEP $\chi = [(Z(t), t \geqslant 0), (T_n,\ n \geqslant 0)]$, we have in this case the equation

$$\lim_{t \to \infty} t^{-1} \int_0^t I_{(\cdot)}(S_u\chi)du = \bar{P}(\bar{\chi} \in (\cdot))\quad P_\nu - a.s.$$

Theorem 3.4 can be obtained as an elementary consequence of theorem 3.3, cf. Streller (1982). Another proof based on the coupling method can be found in Athreya, McDonald and Ney (1978). Theorem 3.3 states that the distribution $\bar{P}$ of a stationary PEP arises as the limiting distribution of the underlying synchronous PEP χ. It can be shown that, vice versa, the distribution P of the synchronous PEP χ arises as the limiting distribution of the stationary PEP $\bar{\chi}$:

$$\lim_{n \to \infty} n^{-1} \sum_{i=1}^n I_{(\cdot)}(S_{\bar{T}_i}\bar{\chi}) = P(\chi \in (\cdot))\quad \bar{P} - a.s.$$

Together with theorem 3.2 this leads to the following interpretation. Consider a certain system in steady state and define the embedded points in a suitable way. If the origin is an arbitrary point of the time axis, then the evolution of the system is described by a stationary PEP $\bar{\chi}$. If the system is viewed by an observer placed at some embedded point (i.e. the origin is an arbitrary embedded point) then the temporal behaviour of the system can be described by the corresponding synchronous PEP χ. This interpretation is confirmed by the following statement : If the occurrence of multiple points is excluded, i.e. (15) is satisfied, then the distribution P of a synchronous PEP χ can be considered as the conditional distribution of the corresponding stationary PEP $\bar{\chi}$ under the condition that in the origin is an embedded point. More precisely,

$$\lim_{h \downarrow 0} \left(\sup_{(\cdot)} \left| P(\chi \in (\cdot)) - \bar{P}(S_{\bar{T}_1} \bar{\chi} \in (\cdot) \mid \bar{T}_1 \leqslant h) \right| \right) = 0.$$

3.5. Remark

For a given system there may be several possibilities to choose the embedded points. Then, for the steady state behaviour of the system, we have the following important statement : Let be $\bar{\chi}_1 = [(\bar{Z}_1(t)),(\bar{T}_{1n})]$ and $\bar{\chi}_2 = [(\bar{Z}_2(t)),(\bar{T}_{2n})]$ two stationary PEP with distribution $\bar{P}_1$ and $\bar{P}_2$, respectively, and assume

$$\bar{P}_1((\bar{Z}_1(t)) \in (\cdot)) = \bar{P}_2((\bar{Z}_2(t)) \in (\cdot)),$$

i.e. the stationary state process $(\bar{Z}(t))$ is in fact the same one in both the cases. If $\chi_1 = [(Z_{(1)}(t)),(T_{1n})]$ and $\chi_2 = [(Z_{(2)}(t)), (T_{2n})]$ are the underlying synchronous PEP with distribution P_1 and P_2, then we have in view of (11) that

$$\bar{P}((\bar{Z}(t),\ t \geqslant 0) \in (\cdot))$$

$$= \mu_1^{-1} \int_0^\infty P_1(D_{10} > u, (Z_{(1)}(t + u),\ t \geqslant 0) \in (\cdot))du$$

$$= \mu_2^{-1} \int_0^\infty P_2(D_{20} > u, (Z_{(2)}(t + u),\ t \geqslant 0) \in (\cdot))du.$$

This means that stationary probabilities can always be calculated by means of formula (11) independent of the choice of the embedded points.

Now we are going to calculate some stationary probabilities for the classes of stochastic processes considered in section 1. (The results are known at least for countable spaces $\mathcal{L}$ and $\mathcal{Z}$). We emphazise that for this aim we do not need any limiting considerations (as they are familiar in the traditional approach). The results are straightforward implications of formula (11) in theorem 3.2.

First, consider a SMP governed by the kernel $p(x,E)$ and distribution functions $F_{xy}(t)$. Additionally we assume the existence of a unique invariant probability measure for the embedded Markov chain $(Z_n, n \geqslant 0)$, i.e. the existence of a unique probability solution $p(\cdot)$ of the equation

$$(18) \quad p(E) = \int_{\mathcal{Z}} p(x,E)p(dx), \quad E \in \mathfrak{z}.$$

With the initial distribution

$$Pr(Z_0 \in E) = p(E)$$

$$Pr(D_0 \leqslant t \mid Z_0 = x, Z_1 = y) = F_{xy}(t)$$

the sequence of cycles $(\xi_n, n \geqslant 0)$, $\xi_n = [D_n, Z_n]$, becomes strictly stationary. Thus, the PEP χ defined by (3) and (4) is synchronous. Using the notations

$$\mu_{xy} = E_P(T_1 \mid Z_0 = x, Z_1 = y) = \int_0^\infty (1 - F_{xy}(t))dt$$

$$\mu_x = E_P(T_1 \mid Z_0 = x) = \int_{\mathcal{Z}} \mu_{xy}p(x,dy)$$

$$\mu = E_P T_1 = \int_{\mathcal{Z}} \int_{\mathcal{Z}} \mu_{xy}p(x,dy)p(dx),$$

we obtain from (13) and (14), respectively,

$$\bar{P}(\bar{T}_1 \leqslant t \mid \bar{Z}_0 = x, \bar{Z}_1 = y) = \mu_{xy}^{-1} \int_0^t (1 - F_{xy}(u))du$$

$$(19) \quad \begin{aligned} \bar{P}(\bar{Z}(0) \in E) &= \mu^{-1} \int_E \int_{\mathcal{Z}} \mu_{xy}p(x,dy)p(dx) \\ &= \mu^{-1} \int_E \mu_x p(dx) \end{aligned}$$

In the case of semi-regenerative processes we assume again the existence of an invariant distribution $\hat{p}(\cdot)$ of the embedded Markov chain $(Z_n(0), n \geqslant 0)$. With the initial distribution

$$Pr(Z_0(0) \in E) = \hat{p}(E)$$

$$Pr(Z_1(0) \in E \mid Z_0(0) = x) = p(x,E)$$

$$Pr(D_0, (Z_0(t), 0 \leqslant t < D_0) \in C \mid Z_0(0) = y) = Q_y(C)$$

the Markov chain $(\xi_n, n \geqslant 0)$ becomes strictly stationary and the PEP χ defined by (3) and (4) is synchronous. For the associated stationary PEP $\bar{\chi}$ we obtain the following equations :

$$\bar{P}(\bar{Z}(0) \in E) = \mu^{-1} \int_0^\infty \int_{\mathcal{L}} Q_x(D_0 > t, Z_0(t) \in E)\hat{p}(dx)dt,$$

where $\mu = E_p D_0$. Using (16) with $f = I_E$, $E \subseteq \mathcal{L}$, we have

$$(20) \quad \bar{P}(\bar{Z}(0) \in E) = \mu^{-1} E_p\left(\int_0^{D_0} I_E(Z(t))dt\right) = \mu^{-1} \int_{\mathcal{L}} \mu_x \hat{p}(dx)$$

where

$$\mu_x = E_p\left(\int_0^{D_0} I_E(Z(t))dt \mid Z(0) = x\right)$$

is the conditional mean sojourn time in E if the process starts at time 0 with state x.

Finally, consider a piecewise Markov process with the parameters $R(y,\cdot)p(x,t,\cdot)$ and $F_x(t)$ · and a unique invariant distribution $p(\cdot)$ of the embedded Markov chain $(Z_n(0), n \geqslant 0)$, i.e. $p(\cdot)$ is the unique probability solution of (14) with $p(x,E)$ given by (9). If the initial distribution is chosen to be

$$Pr(Z_0(0) \in E) = p(E),$$

$$Pr([D_0, (Z_0(t), 0 \leqslant t < D_0)] \in C) = \int_{\mathcal{Z}} Q_x(C)p(dx),$$

where $Q_x(\cdot)$ is defined in section 1, then the Markov chain $(\xi_n, n \geqslant 0)$ is strictly stationary and the PEP χ defined by (3) and (4) is synchronous. For the associated stationary PEP $\bar{\chi}$ we have

$$\bar{P}(\bar{Z}(0) \in E) = \mu^{-1} \int_{\mathcal{Z}} \int_0^\infty p(y,t,E)(1 - F_y(t))du \, p(dy)$$

where

$$\mu = \int_{\mathcal{Z}} \mu_y p(dy), \quad \mu_y = \int_0^\infty (1 - F_y(t))dt = E_P(D_0 \mid Z_0(0) = y).$$

In accordance with (13) we obtain

$$\bar{P}(\bar{T}_1 \leqslant t) = \mu^{-1} \int_{\mathcal{Z}} \int_0^t (1 - F_y(u))du\,p(dy).$$

4. Some formulas for stationary PEP

Sometimes the probabilities $p(E) = P(Z(0) \in E)$ and $\bar{p}(E) = \bar{P}(\bar{Z}(0) \in E)$, $E \in \mathcal{Z}$, are called <u>stationary embedded</u> and <u>time-stationary</u> probabilities, respectively. The relationships between these probabilities and some suitably chosen mean sojourn times are very important in applied probability, in particular in queueing theory, cf. e.g. Franken et al. (1981). Now we derive some of such relationships in general setting; for the interpretation in reliability cf. the next section.

Consider a stochastic system in steady state and assume that the epochs at which the system enters a given subset $E \in \mathcal{Z}$ are well-defined. Moreover, assume that the sequence of embedded points is chosen in such a way that it contains all these epochs. Then we obtain in view of (16) with $f = I_E$

$$(21) \quad \bar{p}(E) = \mu^{-1} \int_E \mu_x p(dx),$$

where

$$\mu_x = E_P \left(\int_0^{D_0} I_E(Z(t)\,dt \mid Z(0) = x) \right), \quad x \in E,$$

denotes the conditional mean sojourn time in E if the system starts at time 0 with state x.

Regardless of different notations, formula (21) coincides with (20) which was derived for semi-regenerative processes and $E \subseteq \mathcal{L}$ using all the epochs of entrance into $\mathcal{L}$ as embedded points. In this case $p(\cdot)$ is the stationary distribution of the embedded

Markov chain, μ is the mean time between two consecutive entrances into $\mathcal{L}$ and μ_x is the conditional mean sojourn time in E.

Now we consider the same system under the assumption that the sequence of embedded points contains all the epochs at which the system enters or leaves the subset E. Using the notations $\hat{Z}$, $\hat{P}$, $\hat{p}$,... instead of Z, P, p,... for this choice of embedded points and assuming the existence of the mean conditional expectations

$$\hat{\mu}_x = E_{\hat{P}}(\hat{D}_0 \mid \hat{Z}(0) = x)$$

for $\hat{P}$-almost all x we get from (16) with $f = I_E$,

$$(22) \quad p(E) = \hat{\mu}^{-1} E_{\hat{P}}\left(\int_0^{\hat{D}_0} I_E(Z(t))dt\right) = \hat{\mu}^{-1} \int_E \hat{\mu}_x \hat{p}(dx),$$

where

$$\hat{\mu} = E_{\hat{P}}\hat{D}_0 = \int_{\mathcal{Z}} \hat{\mu}_x \hat{p}(dx).$$

(Notice, that in view of the given choice of the embedded points the system starting with x, $x \in E$, cannot leave E before $T_1 = D_0$). Observe that (22) formally coincides with (19) which was derived for a semi-Markov process using all jump points as embedded points. In fact, (22) is true for an arbitrary process under the assumptions mentioned above. However, unlike the semi-Markov case, where $p(\cdot)$ and μ_x, $x \in \mathcal{Z}$, are given in some sense, the determination of the right side in (22) may be quite difficult in general.

If the embedded points T_n coincide with the epochs at which the system enters the subset E, then the random variable X with distribution

$$\Pr(X \leqslant x) = P(\inf(t : Z(t) \notin E) \leqslant x)$$

is called <u>sojourn time</u> in E. The corresponding random variable $\bar{X}$ for the process $(\bar{Z}(t))$ is called <u>exit time</u>. Its distribution is

$$\bar{P}(\bar{X} \leqslant x) = \bar{P}(\inf(t : \bar{Z}(u) \notin E) \leqslant x \mid \bar{Z}(0) \in E).$$

Intuitively, $\bar{X}$ denotes the remaining time that the system in steady state spends in E under the condition that it has been in E at time 0.

4.1. Theorem

If the embedded points T_n are chosen according to

$$(23) \quad Z(T_n) \in E \quad \text{and} \quad Z(T_n-) \notin E \quad \text{for all} \quad n \geq 1,$$

then the following formular are valid :

$$(24) \quad \bar{p}(E) = (E_p D_0)^{-1} E_p X,$$

$$(25) \quad \bar{P}(\bar{X} \geq x) = (E_p X)^{-1} \int_x^\infty P(X \geq t)\,dt.$$

Proof

(24) can be proved in the same way as (21). For proving (25) we show that

$$(26) \quad \bar{P}(\bar{Z}(u) \in E \quad \text{for} \quad 0 \leq u \leq x) = (E_p D_0)^{-1} \int_x^\infty P(X \geq t)\,dt.$$

Indeed, in view of (11) we have that

$$\bar{P}(\bar{Z}(u) \in E \quad \text{for} \quad 0 \leq u \leq x)$$

$$= (E_p D_0)^{-1} \int_0^\infty P(D > t, Z(t+u) \in E \quad \text{for} \quad 0 \leq u \leq x)\,dt$$

$$= (E_p D_0)^{-1} \int_0^\infty P(Z(u) \in E \quad \text{for} \quad 0 \leq u \leq x+t)\,dt$$

$$= (E_p D_0)^{-1} \int_0^\infty P(X \geq x+t)\,dt.$$

Thus, (26) is proved and (25) follows from (24). $\square$

Sometimes formula (24) is used to calculate the mean sojourn time $E_p X$ by means of the time stationary probability $\bar{p}(E)$ and the mean cycle length $E_p D_0$. In the case of time-homogeneous Markov processes with finite state space Z, formula (25) was proved in Keilson (1974).

Now we consider stochastic systems with speeds. Assume that being in state $z \in Z$ the system works with speed $c(z) \geq 0$. For a given subset $C \in \mathfrak{z}$ denote by $Y(C)$ the required "sojourn time" in C (for $c(z) = 1$ this is indeed the sojourn time in C). Assume, for example, that the repairs of the system are provided with speeds depending on the system state. In this case, $Y(B)$ is equal to the down-time of the system.

Consider the synchronous PEP $[(Z(t),Y(t)),(T_n)]$ in the real time where T_n are the epochs of entrances into C, $Y(t) = Y_n(C)$ for $T_n \leqslant t < T_{n+1}$ and $Y_n(C)$ are the consecutive required sojourn times in C. Let $[(\bar{Z}(t),\bar{Y}(t)),(\bar{T}_n)]$ be the corresponding stationary PEP. We use the following notations :

$$\mu = E_P T_1, \quad \mu(C) = E_P Y_1(C), \quad G(x) = P(Y_1(C) \leqslant x),$$

$$\tau(C) = \int_0^{T_1} I_C(Z(t))dt, \quad \nu(C) = E_P \tau(C),$$

$$\bar{p}(E) = \bar{P}(\bar{Z}(0) \in E) \quad \text{and} \quad r(C) = \int_C c(z)\bar{p}(dz).$$

Notice that $\tau(C)$ is the real sojourn time in C.

4.2. Theorem

For every $C \in \mathfrak{z}$ with $\mu < \infty$ we have the equations

(27) $\quad \bar{p}(C) = \mu^{-1}\nu(C),$

(28) $\quad r(C) = \mu^{-1}\mu(C).$

Proof

Formula (27) is an analogy of (22). For proving (28) we apply (16) to the PEP $[(Z(t)),(T_n)]$ and $f(z) = c(z)I_C(z)$ and take into consideration that

$$Y(C) = \int_0^{\tau(C)} I_C(Z(t))c(Z(t))dt.$$

For some applications it is important to consider the conditional mean sojourn time in C

$$\nu_x(C) = E_P(\tau(C) \mid Y(C) = x)$$

under the condition that the required sojourn time is x. $\square$

4.3. Theorem

If for every $y \geq 0$ the equation

$$(29) \quad \bar{P}(\bar{Z}(0) \in C, \bar{Y}(0) \geq y) = \bar{p}(C)(\mu(C))^{-1} \int_y^\infty x\,dG(x)$$

is fulfilled, then

$$(30) \quad \nu_x(C) = (r(C))^{-1} \bar{p}(C)x.$$

Proof

We have to show that

$$(31) \quad \int_y^\infty (r(C))^{-1} \bar{p}(C)x\,dG(x) = E_P\left(\int_0^{T_1} I_{[y,\infty)\times C}(Z(t),Y(t))\,dt\right)$$

Applying (16) we get that the right side in (31) is equal to

$$\mu\bar{P}(\bar{Z}(0) \in C, \bar{Y}(0) \geq y).$$

Using the assumption (29) and (27) we obtain (31).

The assumption (29) is fulfilled e.g. for the so-called insensitive generalized semi-Markov processes for which (30) was proved in Barbour and Schassberger (1981) and Jansen (1984).

5. Stationary reliability characteristics

Consider a certain repairable system in steady state. We assume that the state space Z (which is obviously finite) consists of two disjoint sets G and B, G being the set of "good" states (where the system is able to function) and B the set of "bad" (failure) states; $Z = G \cup B$. The temporal behaviour of the system can be described by a suitably chosen synchronous PEP $\chi = [(Z(t), t \geq 0),(T_n, n \geq 0)]$. Usually the state process is given in a natural way while the embedded points have to be chosen. We emphazise that there may be several possibilities of a "suitable" choice of the embedded points and that a general rule for doing so cannot be given. In general, one can only say that the embedded points must be chosen individually for each specific model having regard to what is useful for the calculation. However, we require

(i)   each T with

$$(32)\quad Z(t) \in \mathfrak{C} \quad \text{and} \quad Z(T-) \in \mathfrak{B}$$

is an embedded point,

(ii) for every embedded point T_n we have that $Z(T_n) \in \mathfrak{C}$.

Assumption (ii) ensures that a generic cycle starts with a positive working phase of length X say. In view of (i) this working phase X can be followed at most by one failure phase Y say, $D = X + Y$, $P(X > 0) = 1$, $P(Y > 0) \leqslant 1$. If the embedded points are only the epochs when a system failure is finished (i.e. (32) holds for all the T_n), then $P(Y > 0) = 1$. In this case, however, the cycles may become very long and complicated. In particular, this will be the case for highly reliable systems. Therefore, it is very useful to have the chance to split such long cycles up by choosing additionally other epochs (e.g. some transitions within $\mathfrak{C}$) as embedded points, cf. the example in section 6.

To summarize, for the description of a given system in steady state one can use several sequences of embedded points and then obtains several synchronous PEP. However, the application of equation (11) to each of them leads to the same stationary process $(\bar{Z}(t),\ t \geqslant 0)$, cf. remark 3.5. Now, the stationary system availability and interval reliability can be defined by

$$A = \bar{P}(\bar{Z}(0) \in \mathfrak{C}) \quad \text{and} \quad A_x = \bar{P}(\bar{Z}(t) \in \mathfrak{C} \text{ for } 0 \leqslant t \leqslant x)$$

respectively, independently of the choice of the embedded points.

5.1. Theorem

If the assumptions (i) and (ii) are fulfilled, then the following formulas are valid for every system as described above :

$$(33) \quad A = \frac{E_p X}{E_p D} = \frac{E_p X}{E_p X + E_p Y} \, ;$$

$$(34) \quad A = (E_p D)^{-1} \int_x^\infty P(X \geq t) dt.$$

Proof

(34) is an immediate consequence of (26) and (33) follows from (34)
with $x = 0$. $\square$

5.2. Remark

Formulas (33) and (34) are well-known for the repairable unit
with independent and identically distributed life- and repair.
times. (In this case, $(Z(t))$ and $(\bar{Z}(t))$ are alternating renewal
processes). Theorem 5.1 shows that these formulas are formally
valid without any independence assumptions. Although, the right
sides are not known in che general case in several cases it is
easier to deal with the embedded probabilities and sojourn times.
That makes the formulas useful.

5.3. Remark

In the reliability literature it is familiar to call $E_p X$ and
$E_p Y$ the MTBF (mean time between failures) and the mean repair time,
respectively. It is worth mentioning that these notions are correct
only if $P(Y > 0) = 1$, i.e. if the embedded points are chosen according to (32).

Using the embedded points satisfying (i) and (ii) and $E = \{j\}$
in formula (21) we obtain for the stationary state probabilities

$$(35) \quad \bar{p}_j = \bar{P}(\bar{Z}(0) = j) = \mu^{-1} p_j \mu_j, \quad j \in G,$$

where

$$p_j = P(Z(0) = j), \quad \mu_j = E_p\left(\int_0^{D_0} I_{[j]}(Z(t)\, dt \mid Z(0) = j)\right).$$

Using as embedded points all the jump epochs we obtain in the same way in ivew of (32)

$$(36) \quad \bar{p}_j = \hat{\mu}^{-1} \hat{p}_j \hat{\mu}_j, \quad j \in \mathbb{Z}$$

where the symbol "^" is used for quantities of the synchronous process with these embedded points.

Notice that (36) and (35) are the standard formulas for calculating the stationary state probabilities if the state process is semi-Markov or semi regenerative with $\mathcal{L} = \mathbb{G}$, respectively. The formulas (33)-(36) were derived without any assumptions concerning the independence and concerning the distributions of the life- and repair-times involved. Their application leads to interesting results for coherent (binary and multistate) systems with separately maintained components and for several systems with cold redundancy, cf. Franken and Streller (1978), Beichelt and Franken (1983), Streller (1979), Natvig and Streller (1984). However, for the application of these formulas to systems with warm redundancy and restricted repair facilities some independence assumptions are needed. We have good experiences in calculating stationary reliability characteristics by means of the formulas (33)-(36) in the case that the state process is semi-regenerative with $\mathcal{L} = \mathbb{G}$. (In particular, the temporal behaviour of the system can be described by a semi-Markov process up to the first system failure, cf. the example in section 6).

5.4. Remark

Obviously, the stationary availability A can be calculated by means of (35) or (36) as

$$A = \sum_{j \in \mathbb{G}} \bar{p}_j .$$

However, from our experience, we suggest to use (33) directly, even in the semi-regenerative case.

6. An example : Warm standby

Consider a binary redundant system consisting of two statistically identically components and a single repair facility. If both the components are available one of them is active on-line (with life time X_a having distribution function $F_a(t) = 1 - e^{-\lambda t}$) while the other one is active off-line (with life time X_r having distribution function $F_r(t) = 1 - e^{-c\lambda t}$, $0 \leqslant c \leqslant 1$). If the on-line (off-line) component fails, its repair begins immediately and completely restores it during a repair time Y_a (Y_r) having an arbitrary distribution function $G_a(t)$ (resp. $G_r(t)$), and the other component continues to work on-line. A system failure occurs, if the on-line component fails during the repair of the other component. Using the method described in section 5 we calculate the stationary availability if all life and repair times are independent. For this aim, we construct a semi-regnerative process with the states

1 – both the components are available

2 – one component is active on-line, the other one is under repair; before the on-line component switched into operation, a system failure had been

3 – one component is active on-line, the other one is under repair; before the on-line component switched into operation, both the components has been available

4 – one component is active on-line, a component failed off-line is under repair

5 – both the components are failed; one component is under repair, the other one is waiting for repair.

Obviously, $\mathcal{G} = \{1,2,3,4\}$ and $\mathcal{B} = \{5\}$. As embedded points we choose all the epoches when a repair of a component is finished. Then, $(Z(t))$ is a semi-regenerative process with $\mathcal{L} = \{1,2\}$. For the embedded Markov chain $(Z_n(0))$ we get the following transition probabilities $p_{ij} = P(Z_{n+1}(0) = j \mid Z_n(0) = i)$, $i = 1,2$, $j = 1,2$,

$$P_{11} = P(X_a > Y_a)P(X_a \leq X_r) + P(X_a > Y_r)P(X_a > X_r)$$

$$= g_a(\lambda)(1 + c)^{-1} + cg_r(\lambda)(1 + c)^{-1},$$

where $\displaystyle g(s) = \int_0^\infty e^{-st}dG(t)$ denotes the Laplace–Stieltjes transform of a distribution function $G(t)$ at the point s,

$$P_{12} = (1 + c)^{-1}(1 + c - g_a(\lambda) - cg_r(\lambda)),$$

$$P_{21} = P(X_a > Y_a) = g_a(\lambda),$$

$$P_{22} = 1 - g_a(\lambda).$$

For the stationary state probabilities of $(Z_n(0))$ we get

$$p_1 = p(\{1\}) = g_a(\lambda)(p_{12} + g_a(\lambda))^{-1},$$

$$p_2 = p(\{2\}) = p_{12}(p_{12} + g_a(\lambda))^{-1}.$$

Further,

$$\mu_1 = E\min(X_a,X_r) + E(Y_a \mid X_a \leq X_r)P(X_a \leq X_r)$$

$$+ E(Y_r \mid X_a > X_r)P(X_a > X_r)$$

$$= (1 + c)^{-1}(\lambda^{-1} + EY_a + cEY_r),$$

$$\mu_2 = EY_a,$$

$$\mu = (p_{12} + g_a(\lambda))^{-1}[g_a(\lambda)(1 + c)^{-1}(\lambda^{-1} + EY_a + cEY_r) + p_{12}EY_a].$$

Denote by X_a' another random variable with distribution function $F_a(t)$ independent of X_a. Then, the generic up-time X is equivalent to

$$X_1 = \min(X_a,X_r) + \begin{cases} \min(X_a',Y_a), & \text{if } X_a \leq X_r \\[2ex] \min(X_a',Y_r), & \text{if } X_a > X_r \end{cases}$$

if the cycle starts in 1, and otherwise to

$$X_2 = \min(X_a,Y_a).$$

Together, we obtain

$$EX = (1 + c - cg_r(\lambda)) \, [(1 + c)(p_{12} + g_a(\lambda))]^{-1},$$

and, consequently

$$A = (1 + c - cg_r(\lambda)) \, [g_a(\lambda) + (1 + c)\lambda EY_a$$
$$+ \; c(g_a(\lambda)EY_r - g_r(\lambda)EY_a]^{-1}.$$

If, in particular, $c = 1$ (parallel redundancy) and $G_a = G_r = G$, then

$$A = (2 - g(\lambda))(g(\lambda) + 2\lambda EY)^{-1}.$$

References

Arndt, K. and P.Franken (1979). Construction of a class of stationary processes with applications in reliability, Zast.Matem. 16, 379–393.

Athreya, K.B., D.McDonald, P.Ney (1978). Limit theorems for semi-Markov processes and renewal theory for Markov chains, Ann. Prob. 6, 788–797.

Barbour, A.D. and R.Schassberger (1981). Insensitive average residence times in generalized semi-Markov processes, AAP 13, 846–859.

Beichelt, F. and P.Franken (1983). Zuverlässigkeit und Instandhaltung. Verlag Technik, Berlin.

Çinlar, E. (1975). Introduction to stochastic processes. Prentice Hall, Englewood Cliffs, New York.

Cohen, J.W. (1976). On regenerative processes in queueing theory. Lect.Notes in Econ.Math.Systems, 121, Springer-Verlag, Berlin/ Heidelberg/New York.

Franken, P., D.König, U.Arndt and V.Schmidt (1982). Queues and point processes. Akademie-Verlag, Berlin, 1981, Wiley, New York.

Franken, P. and A.Streller (1978). A general method for calculation of stationary interval reliability of complex systems with repair. Elektron. Informationsverarb.Kybern. (EIK) 14, 283–290.

Franken P. and A.Streller (1979). Stationary generalized regenerative
   processes. Theory Prob.Applic. $\underline{24}$, 78-90

Franken P. and A.Streller (1980). Reliability analysis of complex
   repairable systems by means of marked point processes, JAP, $\underline{17}$,
   154-167.

Jankiewicz, M. and T.Rolski (1977). Piecewise Markov processes on a
   general state space, Zast.Matem., $\underline{15}$, 421-436.

Jansen, U. (1984). Conditional expected sojourn times in insensitive
   queueing systems.

Keilson, J. (1974). Monotonicity and convexity in systems survival
   functions and metabolic disappearance curves. in : Reliability
   and Biometry, SIAM, Philadelphia.

König, D., K.Matthes and K.Nawrotzki (1971). Unempfindlichkeits-
   eigenschaften von Bedienungsprozessen. Supplement to B.W.
   Gnedenko, I.N.Kovalenko, Einführung in die Bedienungstheorie.
   Akademie-Verlag, Berlin.

Kuczura, A. (1973). Piecewise Markov processes, SIAM J.Appl.Math.,
   $\underline{24}$, 169-181.

Natvig, B. and A.Streller (1984). The steady state behaviour of
   multistate monotone systems. (to appear in JAP).

Nawrotzki K. (1975). Markovian random marked sequences and their
   applications in queueing theory (in Russian). Math.Operations-
   forsch.Statist., $\underline{6}$, 445-477.

Port, S.C. and C.J.Stone (1973). Infinite particle systems. Trans.
   Amer.Math.Soc., $\underline{178}$, 307-340.

Pyke, R. and R.Schaufele (1966). The existence and uniqueness of
   stationary measures for Markov renewal processes. Ann.Math.
   Stat., $\underline{37}$, 1439-1462.

Rolski, T. (1981). Stationary random processes associated with
   point processes. Lect.Notes in Statistics, $\underline{5}$, Springer-Verlag,
   New York/Heidelberg/Berlin.

Streller, A. (1979). Stationary interval reliability of a doubled
    system with renewal and prophylactic (in Russian), Izv.Akad.
    Nauk SSSR, Tech.Kybern. 3, 98-103.
Streller, A. (1982). On stochastic processes with an embedded marked
    point process. Math.Operationsforsch.Stat., $\underline{13}$, 561-576.

SECTION VI. SIMULATION AND STATISTICS FOR SEMI-MARKOV PROCESSES

SEMI-MARKOV MODELS FOR MANPOWER PLANNING

Sally McClean

Mathematics Department
New University of Ulster

1. Introduction

In this paper we consider the use of semi-Markov models for
manpower planning. In general, the system consists of a number of
transient states, or grades, and the state of having left, which is
usually absorbing. The individual progresses from one state to another as he is promoted through the company hierarchy. For example,
in a university the states could be lecturer, senior lecturer, professor, and left as described by Young and Almond (1961). Alternatively
the states may correspond to degrees of committment to the firm
(e.g., Herbst, 1963).

The discrete- and continuous-time Markov model has been applied
frequently to such situations (e.g., Young and Almond, 1961; Bartholomew and Forbes, 1981), with considerable success. However, the simple
Markov model has the disadvantage of containing implicitly the assumption that the distribution of duration of stay in a grade is geometric
in the discrete-time case and exponential in the continuous-time case,
i.e., the rate of leaving a grade is constant irrespective of how long
an individual has been there. However, it is well known (Hedberg,
1961) that the length of stay is highly dependent on length of

service, and there is often what has been described by McGinnis (1968)
as a cumulative inertia effect where the probability of leaving a
grade decreases as length of stay increases. Several distributions
have been successfully fitted to the distribution of length of stay
until leaving which take this time dependence into account. Examples
are the lognormal distribution (e.g., Lane and Andrew, 1955 ; Young,
1971) and the mixed exponential distribution (e.g., Bartholomew, 1959;
McClean, 1976). It is also quite likely that the length of stay in a
grade, before promotion occurs, increases up to an optimum length of
service, after which promotion prospects decline as the person is
"passed over" in favour of someone younger. Thus transitions to both
the state of having left and to higher states in the hierarchy are not
adequately described by a simple Markov model. A semi-Markov model
has, therefore, been proposed by Ginsberg (1971), Mehlmann (1980), and
McClean (1978, 1980) which combines the desirable properties and more
general soujourn time distributions.

Some results have been obtained for the distributions of numbers
in each grade, when recruits to each grade occur according to a Poisson
process. A particularly useful special case is the hierarchical model,
where an individual on leaving his present grade is either promoted
to the next one in the hierarchy or leaves. For more general
transitions a practical and tractable distribution for soujourn times
is the phase-type distribution, discussed by Neuts (1982). This lets
the system be regarded as a simple Markov chain while allowing for
more appropriate leaving distributions such as the mixed exponential,
which is a special case of the phase-type. These theoretical results
are summarized in section 2.

However, we shall mainly be concerned with the problem of inference for semi-Markov manpower models. Individuals are subject to
observation, either over some continuous time interval or over a discrete set of points - the latter being referred to as panel data. In
any finite observation period there are incomplete soujourn times at

either end. Their exclusion would bias the analysis towards the smaller lengths of stay. These inadequacies of censoring and grouping of the data thus pose estimation problems which must be solved if we are to fully utilise the data, thus providing a realistic description of the evolution of the firm. The problem of estimation for a semi-Markov model using incomplete data is discussed in section 3, where both a non-parametric and a parametric approach are described.

Finally we apply these methods to data for movements within the nursing service, where typical grades are student nurse, staff nurse and sister, and all grades are subject to leaving. In some cases, nurses may leave service to pass into a limbo state from which they have a certain probability of returning to service after an appro-priate delay.

2. Definitions and Results

We consider a system with states S_1, ..., S_k and semi-Markov transition probabilities α_{ij}, i=1, ..., k, j=1, ..., k+1 of moving from S_i to S_j, where S_{k+1} is the state of having left. The length of stay in state S_i before making a transition to S_j has p.d.f. $f_{ij}(t)$ and d.f. $F_{ij}(t)$.

Let, $P_{ij}(t) = \Pr\{$individual in S_j at time t $\mid$ he is in state i at time 0$\}$. Then $\underset{\sim}{P}(t) = \{P_{ij}(t)\}$ may be obtained from

$$\underset{\sim}{P}^\star(s) = (\underset{\sim}{I} - \underset{\sim}{g}^\star(s))^{-1} \underset{\sim}{H}^\star(s)$$

where

$$\underset{\sim}{g}(t) = \{\alpha_{ij}\, f_{ij}(t)\}$$

$$\underset{\sim}{H}(t) = \mathrm{diag}\left\{\sum_{r=1}^{k+1} \alpha_{ir}(1-F_{ir}(t))\right\}$$

and $\underset{\sim}{P}^\star(s)$, $\underset{\sim}{g}^\star(s)$, $\underset{\sim}{H}^\star(s)$ denote the Laplace-Stieltjes transforms of $\underset{\sim}{P}(t)$, $\underset{\sim}{g}(t)$ and $\underset{\sim}{H}(t)$ respectively. This equation is a modification of a standard result for semi-Markov processes given in Pyke (1961a, 1961b) for the special case of one absorbing state S_{k+1}.

In the particular case when transitions are <u>hierarchical</u>, e.g.
a manpower planning model with no demotions, an individual either
stays where he is, enters the absorbing state or is promoted to a
higher grade. So $\alpha_{ij} = 0$ for $j < i$. In this case

$$P_{ii}(t) = \sum_{r=i+1}^{k} \alpha_{ir}(1-F_{ir}(t)) = \Pr\{\text{does not leave in }(0,t)\} \text{ and}$$

$$P_{ij}(t) = \{(\alpha_{i,i+1}f_{i,i+1}(t)) \star (\alpha_{i+1,i+2}f_{i+1,i+2}(t)) \cdots$$

$$\star (\sum_{r=j+1}^{k+1} \alpha_{jr}(1-F^{\star}_{jr}(t)))\}$$

for $i < j$, where $\star$ denotes convolution. So $P_{ij}(t)$ is the convolution
of completed lengths of stay in grade S_i before promotion to S_j, with
length of stay in S_j until t. The special case of exponential promotions and mixed exponential length of stay until leaving was evaluated by McClean (1978) and compared with the simple Markov model for
several data sets.

However, for manpower planning we are often concerned with predicting future numbers in each grade, as well as evaluation of the individual's career prospects. To this end, we assume recruitment occurs
at a Poisson rate λ_i to grade S_i for $i = 1, \ldots, k$. Let $N_i(t)$ be the
numbers of staff in grade S_i at time t and ν_i the initial numbers in
grade S_i. Then the joint p.g.f. of the $N_i(t)$'s at time t is given by

$$G(\underset{\sim}{z};t) = \prod_{i=1}^{k}(1+\underset{\sim i}{R}\ \underset{\sim}{P}(t)\underset{\sim}{z})^{\nu_i}\{\exp\{\underset{\sim}{\Lambda}\underset{\sim}{r}(t)\underset{\sim}{z}\}$$

where $\underset{\sim i}{R}$ is a vector with 1 in the ith position and 0's elsewhere,
$\underset{\sim}{z} = (z_1 - 1, \ldots, z_k - 1), \Lambda = (\lambda_1, \ldots, \lambda_k)$ and $\underset{\sim}{r}(t) = \int_0^t \underset{\sim}{P}(u)du$.

This expression is the convolution of a multinomial distribution
corresponding to the redistribution of the initial staff and a Poisson
distribution corresponding to the distribution of recruits who join
in $(0,t)$. Some special cases of time-dependent Poisson arrivals to
such a system are discussed in McClean (1980).

For a general distribution of stay in a grade before making a transition it may prove difficult to evaluate these expressions. However, one distribution which is both tractable and practical is the phase type distribution described by Neuts (1982). Consider a n+1 state continuous-time Markov process with n transient states and one absorbing state. Its infinitesimal generator, i.e., matrix of instantaneous transition probabilities, is of the form

$$
\underset{\sim}{Q} = \begin{bmatrix} \underset{\sim}{T} & \underset{\sim}{T}^{O} \\ \\ \underset{\sim}{0} & 0 \end{bmatrix}
$$

where $\underset{\sim}{T} = \{t_{ij}\}$ is an n x n square matrix with t_{ij} dt the probability of going from S_i to S_j in (t,t+dt) given an individual is in S_i at t, and

$$
t_{ii\cdot} = - \sum_{j=1}^{n+1} t_{ij} < 0.
$$

The vector $\underset{\sim}{T}^{O} = \{t_{i,n+1}\}$ and is equal to $- \underset{\sim}{T}.\underset{\sim}{e}$, where $\underset{\sim}{e} = (1,1,\ldots1)'$. The vector of initial probabilities is denoted by $(\underset{\sim}{\beta},\beta_{n+1})$ and satisfies $\underset{\sim}{\beta}\underset{\sim}{e} + \beta_{n+1} = 1$, where β_{n+1} is normally zero.

Then, the distribution function F of the time until absorption in state S_{n+1} is given by $F(x) = 1 - \underset{\sim}{\beta}\exp\{\underset{\sim}{T} x\} \underset{\sim}{e}$ for x $\geqslant$ 0, and F is said to be of phase type, with representation $(\underset{\sim}{\beta},\underset{\sim}{T})$.

An example of the phase-type distribution is the mixed exponential, which is well known to give a good fit to length of service before leaving a firm. If we hypothesise that, before leaving, an individual moves through a number of stages representing degrees of committment to the firm, (e.g., Herbst, 1963, McClean, 1976), then the distribution of time until leaving is also phase-type. We may, therefore, use phase-type distributions to describe the duration of length of stay in grade S_i before transition to S_j occurs.

Let $\underset{\sim}{A} = \{a_{ij}\}$ be the transition matrix of the underlying Markov chain. Then, the length of stay in grade S_i before making the transition to S_j has a phase-type distribution $F_{ij}(.)$ with representation $(\underset{\sim}{\beta}_{ij}, \underset{\sim}{T}_{ij})$ for $i = 1, \ldots, k$; $j = 1, \ldots, k+1$, $i \neq j$. There are n_{ij} phases in the transition from S_i to S_j, i.e., $\underset{\sim}{T}_{ij}$ is a $n_{ij} \times n_{ij}$ matrix. This is a similar situation to that described by Latouche (1980), except in his case $F_{ij} = F_i$ with representation $(\underset{\sim}{\alpha}_i, \underset{\sim}{T}_i)$ for all j.

With these assumptions we may represent our semi-Markov process as a continuous-time Markov process with each phase of each state of the semi-Markov process corresponding to a state of the continuous-time Markov process. So, if $\underset{\sim}{Q}$ is the appropriate generator of the continuous-time Markov process we obtain the standard result $\underset{\sim}{P}(t) = \underset{\sim}{P}(0)$ exp $\{Qt\}$ which may be evaluated by numerical integration of the Chapman-Kolmogorov equation $\underset{\sim}{P}'(t) = \underset{\sim}{P}(t) \underset{\sim}{Q}$ using a standard procedure such as the Runge-Kutta method. The particular structure of the matrix $\underset{\sim}{Q}$ may be taken into account since it is relatively sparse.

The distribution of the number of events in the time interval $(0,t)$ may also be obtained as a special case of Neuts (1979) and the distributions of the numbers in each grade when arrivals are Poisson is easily determined using our previous result evaluated for the special case of a continuous-time Markov process.

These results for phase-type soujourn distributions are discussed further by McClean (1982). Since they may serve as a numerical approximation to other useful families of distributions, we thus extend the class of distributions available for describing length of stay in a grade, in a manner which is computationally convenient.

3. Estimation

In manpower planning, data is generally available over a period

of years. Management begins collecting suitable information at a
particular point in time and continues up to the present. Consequently there will be staff already in service when data collection commences, with corresponding left truncated soujourn time, and staff
still in service when data collection ceases, with corresponding right
censored soujourn times. The problem of both non-parametric and parametric estimation of failure distributions, subject to right censoring, has been discussed by a number of authors, (e.g. Kalbfleisch and
Prentice, 1980).

This situation is of particular relevance to medical applications
where, typically, subjects are observed from the onset of illness
until death occurs, and those still alive at the end of the study are
considered to have right censored failure times. The problem of estimation for a semi-Markov model with right censoring is discussed by
Lagakos et al. (1978) who obtained the non-parametric maximum likelihood estimators of the model.

However, for manpower planning, we wish to include left truncated
as well as right censored data. By fully utilizing such incomplete
data we may estimate more effectively by using observations for only
a small number of years, thus reducing the effect of time inhomogeneity. The problem of estimation for left truncated and right censored
data was discussed by McClean and Gribbin (1982) for both non-parametric and parametric estimation and applied to data for the Northern
Ireland nursing service. We propose to extend these methods to a
semi-Markov model where there are an arbitrary number of states as
well as left truncated and right censored observations. In this
section we extend the results of Lagakos et al. (1978) to such a
model to give non-parametric estimators and also consider the case
of parametric estimation which is discussed by Bartholomew (1977)
for particular soujourn time distributions.

3.1. Non-parametric estimation

We assume that there are k states $S_1, \ldots, S_k$. A subject is first observed in a particular state z_o and last observed in the state z_m. Transitions between states correspond to promotion, demotion or entry into limbo or left.

Let z_m denote that state corresponding to the subjects' nth epoch and z_o denote the initial state. Then T_n represents the sojourn time between the (n-1)th and nth epochs. However, the data may be left truncated, in which case the subject is first observed when he has already been in state z_o for a time T_o and transfers to z_1, when he has been there for a time T_1. If the data are right censored then, when the subject has been in the last state z_{m-1} for a time T_m, observation ceases, and this is represented by putting $z_m = k+1$. The complete history of the process is therefore,

$$H_m = \{T_o, z_o, T_1, z_1, \ldots, z_{m-1}, T_m, z_m\}$$

where $T_o > 0$ for left truncated data and $z_m = k+1$ for right cansored data. Then

$$\alpha_{ij} = P\{z_{n+1} = j \mid z_n = i\} \text{ and}$$

$$\bar{F}_{ij}(t) = P\{T_{n+1} > t \mid z_n = i, z_{n+1} = j\} = 1 - F_{ij}(t)$$

$$r_{ij}(t)dt = P\{S_i \text{ to } S_j \text{ in } (t, t+dt)\}.$$

Then the probability element for H_m is

$$\left\{\sum_{j=1}^{k} \alpha_{z_o,j} \bar{F}_{z_o,j}(t_o)\right\}^{-u(t_o)} \left\{\prod_{n=1}^{m-1} \{\alpha_{z_{n-1},z_n} f_{z_{n-1},z_n}(t_n)\}\right\}$$

$$\times \{\alpha_{z_{m-1},z_m} f_{z_{m-1},z_m}(t_m)\}^{u(k+1-z_m)}$$

$$\times \left\{\sum_{j=1}^{k} \alpha_{z_{m-1},j} \bar{F}_{z_{m-1},j}(t_m)\right\}^{u(z_m-k)}$$

where

$$u(i) = \begin{cases} 1 & i>0 \\ 0 & \text{otherwise} \end{cases}$$

In order to facilitate estimation of F and α we define new function $G(t;i,j)$ as in Lagakos (et al 1978) where $G(t;i,j) = \int_t^\infty r_{ij}(x)\,dx$ may be regarded as the survivor function of the random variable W_j where $T_{n+1} = \min(W_1,\ldots,W_k)$ and $z_{n+1}=j$ if and only if $W_r>W_j$ for $r<j$ and $W_r \geqslant W_j$ for r_j. So the W_j may be regarded as the latent soujourn times from state i into state j. We may then write the likelihood function as $L = \Pi L_{ij}$ where

$$\log L_{ij} = \sum_{r=1}^{M} \left\{ \sum_{\ell=j+1}^{k+1} m_{ilr} \log G(v_r+0;i,j) \right.$$
$$+ \sum_{1=1}^{j-1} m_{ilr} \log G(v_r;i,j) + m_{ijr} \log dG(v_r+0;u,j)$$
$$\left. - \sum_{\ell=1}^{r} \sum_{u=1}^{\gamma_{ijr\ell}} s_{ijr\ell} \, G(w_{ijr\ell}+0;i,j) \right\}$$

where $v_1<\ldots<v_M$ are the soujourn times from state i to state j and m_{ijr} are the number of soujourn times state i into state j of length v_r. $S_{ijr\ell u}$ are the number who go from i to j at N_r and are left truncated at $w_{ijr\ell u}$

where $v_{\ell-1} \leqslant w_{ijr\ell u} < v_\ell$ and $u=1,\ldots,\gamma_{ijr\ell}$.

So, $\gamma_{ijr\ell}$ are the number of truncation points of individuals who are left truncated in $[v_{\ell-1},v_\ell)$ and go from i to j at v_r, and $w_{ijr\ell1}$, $w_{ijr\ell2}$, $\ldots$ are the truncation times.

From Kaplan and Meier (1958) and McClean and Gribbin (1982) it follows that the maximum likelihood solution is a discrete distribution concentrated at v_1, $\ldots$, v_m. So

$$G(v_r;i,j)=G(v_r+0;i,j) \quad \text{and} \quad dG(v_r;i,j)=G(v_{r-1};i,j)-G(v_r;i,j).$$

Define :

$$\lambda_{ijr} = G(v_r;i,j)/G(v_{r-1};i,j)$$

so that

$$G(v_r;i,j) = \prod_{\ell=1}^{r} \lambda_{ij\ell}.$$

The likelihood component becomes

$$\sum_{r=1}^{M} (N_{ijr} - m_{ijr}) \log \lambda_{ijr} + m_{ijr} \log(1-\lambda_{ijr})$$

where

$$N_{ijr} = \sum_{1=j}^{k+1} m_{ilr} + \sum_{1=1}^{k+1} \sum_{u=r+1}^{M} (m_{ilu} - \sum_{b=r+1}^{u} s_{ilub})$$

is the risk set at time v_r given that a transition eventually occurs from i to j. From this we have removed cases which are left truncated in $[v_{b-1}, v_b)$ where $v_b > v_r$ since they have not yet been observed at v_r. We note that the first term in the expression for N_{ijr} allows for the possibility of identical soujourn times for different transitions. This situation will not normally occur for manpower data, in which case the term is zero. The maximum likelihood estimate of λ_{ijr} is given by

$$\hat{\lambda}_{ijr} = 1 - m_{ijr} / N_{ijr}$$

and defining the probability of going from S_i to S_j at v_r as $p(i,j;v_r)$ we obtain

$$\hat{p}(i,j;v_r) = (1-\hat{\lambda}_{ijr}) \prod_{1=1}^{j-1} \hat{\lambda}_{ilr} \prod_{1=1}^{k} \prod_{b=1}^{r-1} \hat{\lambda}_{ilb}$$

The transition probabilities are estimated by

$$\hat{\alpha}_{ij} = \sum_{r=1}^{M} \hat{p}(i,j;v_r)$$

and

$$\hat{F}_{ij}(t) = \{\hat{\alpha}_{ij}\}^{-1} \sum_{1=r+1}^{M} \hat{p}(i,j;v_1)$$

where $v_r < t \leqslant v_{r+1}$

The proof of Lagakos et al. (1978) that $\hat{\lambda}_{ijr}$ is unbiased, and $\hat{p}(i,j;v_r)$ is approximately unbiased, still holds in this left truncated case as do the expressions which they derive for the variance of $\hat{\alpha}_{ij}$ and $\hat{F}_{ij}(v_r)$.

We have thus succeeded in fitting a semi-Markov model to data which are both left truncated and right censored. The estimates of the various parameters have the advantage of being maximum likelihood, along with other desirable statistical properties, and are reasonably easy to compute. The non-parametric semi-Markov model, once fitted, may be used to test the suitability of various parametric models.

3.2. Parametric estimation

Bartholomew (1977) discusses a method which is essentially the same as the actuarial approach to competing risks. He assumes

$$r_{ij}(t) = r_{ij}(m) \text{ for } t_{m-1} < t < t_m$$

which is reasonable over small intervals. The hazard function is, therefore, piecewise exponential and the maximum likelihood estimate is

$$\hat{r}_{ij}(m) = n_{ij}(m)/T_i(m)$$

where $n_{ij}(m)$ is the number making the transition from i to j in (t_{m-1}, t_m) and $T_i(m)$ is the total time at risk. If the data is left truncated then the contribution to $T_i(m)$ is suitably adjusted i.e. if an individual is truncated at s_1 and leaves at s_2 where

$$t_{m-1} < s_1 < t_m < \ldots < t_{p-1} < s_2 < t_p$$

then he contribues $t_m - s_1$ to $T_i(m)$, $t_{m+1} - t_m$ to $T_i(m+1), \ldots$, and $s_2 - t_{p-1}$ to $T_i(p)$.

This follows from McClean and Gribbin (1982).

To estimate α_{ij} we define $\theta(i,j;m) = P\{S_i \text{ to } S_j \text{ in } (t_{m-1}, t_m)\}$. Then

$$\theta(i,j;m) = P\{\text{survives in } S_i \text{ to } t_{m-1}\}.P\{S_i \text{ to } S_j \text{ in } (t_{m-1}, t_m) \mid \text{survives to } t_{m-1}\}$$

$$= \exp\{-\sum_{k=1}^{m-1} \hat{r}_i(k)(t_k - t_{k-1})\}.\frac{\{\hat{r}_{ij}(m)\}}{\hat{r}_i(m)}.\{1 - e^{-\hat{r}_i(t_m - t_{m-1})}\}$$

Kuhn et al.(1973) consider a similar situation where all holding time distributions are exponential and censoring has a seperate exponential distribution in each interval.

We may fit other distributions which are known to provide good descriptions of manpower data, such as the lognormal and mixed exponential, by using the non-parametric estimates of the survivor function. This approach is discussed in McClean and Gribbin (1982). If the data are collected at a discrete set of time points, i.e., panel data, and we do not know the exact time at which events occur, only the interval, then we may assume that the arrivals are unformly distributed on the interval between observations (McClean, 1983) or use the equilibrium distribution (Thompson, 1981).

Once a semi-Markov model has been fitted, the goodness of fit may be tested by using a chi-squared test. The model is fitted using most, but not all, of the data and, knowing the numbers in each grade at time t_1, we may use the model to predict the numbers transferring to each grade by t_2, and compare our predictions with the observed numbers.

3.3. Results

A semi-Markov model was fitted to data for the movements of nurses. In this case there are the state registered grades - (7) student (5) state registered nurse (S.R.N.) and (4) sister or above, and the state enrolled grades - (8) pupil and (6) state enrolled nurse. In addition there is the unskilled grade (9) auxiliary. A state registered nurse may be promoted from (7) student to (5) S.N.R. to (4) sister or above. However, a student (7) may be demoted to a pupil (8). A S.R.N. (5) or a sister or above (4) may return to being a student (7). A pupil (8) may qualify to become a S.E.N. (6) o. be demoted to an auxiliary (9) and a S.E.N. (6) may become a student (7) or return to being a pupil (8). An auxiliary (9) may become a pupil (8).

All grades are subject to leaving. These transitions are described
in figure 1.

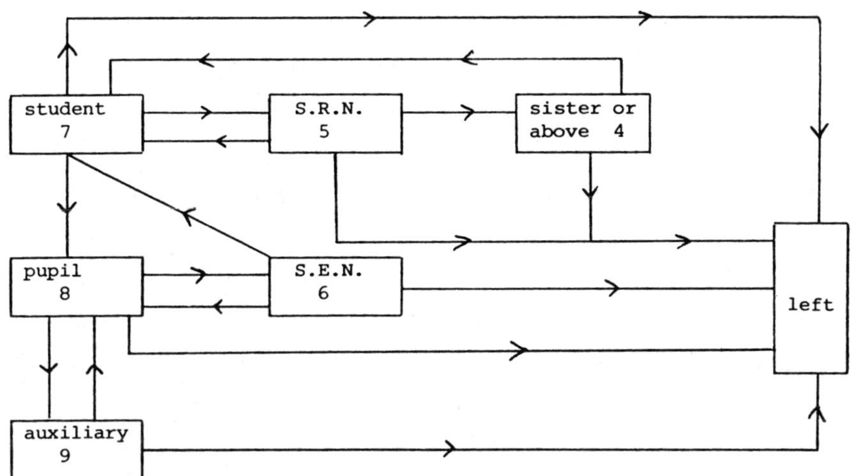

fig. 1.Possible transitions within the nursing service.

        Data are available for the years 1977 to 1984. However, it is
not desirable to use all the data for estimation since there have
been considerable changes in the economic climate during this period
and it is therefore likely that the data are not time homogeneous.
We therefore use a small number of years'data to estimate the parameters of the model. These estimates are then used to calculate the
expected number who move to each grade between the last time point
used in the estimation and some future date. If the estimation period
is too long, then the fit is bad, due to time inhomogeneity. However
we want to use as much information as possible, otherwise our estimation will be worse than it need be. We have therefore considered a
number of estimation periods and selected the best. In general, the
shorter the prediction period, the better the estimates. However,
some of the data is seasonal, which means that a better prediction
may be obtained by projecting a year ahead rather than the shorter

periods of three or six months. For each grade there is a chi-squared
figure associated with the possible movements of staff, from that
grade, during the prediction period. These chi-squared values are independent, so to obtain an overall goodness of fit test we sum the
various chi-squared values to obtain a chi-squared goodness of fit
test for the model. These results are given in table 1, for different
prediction periods using non-parametric estimation.

Looking at the results in table 1, we see from the total chi-
squared figures that the overall fit of the model is acceptable at the
5% level, for a prediction period of three months, and deteriorates as
the prediction period increases, as would be expected. The exact
results, for the best prediction period of a quarter, are given in
table 2.

Table 1   Goodness-of-fit of the non-parametric model.

| | Prediction period | | | | |
	3 months	6 months	1 year	2 years	Degrees of freedom
Grade 4	0.4	0.2	2.1	4.0	2
Grade 5	11.5	16.1	27.2	61.5	3
Grade 6	0.4	12.6	3.1	12.1	3
Grade 7	0.7	8.7	69.7	96.3	3
Grade 8	10.1	13.6	0.4	12.6	3
Grade 9	0.5	3.2	0.9	7.0	2
Total	23.6	54.4	103.4	193.5	16

Table 2   Transitions over a three months period.

from grade 4 (sister)			from grade 5 (S.R.N.)			from grade 6 (S.E.N.)		
to	obs	expt	to	obs	expt	to	obs	expt
student	2	3.4	sister	14	35.9	student	0	1.3
left	53	49.7	student	36	28.7	pupil	8	7.9
stayed	3057	3058.9	left	244	225.0	left	89	83.4
			stayed	5761	5765.5	stayed	3117	3121.4

from grade 7 (student)			from grade 8 (pupil)			from grade 9 (auxiliary)		
to	obs	expt	to	obs	expt	to	obs	expt
S.R.N.	180	173.1	S.E.N.	24	51.3	pupil	7	4.6
pupil	17	20.5	auxiliary	2	3.3	left	72	77.3
left	81	80.8	left	59	45.1	stayed	3386	3383.1
stayed	3392	3395.6	stayed	785	770.3			

We see from table 2 that for both grade 5 (S.R.N.) and grade 8 (pupil), the number of nurses promoted is over-estimated while the number leaving is underestimated. This effect could have been caused by worsening economic conditions where jobs are more scare so leaving decreases and cutbacks result in less promotions.

The non-parametric semi-Markov model has therefore been fitted to the movements of nurses and shown, in most cases, to give reasonable short term predictions of the numbers in each grade. A parametric model would have the advantage of requiring only a few parameters to estimate leaving probabilities rather than the large amount of information which must be stored for a non-parametric model. Further-more, it has been shown by McClean and Gribbin (1982) that parametric models, such as the lognormal and mixed exponential models of leaving, may, in some cases give as good or better fits than their non-parametric counterpart due to their smoothing effect on the data.

However, it may be the case that the length of stay in a grade before
making a transition is not readily described by a parametric distribution, in which case the non-parametric model provides an easily
calculated method of estimation future trends.

Acknowledgements

I am greatly indebted to Owen Gribbin and Bob Devine for doing
the necessary programming. Thanks are also due to the D.H.S.S.(N.I.)
for permission to use the data.

References

D.J. Bartholomew (1959), "Note on the measurement and prediction of
   labour turnover", J.R. Statist. Soc. A 122, 232-239.

D.J. Bartholomew (1977), "The analysis of data arising from stochastic
   processes", 145-174, in "The analysis of survey data, Vol. 2,
   model fitting" ed. C.A. O'Murcheartaigh and P. Payne, Wiley.

D.J. Bartholomew and A.F. Forbes (1979), "Statistical techniques for
   manpower planning", Wiley.

R.B. Ginsberg (1971), "Semi-Markov processes and mobility", J. Math.
   Sociology, 1, 233-262.

P.G. Herbst (1963), "Organizational committment:a decision model"
   Acta Sociologica 7, 34-35.

M. Hedberg (1961), "The turnover of labour in industry, an actuarial
   study", Acta Sociologica 5, 129-143.

J.D. Kalbfleisch and R.L. Prentice, "The statistical analysis of
   failure time data", New York, Wiley.

E.L. Kaplan and P. Meier (1958), "Non-parametric estimation from
   incomplete observations", J. Am. Stat. Assoc., 53, 457-481.

S. Lagakos, S.J. Sommer and M. Zelen (1978), "Semi-Markov models for partially censored data", Biometrika 65, 311-317.

K.F. Lane and J.E. Andrew (1955), "A method of labour turnover analysis", J.R. Statist. Soc. A118, 296-323.

G. Latouche (1982), "A phase type, semi-Markov point process", Siam J. Alg. Disc. Math. 3, 77-90.

S.I. McClean (1976), "The two stage model of personnel behaviour", J.R. Statis. Soc. A139, 205-217.

S.I. McClean (1978), "Continuous-time stochastic models for a multi-grade population", J. Appl. Prob. 15, 26-32.

S.I. McClean (1980), "A semi-Markov model for a multigrade population with Poisson recruitment", J. Appl. Prob. 17, 846-852.

S.I. McClean (1982), "Some results for a semi-Markov model of mobility" submitted to J. of Math. Soc.

S.I. McClean (1983), "Estimation for failure time distribution using grouped follow-up data" submitted to Applied Statistics.

S.I. McClean and J.O. Gribbin (1982), "Statistical estimation for failure time data which is right censored and left truncated" presented to the R.S.S. conference in York, March, 1982.

R. McGinnis (1968), "A stochastic model of social mobility" Amer.Soc. Rev. 33, 712-721.

A. Mehlmann (1980), "Semi-Markovian models in continuous time", J. Appl. Prob. 16, 416-422.

M.F. Neuts (1979), "A versatile Markovian point process", J. Appl. Prob. 16, 764-779.

M.F. Neuts (1982), "Matrix - geometric solutions in stochastic models" John Hopkins.

R. Pyke (1961a), "Markov renewal processes : definitions and preliminary properties", Ann. Math. Stat. 32, 1231-42.

R. Pyke (1961b), "Markov renewal processes with finitely many states" Ann. Math.Stat. 32, 1243-59.

M.E. Thompson (1981), "Estimation of parameters of a semi-Markov process from censored records", A.A. Prob. 13, 804-825.

A. Young (1971), "Demographic and ecological modelling for manpower
    planning", in D.J. Bartholomew and B.R. Morris (1971) "Aspects
    of manpower planning", English Universities Press.
A. Young and G. Almond (1961), "Predicting distributions of staff",
    Computer J., 3, 246-250.

STATISTICAL ANALYSIS OF SEMI-MARKOV PROCESSES BASED ON THE
THEORY OF COUNTING PROCESSES

Niels Keiding

Statistical Research Unit, University of Copenhagen

1. Introduction

Many applications of stochastic process models in biostatistics
involve several time-scales simultaneously.

Thus in an epidemiological study of the mortality of diabetics
(Andersen et al., 1986), the death intensity was related primarily to age
of the individual. In addition to this, time since diagnosis was
important; we may call this a duration variable. Finally calendar
time played a rôle that may be called concomitant.

In many cancer clinical trials with death as endpoint the
primary time variable is time since admission (which usually happens
either at diagnosis or at primary (e.g. surgical) treatment). However, for patients who have relapsed the duration in relapse may
well be important, and it is usually important to take age into
account as a concomitant time variable. For a third example, consider the analysis by Andersen & Rasmussen (1986) of psychiatric admissions of women giving birth. The primary time variable is here
time relative to birth, a duration variable is time since last discharge from psychiatric care, and as concomitant time variable we
again have age. Also parity (the number of previous births) ought

to be taken into account, albeit time-discrete.

These examples indicate the rather common situation in biostatistics that the _duration_ of an individual in a state will be expected to influence the intensity of departure from that state. It is
obvious that semi-Markov processes suggest themselves as models for
such phenomena.

The purpose of this paper is to review and discuss statistical
analysis of (continuous-time) semi-Markov processes within the
framework of the theory of statistical models based on counting
processes. For a review of other aspects of the use of semi-Markov
processes in medical statistics, reference may be made to Cox (1986).

2. Multivariate counting processes and the multiplicative intensity model

The systematic use of concepts and tools from the theory of
martingales and stochastic integrals was initiated by Aalen (1978)
and has recently been reviewed by Andersen et al. (1982), Andersen
and Borgan (1985) and others.

Briefly, one considers a complete probability space $(\Omega, \mathcal{F}, P)$
with a filtration

$$(\mathcal{F}_t, \ t \in [0,1]), \ \mathcal{F}_u \subseteq \mathcal{F}_v \subseteq \mathcal{F} \ \text{ for } \ u < v.$$

Let $N(t) = (N_1(t), \ldots, N_k(t))$, $t \in [0,1]$ be a _multivariate counting
process_, i.e. the sample function of each $N_i(t)$ is a right continuous, increasing step function with steps of size 1 and adapted to
$(\mathcal{F}_t)$. The regularity condition $E(N_i(1)) < \infty$ for all i is convenient, although most results hold under the weaker assumption
$N_i(1) < \infty$ a.s. (all i), provided local martingale techniques are
used.

It is assumed that the Doob-Meyer decomposition $N_i = \Psi_i + M_i$
of each component into the compensator Ψ_i (a predictable increasing process) and a martingale M_i has the special feature that Ψ_i

is a.s. absolutely continuous :

$$\Psi_i(t) = \int_0^t \Lambda_i(s)ds$$

where the _intensity process_ $\Lambda_i(t)$ is adapted to $(\mathcal{F}_t)$ and is assumed a.s. left continuous with right-hand limits. It is also assumed that the martingales M_i are _orthogonal_ in the sense that $E(M_i M_j) = 0$, $i \neq j$.

The compensator of the submartingale M_i^2 is called the _pre-dictable variation process_ and denoted $<M_i>(t)$; in fact

$$<M_i>(t) = \int_0^t \Lambda_i(s)ds.$$

The _multiplicative intensity model_ is a family of probabilities $\mathcal{P}$ specified by assuming that

$$\Lambda_i(t) = \alpha_i(t)Y_i(t), \quad t \in [0,1]$$

for all i, where the stochastic process $Y_i(t)$ has a.s. left-continuous sample functions with right-hand limits, whereas $\mathcal{P}$ is spanned by the different deterministic non-negative functions $\alpha_i(t)$ (assumed left-continuous w.r.h.l.).

3. Censoring

A distinctive feature of the counting process approach is the ease with which most cases of _censoring_ is handled. Let $(C_i(t))$ specify an $(\mathcal{F}_t)$-adapted multivariate indicator process, assumed left continuous to avoid unnecessary technicalities, and assume that the original process $N_i(t)$ may only be observed when $C_i(t) = 1$. Since the observable ("censored") process $N_i^C(t)$ is given by

$$N_i^C(t) = \int_0^t C_i(s)dN_i(s)$$

it is seen that $N_i^C(t)$ has intensity process given by $C_i(t)Y_i(t)\alpha_i(t)$.

It follows that inference in the censored process may proceed in a similar fashion as in the original process by replacing Y_i by $C_i Y_i$. Censoring is not our main interest in this report, so we shall not make further explicit reference to it but rather mention the discussion by Aalen & Johansen (1978) and Andersen et al. (1982).

4. Estimation

The Doob-Meyer decomposition

$$N_i(t) = \int_0^t \alpha_i(s) Y_i(s) ds + M_i(t)$$

may informally be written

$$dN_i(t) = \alpha_i(t) Y_i(t) dt + dM_i(t)$$

and since $dM_i(t)$ represents a zero-mean "innovation" component, a natural solution of the problem of estimating the unknown function $\alpha_i(t)$ is to use

$$\hat{\alpha}_i(t) dt = dN_i(t)/Y_i(t).$$

Formally, it is necessary to take care of the possibility that $Y_i(t) = 0$ for some t. Thus define

$$J_i(t) = I\{Y_i(t) > 0\},$$

$$A_i(t) = \int_0^t \alpha_i(s) ds$$

and

$$A_i^*(t) = \int_0^t \alpha_i(s) J_i(s) ds;$$

then the suggested estimator (the Nelson-Aalen estimator) is

$$\hat{A}_i(t) = \int_0^t \frac{J_i(s)}{Y_i(s)} dN_i(s).$$

The main results about the Nelson-Aalen estimator are given in the following theorem.

Estimation : Exact properties

$\hat{A}_i(t) - A_i^*(t)$, $i = 1,\ldots,k$, are orthogonal square integrable martingales. (This implies the unbiasedness property that for all stopping times T w.r.t. $(\mathcal{F}_t)$, $E[\hat{A}_i(T) - A_i^*(T)] = 0$).

Furthermore

$$< \hat{A}_i - A_i^* > (t) = \int_0^t \alpha_i(s) \frac{J_i(s)}{Y_i(s)} \, ds$$

which defines the predictable variation of $\hat{A}_i - A_i^*$. An "estimator" of $< \hat{A}_i - A_i^* >$ is given by

$$\hat{\tau}_i(t) = \int_0^t \frac{J_i(s)}{Y_i(s)^2} \, dN_i(s)$$

which may also be viewed as an estimator of the mean squared error function

$$\tau_i(t) = E[\{\hat{A}_i(t) - A_i^*(t)\}^2].$$

Asymptotic results may also be derived for a sequence of statistical problems indexed by n :

Consistency

If $\tau_i^{(n)}(1) \to 0$ as $n \to \infty$ then

$$E[\sup\{\hat{A}_i^{(n)}(t) - A_i^{*(n)}(t)\}^2] \to 0.$$

Asymptotic normality

Assume that there exists a sequence of positive constants $\{a_n\}$, increasing to infinity as $n \to \infty$, and non-negative square integrable functions g_h, $h = 1,2,\ldots,k$, defined on $[0,1]$, such that

A)   For each $t \in [0,1]$ and $h = 1,2,\ldots,k$

$$a_n^2 \int_0^t \frac{J_h^{(n)}(s)}{Y_h^{(n)}(s)} \alpha_h(s) ds \xrightarrow{P} \int_0^t g_h^2(s) ds \quad \text{as} \quad n \to \infty.$$

B)   For all h and $\varepsilon > 0$

$$a_n^2 \int_0^t \frac{J_h^{(n)}(s)}{Y_h^{(n)}(s)} \alpha_h(s) I\left(\left| a_n \frac{J_h^{(n)}(s)}{Y_h^{(n)}(s)} \right| > \varepsilon\right) ds \xrightarrow{P} 0 \quad \text{as} \quad n \to \infty.$$

Then

$$a_n \begin{pmatrix} \hat{A}_1^{(n)} - A_i^{*(n)} \\ \vdots \\ \hat{A}_k^{(n)} - A_k^{*(n)} \end{pmatrix} \xrightarrow{D} \begin{pmatrix} X_1 \\ \vdots \\ X_k \end{pmatrix}$$

where $X_1, X_2, \ldots, X_k$ are independent Gaussian martingales with
$X_h(0) = 0$ and $\mathrm{Cov}(X_h(s), X_h(t)) = \int_0^{t \wedge s} g_h^2(u) du.$

In practice the verification of the conditions A and B is not always
so direct, and it is useful to have alternative and more easily
verifiable set of conditions.   A simple set of conditions, sufficient
for A and B to hold true, and which is often fulfilled in practice,
is

A')   For $h = 1, 2, \ldots, k$

$$\sup_{t \in [0,1]} \left| a_n^2 J_h^{(n)}(t) [Y_h^{(n)}(t)]^{-1} \alpha_h(t) - g_h^2(t) \right| \xrightarrow{P} 0 \quad \text{as} \quad n \to \infty.$$

B')   For $h = 1, 2, \ldots, k$

$$\sup_{t \in [0,1]} \left| a_n J_h^{(n)}(t) [Y_h^{(n)}(t)]^{-1} \right| \xrightarrow{P} 0 \quad \text{as} \quad n \to \infty.$$

To apply the weak convergence result in practice, one must be able
to estimate the covariance function of the limiting Gaussian martingale.   If conditions A and B hold true, then for all   h   and
$t \in [0,1]$

$$a_n^2 \hat{\tau}_h^{(n)}(t) \xrightarrow{P} \int_0^t g_h^2(u) du.$$

Hypothesis testing was reviewed by Andersen et al. (1982), and two
approaches to <u>regression analysis</u> were presented by Aalen (1980)

and Andersen & Gill (1982), respectively. The latter authors discussed <u>Cox regression</u>, and we shall briefly review this theory.

5. Cox regression

Assume that the multivariate counting process $(N_1(t),\ldots,N_k(t))$ has intensity process specified by

$$dN_i(t) = Y_i(t)\lambda_0(t)e^{\beta' z_i(t)} dt + dM_i(t)$$

where we assume $Y_i(t)$ to be a random indicator process with a.s. left continuous sample functions common to all parameters, and where the statistical model is specified by the "unknown underlying intensity" $\lambda_0(t)$ (left-continuous, non-negative) and the regression coefficients $\beta = (\beta_1,\ldots,\beta_p)'$ which operate on the (possibly time-dependent) covariates $z_i(t) = (z_i^{(1)}(t),\ldots,z_i^{(p)}(t))$.

The estimation problem concerns β and λ_0. For β, one may use the Cox likelihood ("partial likelihood")

$$\prod_{T_{ij} \leqslant t} \frac{e^{\beta' z_i(T_{ij})}}{\sum_{h=1}^{k} Y_h(s)e^{\beta' z_h(T_{ij})}}$$

where T_{ij} is the time of the j'th jump of the i'th counting process N_i. Intuitively, at each jump, the likelihood is multiplied by the conditional probability that the i'th process jumps, given one of them jumps.

To estimate the integrated underlying intensity

$$\Lambda_0(t) = \int_0^t \lambda_0(s)ds$$

one may use a modified Nelson-Aalen estimator

$$\hat{\Lambda}_0(t) = \int_0^t \frac{d\bar{N}(s)}{\sum_{i=1}^{k} Y_i(s)e^{\hat{\beta}' z_i(s)}}$$

with $\bar{N} = N_1 + \ldots + N_k$.

Consistency and asymptotic normality of $\hat{\beta}$ and $\hat{\Lambda}_0$ were given by Andersen & Gill (1982).

6. Markov processes

Let Γ be a finite state space and consider a time-inhomogeneous Markov process $U(t)$, $t \in [0,1]$, on Γ with transition intensities $\alpha_{ij}(t)$, $i, j \in \Gamma$, $i \neq j$, which are assumed left continuous with right-hand limits. A multivariate counting process is then specified by the transition counts : Let $N_{ij}(t)$ be the number of direct transition from i to j in $[0,t]$; it is then readily seen that defining $Y_i(t) = I\{U(t-) = i\}$, the Doob-Meyer decomposition of $N_{ij}(t)$ may be written

$$dN_{ij}(t) = \alpha_{ij}(t)Y_i(t)dt + dM_{ij}(t).$$

For a set of independent Markov processes $U_1(t),\ldots,U_n(t)$ all driven by the transition intensities $\alpha_{ij}(t)$ but with possibly different initial distributions it is natural to restrict attention to the <u>aggregate</u> processes (in obvious notation) :

$$\bar{N}_{ij}(t) = \sum_{\nu=1}^{n} N_{ij}^{(\nu)}(t) = \text{the number of direct transitions from } i \text{ to } j \text{ in } [0,t] \text{ by all particles}$$

$$\bar{Y}_i(t) = \sum_{\nu=1}^{n} Y_i^{(\nu)}(t) = \text{the number of particles in state } i \text{ at time } t-.$$

In the sense specified above the transition intensities $\alpha_{ij}(t)$ may then be estimated by $d\bar{N}_{ij}(t)/\bar{Y}_i(t)$: Let $J_i(t) = I\{\bar{Y}_i(t) > 0\}$ and define

$$A_{ij}^{*}(t) = \int_0^t \alpha_{ij}(s)J_i(s)ds;$$

then this is "estimated" by

$$\hat{A}_{ij}(t) = \int_0^t \frac{J_i(s)}{\bar{Y}_i(s)} dN_{ij}(s).$$

Properties of the estimators $\hat{A}_{ij}(t)$ follow directly from the general theory above. The non-trivial extension to estimation of the transition <u>probabilities</u> $P_{ij}(s,t) = P\{U(t) = j \mid U(s) = i\}$ was studied by Aalen & Johansen (1978) and Fleming (1978). Aalen & Johansen showed that

$$\hat{P}(s,t) = \prod_{(s,t]} (I + d\hat{A})$$

in product integral notation, with $A = (A_{ij})$. These authors proved exact and asymptotic properties of this estimator, which in a sense may be regard a generalization of the Kaplan-Meier estimator for censored survival data.

7. Semi-Markov processes

An approach to nonparametric statistical models based on semi-Markov processes was proposed by Lagakos et al. (1978) and further developed by Gill (1980). For ease of reference we may implement the change from time-inhomogeneous Markov processes to time-homogeneous (but duration-inhomogeneous) semi-Markov processes by keeping the notation as before, but instead of letting the transition intensities α_{ij} depend on actual "calendar" time it should depend on <u>duration</u> in state i. That is, we define $\alpha_{ij}(s)$ as the infinitesimal transition probability to j for a particle that has spent time s in state i. Define $J_1, J_2, \ldots$ as that state (in Γ) which is occupied before jump $1, 2, \ldots$, and let, as before $N(t) = \sum_{i,j} N_{ij}(t)$ be the number of jumps in $(0,t]$; then $Z(t) = J_{N(t)}$ is a semi-Markov process. If X_i is the i'th sojourn time, $S_0 = 0$, $S_i = X_1 + \ldots + X_i$, $L(t) = t - S_{N(t-)}$ is a left-continuous version of the backward recurrence time. The counting process $N_{ij}(t)$ has Doob-Meyer decomposition given by

$$dN_{ij}(t) = I(Z(t-) = i)\alpha_{ij}(L(t))dt + M_{ij}(t)$$

with $M_{ij}(t)$ a martingale. But this is not useful as estimating equation because of the complicated indirect dependence on time t.

Rather, one should exploit the intrinsic time-scale of this process, which is duration rather than calendar time. Define

$K_{ij}(s)$ = the number of sojourn times in i observed to take on a value $\leq s$ and to be followed by a jump to j

$Y_{ij}(s)$ = the number of sojourn times in i observed to take on a value $\geq s$.

Notice that there is no filtration making $K_{ij}(s)$ a counting process with intensity process proportion to Y_i. This is because the relevant sojourn times may be spent at many different calendar times, so that additional information on early sojourn times may derive from late calendar times. In the representation indicated by

$$dK_{ij}(s) = Y_i(s)\alpha_{ij}(s)ds + dH_{ij}(s)$$

$H_{ij}(s)$ is therefore not a martingale. However, it does have zero expectation and explicitly representable variance, and as demonstrated by Gill (1980) for the transition probabilities, similar asymptotic results concerning the usual estimators $\hat{\alpha}_{ij}(s)ds = dK_{ij}(s)/Y_i(s)$ may be derived. The difference is that one now has to start from first principles, because the master theorem above is no longer applicable, being dependent on <u>martingale</u> central limit theory.

Gill (1983) has recently indicated that a further parallel between the Markov and semi-Markov process situations is that likehood functions look the same. Gill conjectured that the similarity of the limiting results might be proved directly from this representation.

8. <u>Hierarchical semi-Markov processes and the probability-of-being-</u>
<u>in-response function</u>

Consider the particular example of a semi-Markov process where the flow through the states is unidirectional in the sense that if the transition $i \rightarrow j$ is possible, then $j \rightarrow i$ is not. For such <u>hierarchical</u> semi-Markov processes Aalen (1975) remarked that a random time transformation would allow the definition of a suitable counting process with associated martingale, the idea being to reserve separate parts of the new time axis for the sojourns in the different states. This way the study of the process is really reduced to simple survival analysis (allowing left-censored data). Voelkel & Crowley (1984) developed this approach and obtained useful results on the "probability of being in response function" (PBRF) which was suggested by Temkin (1978) as a descriptor in cancer clinical trials.

To explain the latter, assume that from an initial state 1, a cancer patient may enter either a response state 2 or a progression state 3, and from the response state, a relapse state 4 may be entered. Thus the only possible direct transitions are $1 \rightarrow 2$, $1 \rightarrow 3$ and $2 \rightarrow 4$. A Markov process description of this problem would specify transition probabilities $\alpha_{12}(t)$, $\alpha_{13}(t)$ and $\alpha_{24}(t)$, with t = time since entry, and if $S_i(t)$ is the probability of staying in state i from 0 to t, we have

$$\text{PBRF}(t) = \int_0^t S_1(s)\alpha_{12}(s)[S_1(t)/S_2(s)]ds$$

which may be estimated using the theory by Aalen & Johansen (1978). If the transition $2 \rightarrow 4$ depends on duration in state 2 rather than time t since entry in state 1, a semi-Markov model is appropriate, and we get

$$\text{PBRF}(t) = \int_0^t S_1(s)\alpha_{12}(s)S_2(t - s)ds.$$

Obviously the two are equal when $\alpha_{24}(t)$ is constant. The semi-Markov version of the PBRF may be estimated by

$$\widehat{\text{PBRF}}(t) = \int_0^t \hat{S}_1(s-) \frac{J_1(s)}{Y_1(s)} \hat{S}_2(t - s)dN_{12}(s)$$

where $\hat{S}_i$ may be chosen either as the Kaplan–Meier estimates or as

$$\hat{S}_1 = \exp(-\hat{A}_{12} - \hat{A}_{13}), \quad \hat{S}_2 = \exp(-\hat{A}_{24})$$

Voelkel & Crowley showed how to obtain large sample properties of PBRF(t).

9. Semi-Markov processes and Cox regression

It was noted above that the completely nonparametric specification of the semi-Markov process transition intensity required extra effort in the statistical inference. We want here to remark, by way of an example, how even a modest restriction in the form of a parametric model might often be useful. One feature of this approach is the possibility of allowing for (calendar) time-inhomogeneous semi-Markov processes. (It came to my attention during the conference that Janssen & de Dominicis (1984) have recently studied a class of such processes).

For a particle in state i at time t which has spent time s already in i, denote the intensity of a transition to state j by $\alpha_{ij}(t,s)$. Assume the statistical model given by

$$\alpha_{ij}(t,s) = \gamma_{ij}(t)e^{\beta_{ij}\phi(s)}$$

with $\gamma_{ij}(t)$ an unspecified nonnegative function and $\phi(s)$ known. Since $s = L_i(t)$ (the backward recurrence time specified above) we then have the representation

$$dN_{ij}(t) = Y_i(t)\gamma_{ij}(t)e^{\beta_{ij}\phi(L_i(t))} dt + dM_{ij}(t)$$

with $M_{ij}(t)$ a martingale. Statistical inference may be performed using Cox regression with time-dependent covariates, as discussed above. For an interesting application, we refer to Andersen & Rasmussen's (1986) study of psychiatric admissions for women giving birth. As briefly mentioned in the introduction, t is here time relative to birth (the interval (- 9 months, 12 months) being studied) and s is time since latest discharge. Not surprisingly, the latter variable has a highly significant impact upon the admission intensity. The Cox regression analysis enabled the analysis to simultaneously account for other covariates, the more important ones being marital status, age, parity and urbanization.

10. The Stanford heart transplant programme

After admittance to the project, patients are put on a waiting list for a suitable donor heart; some die while waiting, others receive a transplant and are then followed until death or end of study.

Several statisticians have studied the possible effects of transplantation on the death intensity, often using Cox regression techniques, which allow for the inclusion of additional covariates not discussed below. Thus Crowley & Hu (1977) postulated a pre-transplant death intensity of $\lambda_0(t)$ and a post-transplant intensity of $\theta\lambda_0(t)$. Cox & Oakes (1984, p.129 f.f.) suggested a "transient effect" by generalizing to a post-transplant intensity of $\theta\psi^{-v}\lambda_0(t)$, where v is duration since transplant.

These models further illustrate the considerable potential of the regression methodology just described. However, these and other models such as those studied in the discussion paper by Aitkin et al. (1983) also indicate that although some may naturally be called semi-Markov processes, others may not. Hence this experience agrees with the remark by Çinlar at the conference that perhaps semi-Markov processes do not form a very natural framework as statistical models.

References

Aalen, O.O. (1975). Statistical inference for a family of counting processes. Ph.D.Dissertation, Dept.of Stat., Univ.of Calif., Berkeley.

Aalen, O.O. (1978). Nonparametric inference for a family of counting processes. Ann.Statist. **6**, 701-726.

Aalen, O.O. (1980). A model for nonparametric regression analysis of counting processes. Springer Lect.Notes in Statist. **2**, 1-25.

Aitkin, M.,N.Laird & B.Francis (1983). A reanalysis of the Stanford heart transplant data (with discussion). J.Amer.Statist.Assoc. **78**, 264-292.

Andersen, P.K., K.Borch-Johnsen, T.Deckert, A.Green, P.Hougaard,
     N.Keiding & S.Kreiner (1986). A Cox regression model for the
     relative mortality and its application to diabetes mellitus
     survival data. Biometrics 42 (to appear).

Andersen, P.K. & Ø.Borgan (1985). Counting process models for life
     history data : A review (with discussion). Scand.J.Statist.
     12, 97-158.

Andersen, P.J., Ø.Borgan, R.D.Gill & N.Keiding (1982). Linear non-
     parametric tests for comparison of countig processes, with
     applications to censored survival data (with discussion).
     Int.Statist.Rev. 50, 219-258.

Andersen, P.K. & R.D.Gill (1982). Cox's regression model for counting
     processes : a large sample study. Ann.Statist.10, 1100-1120.

Andersen, P.K. & N.K.Rasmussen (1986). Psychiatric admissions and
     choice of abortion. Statist.in Medicine 5 (to appear).

Cox, D.R. (1986). Some remarks on semi-Markov processes in medical
     statistics. This volume.

Cox, D.R. & D.Oakes (1984). Analysis of survival data. Chapman and
     Hall, London.

Crowley, J. & M.Hu (1977). Covariance analysis of heart transplant
     survival data. J.Amer.Statist.Assoc. 73, 27-36.

Fleming, T.R. (1978). Nonparametric estimation for nonhomogeneous
     Markov processes in the problem of competing risks. Ann.
     Statist. 6, 1057-1070.

Gill, R.D. (1980). Nonparametric estimation based on censored
     observations of a Markov renewal process. Z.Wahrscheinlich-
     keitsthe.verw.Geb. 53, 97-116.

Gill, R.D. (1983). Discussion of the papers by Helland and Kurtz.
     Bull.Internat.Statist.Inst. 50(3), 239-243.

Janssen, J. & R.de Dominicis (1984). Finite non-homogeneous semi-
     Markov processes : Theoretical and computational aspects.
     Insurance : Math.and Econ. (To appear).

Lagakos, S.W., C.J.Sommer & M.Zelen (1978). Semi-Markov models for
    partially censored data. Biometrika 65, 311-317.
Temkin, N. (1978). An analysis for transient states with application
    to tumor shrinkage. Biometrics 34, 571-580.
Voelkel, J.G. & J.Crowley (1984). Nonparametric inference for a
    class of semi-Markov processes with censored observations.
    Ann.Statist. 12, 142-160.

SECTION VII. SEMI-MARKOV PROCESSES AND QUEUEING THEORY

APPROXIMATION OF SOME STOCHASTIC MODELS

V.V. Kalashnikov

Institute for Systems Studies

9, Prospect 60 let Oktjabrja, 117312 Moscow

1. Introduction

Approximation problems are of great importance in applied
mathematics. The approximation itself consists in changing the
initial mathematical model to another one which is more preferable
to consider (e.g. from a computational point of view). At most the
arising problems are implied by the problem of receiving quantitative estimates of the approximation accuracy.

In this paper we review some results on probabilistic models
approximation in different senses. The central accuracy problem
is solved in terms of probability metrics, see Zolotarev (1976,
1984). Many of the approximation results are based on continuity
theory of stochastic models, see Kalashnikov (1978a, 1981a, 1983a,
1983b), Zolotarev (1977).

The general approximation scheme considered below can be
formulated as follows.

Let us have a model F mapping some initial data U from
the set $\mathfrak{U}$ to output data V from the set $\mathfrak{V}$, i.e. $F : \mathfrak{U} \to \mathfrak{V}$.
Usually (below) the sets $\mathfrak{U}$ and $\mathfrak{V}$ are some sets of random elements
(variables, processes etc.) or their probability distributions.

Suppose that some notions of convergence are inserted in $\mathfrak{U}$ and $\mathfrak{V}$. Having in mind not only the problem of receiving qualitative assertions but quantitative estimates we shall consider these sets to be metric spaces : $(\mathfrak{U}, h_{\mathfrak{U}})$, $(\mathfrak{V}, h_{\mathfrak{V}})$.

In order to define an approximation model we need to choose a subset $\mathfrak{U}_A \subset \mathfrak{U}$ of appropriate input data and define a mapping. $F_A : \mathfrak{U}_A \to \mathfrak{V}$ (in reality F_A can map the subset $\mathfrak{U}_A$ into some subset $\mathfrak{V}_A \subset \mathfrak{V}$). The main problem which is solved here is receiving estimates of quantities $h_{\mathfrak{V}}(FU, F_A U_A)$ for $U \in \mathfrak{U}$, $U_A \in \mathfrak{U}_A$ accumulating the information about the approximation accuracy.

Usually we choose the subset $\mathfrak{U}_A$ to be $h_{\mathfrak{U}}$-dense in $\mathfrak{U}$. In this case the existence of a dense sequence of approximation models $(F_A^{(n)}, \mathfrak{U}_A^{(n)})$ is of interest. It means that for the considered input data $U \in \mathfrak{U}$ and any $n \geqslant 0$ there exist data $U_A^{(n)} \in \mathfrak{U}_A^{(n)}$ such that

$$(1) \quad \lim_{n \to \infty} h_{\mathfrak{V}}(FU, F_A^{(n)} U_A^{(n)}) = 0.$$

It is natural to solve this problem in frames of continuity analysis. Namely, let us suppose for simplicity that $F_A = F$ (the general case can be reduced to the above mentioned one). At first we find a subset $\mathfrak{U}^* \subset \mathfrak{U}$ (a continuity subset) having the following property : $h_{\mathfrak{U}}(U, U') \to 0$, $U, U' \in \mathfrak{U}^* \Rightarrow h_{\mathfrak{V}}(FU, FU') \to 0$. If the sequence of subsets $\mathfrak{U}_A^{(n)}$ has property :

$$\lim_{n \to \infty} h_{\mathfrak{U}}(U, \mathfrak{U}_A^{(n)} \cap \mathfrak{U}^*) = 0$$

then the model F having the initial data U can be approximated by the sequence $(F, \mathfrak{U}_A^{(n)})$ with any necessary accuracy in the sense of relation (1).

This type of approximation problem contains for example the approximation of Markov chains with denumerable or general state space by finite chains. It also contains the problem of state enlarging for stochastic processes. We can consider an approximation of queueing models with semi-Markov input or other simple ones as

problems of this type too. Finally the assertions of some limit
theorems can be treated as approximation problems.

2. Notations. Hyper-Erlangen approximation of distribution functions

Below for the definiteness all considered random elements
are supposed to be defined on the same triplet $(\Omega, \mathcal{F}, P)$. Let X, Y
be random elements taking values in complete separable metric space
$(\mathcal{X}, m)$, their distributions being denoted $P_X(\cdot) = P(X \in \cdot)$ and
$P_Y(\cdot) = P(Y \in \cdot)$ respectively. Let us denote $\Pi(X,Y) = \Pi(P_X, P_Y)$
the Levi-Prohorov distance between X and Y, $L(X,Y)$ the Levi
distance between them (when $\mathcal{X} = R^1$),

$$\zeta_1(X,Y) = \sup\{|Ef(X) - Ef(Y)| : f \in \mathcal{F}_1\},$$

$$d(X,Y) = \sup\{|Ef(X) - Ef(Y)| : f \in \mathcal{F}_0\},$$

where

$$\mathcal{F}_1 = \{f : \mathcal{X} \to R^1, |f(x) - f(y)| \leqslant m(x,y) \quad \forall x,y \in \mathcal{X}\},$$

$$\mathcal{F}_0 = \mathcal{F}_1 \cap \{f : |f| \leqslant 1\},$$

In the case $\mathcal{X} = R^1$, $m(x,y) = |x - y|$ the following equality is
well-known :

$$\zeta_1(X,Y) = \int_{-\infty}^{\infty} |P(X \leqslant x) - P(Y \leqslant x)| dx.$$

Metrics Π, L, d metrize the weak convergence, metric ζ_1
metrizes the weak convergence together with the convergence of the
expectations.

Some further properties of the above-mentioned metrics and
relations between them can be found in Zolotarev (1976) and
Kalashnikov (1978a).

Denote some special distributions functions (d.f.) :

$$D_a(x) = \begin{cases} 0, & x \leqslant a, \\ 1, & x > a, \end{cases} \quad \text{- degenerate d.f.,}$$

$$\text{Exp}_{\lambda}\ (x) = 1 - \exp(-\lambda x) \qquad\qquad - \text{ exponential d.f.,}$$

$$E_{n,\lambda}(x) = 1 - e^{-\lambda x}\sum_{k=0}^{n-1}(\lambda x)^k/k! \qquad - \text{ Erlang d.f.}$$

One of the simplest approximation problems consist in the
approximation of arbitrary d.f. by d.f.'s of sufficiently simple
structure. In queueing theory we often take hyper-Erlang d.f.'s

$$\sum_{i=1}^{N} p_i E_{n_i,\lambda_i}(x),\ N \geqslant 1,\ p_i > 0,\ \sum_{i=1}^{N} p_i = 1,$$

as approximating ones.

Usually such an approximation is conducted in two states :
(1) an approximation of an arbitrary d.f. by an atomic one; (2) an
approximation of an atomic d.f. by a hyper-Erlang one. Hence it is
natural to consider at first an approximation of degenerate d.f. by
hyper-Erlang one.

Theorem 1 (Anichkin (1983))[*]

For any $a > 0$ and $\gamma < 1$ there exists a number $N = N(a,\gamma)$ such
that for all $n \geqslant N$ the following inequality is fulfilled

$$\gamma a \sqrt{\frac{\ln n}{n}} < \Pi(E_{n,n/a}, D_a) \leqslant a\sqrt{\frac{\ln n}{n}}$$

The same inequality is valid for metric L.

We note that for metrics $\mu = \zeta_1$ and $\mu = d$ the following
assertion takes place, see Anichkin (1983) :

$$\lim_{n \to \infty} \left| \frac{\mu(E_{n,n/a}, D_a)}{a}\sqrt{\frac{\ln n}{2}} - 1 \right| = 0.$$

The estimates from Theorem 1 can be used for approximation
estimates in the general case. Namely, let H(x) ce an arbitrary d.f.
Given $\varepsilon > 0$ it is easy to build the following atomic approximation
of this d.f. in terms of some metric μ :

---

[*] Some new estimates one can find in V.Kalashnikov "Hyper-Erlang approximation functions", Lect.Notes in Math., 1155, 1985, Springer-
Verlag.

$$\widetilde{H}(x) = \sum_{i=1}^{N} p_i D_{a_i}(x), \quad \mu(\widetilde{H},H) \leqslant \varepsilon,$$

where $p_i > 0$, $\sum_{i=1}^{N} p_i = 1$.

Further we construct for each degenerate d.f. D_{a_i} a hyper-Erlang approximation $E_{n_i,n_i/a_i}$ using estimates from Theorem 1. Denote

$$H_A(x) = \sum_{i=1}^{N} p_i E_{n_i,n_i/a_i}(x),$$

where the following inequalities are valid for $\gamma_i < 1$:

$$\gamma_i \varepsilon_i < \mu(E_{n_i,n_i/a_i}, D_{a_i}) \leqslant \varepsilon_i.$$

Then for $\mu = \Pi$ or $\mu = L$ we obtain

$$\inf_i \gamma_i \varepsilon_i < \mu(H_A, \widetilde{H}) \leqslant \sup_i \varepsilon_i$$

and for $\mu = d$ or $\mu = \zeta_1$

$$\sum_{i=1}^{N} \gamma_i p_i \varepsilon_i < \mu(H_A, \widetilde{H}) \leqslant \sum_{i=1}^{N} p_i \varepsilon_i.$$

3. Approximation of Markov chains

Consider the results concerning an approximation of denumerable chains and ones with general state space by finite Markov chains. Similar problems are of interest, e.g. in numerical analysis of Markov chains.

3.1. Approximation of denumerable Markov chains

Let us suppose that all Markov chains below are aperiodic, contain the single ergodic class and may be a set of transient states.

Let input data be a transition matrix $U = (u_{ij})$, $i,j = 0,1,\ldots$ of the initial Markov chain. We introduce a metric h_u in the space $\mathfrak{U}$ of all denumerable stochastic matrices which metrizes a convergence of each element of these matrices, e.g. $h_u(U,U') = \sum_{i,j} |u_{ij} - u'_{ij}| 2^{-\max(i,j)}$. Let ν_t, $t = 0,1,\ldots$ be the state at time t of the chain with matrix U. In order not to introduce notations connected with initial state we suppose that both the chain under consideration and all other Markov chains below meet the condition $P(\nu_o = 0) = 1$, i.e. they are in state 0 at time 0. Let $q = (q_1, q_2, \ldots)$ be a stationary distribution of the chain : $q_k = \lim_{t \to \infty} P(\nu_t = k)$. We choose a random sequence $V = \{\nu_t\}$, $t \geqslant 0$, to be output data and define the following metric

$$h_\nu(\{\nu_t\}, \{\nu'_t\}) = \sup_t d(\nu_t, \nu'_t)$$

in the space of random sequences their elements taking values 0,1, 2,... . Such a choice of metric h_ν implies (when $h_\nu(V,V')$ is small) the uniform closeness of all one-dimensional distributions of the chains under comparison.

If we are interested in stationary characteristics only then it is natural to choose a vector $\tilde{V} = q$ as output data and consider the metric

$$h_{\tilde{\nu}}(q,q') = \sum_{k=0}^{\infty} 2^{-k} |q_k - q'_k|$$

which implies the convergence of each element of the sequences.

Let now $\mathfrak{U}_A$ be a subset of $\mathfrak{U}$ consisting of all finite stochastic matrices (non-reducing and aperiodic). It is obvious that $\mathfrak{U}_A$ is h_u-dense in $\mathfrak{U}$. However, we can easily see that if a denumerable metric U is approximated by a sequence of finite matrices $U^{(n)} \in \mathfrak{U}_A$ then it is not sufficient for the closeness of corresponding output characteristics according with the above metrics. More precisely, if

$$h_u(U,U^{(n)}) \underset{n \to \infty}{\to} 0, \quad U \in \mathfrak{U}, \quad U^{(n)} \in \mathfrak{U}_A,$$

then the following relations

$$(2) \quad h_\nu(V,V^{(n)}) \underset{n \to \infty}{\to} 0, \quad h_{\tilde{\nu}}(\tilde{V}, \tilde{V}^{(n)}) \underset{n \to \infty}{\to} 0$$

are not valid in general.

In order to guarantee the validity of (2) we need to choose an approximating chain $U^{(n)}$ from the subset $\mathfrak{U}^* \cap \mathfrak{U}_A$ where $\mathfrak{U}^*$ is a stability subset for denumerable Markov chains. The correct assertions are contained in the following theorems.

<u>Theorem 2</u> (Kalashnikov (1978b))

For any matrix $U \in \mathfrak{U}$ (given that there exists only the stationary distribution q) we can find a sequence of $(n \times n)$-matrices $U^{(n)} \in \mathfrak{U}_A$ such that

$$\lim_{n \to \infty} h_{\mathfrak{u}}(U, U^{(n)}) = 0, \quad \lim_{n \to \infty} h_{\mathcal{v}}(V, V^{(n)}) = 0, \quad \lim_{n \to \infty} h_{\widetilde{\mathcal{v}}}(q, q^{(n)}) = 0.$$

The proof of the theorem is constructive. We can show in particular that it is possible to choose the matrices $U^{(n)}$ having the following elements :

$$(3) \quad \begin{aligned} u_{ij}^{(n)} &= u_{ij}, \quad 0 \leqslant i \leqslant n - 1, \ 1 \leqslant j \leqslant n - 1, \\ u_{io}^{(n)} &= u_{io} + \sum_{k=n}^{\infty} u_{ik} \end{aligned}$$

Moreover we can obtain the following estimates and conclusions from the proof.

As the matrix U has the only stationary distribution there exist constants $\gamma \geqslant 1$, $0 < a < \infty$ $\alpha >$ and convex (up) nonnegative function $g(x) \to \infty$ such that $x \to \infty$

$$(4) \quad E(\tau_o^{\gamma} g(\tau_o)) \leqslant a,$$

or

$$(5) \quad Ee^{\alpha \tau_o} \leqslant a,$$

where τ_o is a recurrence time for the chain $\{\nu_t\}$ to the state 0. The following estimates are valid under conditions (4) and (5).

$$(6) \quad \sum_{k=0}^{\infty} |P(\nu_t = k) - q_k| \leqslant \varepsilon(t),$$

where

$$(7) \quad \varepsilon(t) = \begin{cases} c/t^{\gamma-1} g(t), & \text{in the case (4)}, \\ \\ c_1 e^{-\beta t}, & \text{in the case (5)}, \end{cases}$$

and constants c, c_1 and β are defined by γ, a, α and function g.

If matrices $U^{(n)}$ are built by rule (3) then estimates (6), (7) are preserved :

$$\sup_n \sum_{k=0}^{n-1} |P(\nu_t^{(n)} = k) - q_k^{(n)}| \leqslant \varepsilon(t).$$

Let

$$\alpha_t^{(n)} = \max_{0 \leqslant k \leqslant t} d(\nu_t, \nu_t^{(n)}).$$

It is easy to see that the relation $h_u(U, U^{(n)}) \to 0$ follows $\alpha_t^{(n)} \to 0$ for any t. Denote

$$\theta = \min\{t : \alpha_t^{(n)} \geqslant \varepsilon(t)\}.$$

Then

$$(8) \quad h_\nu(V, V^{(n)}) = \sup_t d(\nu_t, \nu_t^{(n)}) \leqslant \alpha_\theta^{(n)}$$

It is obvious that

$$h_{\tilde{\nu}}(q, q^{(n)}) \leqslant \alpha_\theta^{(n)}.$$

At last we give the following two assertions.

Theorem 3 (Kalashnikov (1978b))

Let $U^{(n)}$ be a sequence of $(n \times n)$ stochastic matrices such that

$$h_u(U^{(n)}, U) \to 0 \ , \ h_{\tilde{\nu}}(q^{(n)}, q) \to 0$$

for some stochastic vector q. Then q is a stationary distribution for U and relation

$$(9) \quad \sup_t d(\nu_t, \nu_t^{(n)}) \to 0$$

is valid.

<u>Corollary</u>

If $U^{(n)}$ is constructed in accordance with (3) then matrix U has the only stationary distribution (and relation (9) is valid) if and only if the sequence converges to some probability distribution.

3.2. Approximation of general Markov chains

The results given in this paragraph are obtained in Anichkin and Kalashnikov (1981). They are some generalization of the above-mentioned results. Further development can be found in Anichkin (1983).

We restrict ourselves to the only theorem illustrating the character of the obtained results.

Let $\{Y_t\}$, $t = 0,1,2,\ldots$ be a time-homogeneous Markov chain, $(\mathcal{Y},r)$ its state space which is a complete, separable and metric one.

Denote a transition kernel of this chain by $K(y,\cdot)$. Let $\{\Gamma_N\}$ be a system of compact subets of the space $(\mathcal{Y},r)$ such that $\Gamma_1 \subset \Gamma_2 \subset \ldots \subset \Gamma_N \subset \ldots$, $\lim\limits_{N\to\infty} \Gamma_N = \mathcal{Y}$, and $\{S_\varepsilon\}_{\varepsilon>0}$ – a centred system of subsets of Γ_1 such that $\sup\limits_{y,y'\in S_\varepsilon} r(y,y') = \varepsilon$,

$$\tau(y,\varepsilon) = \min\{t : Y_t \in S_\varepsilon,\ Y_o = y\}.$$

Let $\{H_i(N,\delta)\}$ be a finite covering of the compact Γ_N consisting of disjoint subsets their diameter not exceeding $\delta/2$ and $Z(N,\delta)$ be a collection of some representatives of these subsets which is of course a δ-net in the compact Γ_N – let α_i be a representative of the subset $H_i(N,\delta)$.

We construct a finite Markov chain $\{Y_t^{N,\delta}\}$, $t = 0,1,\ldots$ having the mentioned δ-net in Γ_N as its state space. The following construction is analog to relation (3).

$$K^N(y,B) = K(y,B), \quad \text{if } B \subset \Gamma_N \setminus S_\varepsilon,\ y \in \Gamma_N,$$

$$K^N(y,S_\varepsilon) = K(y,S_\varepsilon) + K(y,\mathcal{Y} \setminus \Gamma_N),$$

the form of $K^N(y,\cdot)$ on subsets of S_ε is arbitrary.

Define the transition matrix of the finite chain $\{Y_t^{N,\delta}\}$ by equality.

$$K^{N,\delta}(\alpha_i,\alpha_j) = K^N(\alpha_i,H_j(N,\delta)).$$

Theorem 4 (Anichkin, Kalashnikov (1981))

Let

(1) $\mathrm{Var}\,[K(y,\cdot),K(y',\cdot)] \to 0$ when $r(y,y') \to 0$;

(2)   there exist constants $\gamma > 1$, $L > 0$, $a > 0$ such that for any N, ε

$$\sup_{y\in\Gamma_N} E\,[\tau(y,\varepsilon)]^\gamma < \infty$$

and

$$GCD\{n \,:\, n \leqslant L,\; P(\tau(y,\varepsilon) = n) \geqslant a,\; y \in S_\varepsilon\} = 1.$$

Then

$$\sup\{d(Y_t,Y_t^{N,\delta}) \,:\, t \geqslant 0,\; Y_o \in S_\varepsilon,\; Y_o' \in S_\varepsilon \cap Z(N,\delta)\} \to 0$$

if $N \to \infty$, $\delta \to 0$, $\varepsilon \to 0$.

Some of the conditions of this theorem can be relaxed, e.g. condition (2). However we won't consider these question here. We only mention a rather effective way of considering approxima- tion problems for general Markov chains, namely, the representation of these chains as regenerative processes, using the Harris-recurrence property. The matter is that for regenerative processes effective continuity conditions are known (see Kalashnikov (1978a, 1981a)) which are key for approximation.

4. Asymptotic enlarging of Markov chains

Another sort of Markov chains approximation connected with the asymptotic enlarging of their states when we compare the initial chain and another one which is obtained from the initial one by lumping together some of its states.

Let us consider a parametric family of finite Markov chains $\{\xi_t^\varepsilon\}$, $t = 0,1,2,\ldots$ ($\varepsilon > 0$ – the parameter). It is supposed that for any $\varepsilon > 0$ the chain $\{\xi_t^\varepsilon\}$ is aperiodic, irreducible, has the same state space N and $P^\varepsilon = (p_{\alpha\beta}^\varepsilon)$, $\alpha,\ \beta \in N$ – its transition matrix.

For any $\varepsilon > 0$ we divide N into $m(\varepsilon)$ subsets (classes) : $N = \bigcup_{i=1}^{m(\varepsilon)} N^\varepsilon(i)$ and denote $\lfloor\alpha\rfloor^\varepsilon$ the number of the class in which lays state α.

Suppose that the above-mentioned division meets the following condition of limit noncommunication between the classes :

$$\max_{\alpha \in N} \bar{p}^\varepsilon(\alpha) \to 0, \quad \varepsilon \to 0$$

where

$$(10) \quad \bar{p}^\varepsilon(\alpha) = \sum_{\gamma:\ \lfloor\gamma\rfloor^\varepsilon \neq \lfloor\alpha\rfloor^\varepsilon} p_{\alpha\gamma}^\varepsilon.$$

Let $q^\varepsilon(i) = (q_\alpha^\varepsilon(i),\ \alpha \in N^\varepsilon(i))$, $i = 1,\ldots,m(\varepsilon)$, be the eigenvectors of matrices

$$P^\varepsilon(i) = (p_{\alpha\beta}^\varepsilon)_{\alpha,\beta \in N^\varepsilon(i)}$$

corresponding to the maximal eigenvalue $\sum_\alpha q_\alpha^\varepsilon(i) = 1$. Define for every $\varepsilon > 0$ the auxiliary Markov chain $\{\nu_n^\varepsilon\}$ with state space $\{1,\ldots,m(\varepsilon)\}$ and transition probabilities

$$u_{ij}^\varepsilon = \sum_{\alpha \in N^\varepsilon(i)} q_\alpha^\varepsilon(i)[\bar{p}^\varepsilon(\alpha)]^{-1} \sum_{\beta \in N^\varepsilon(j)} p_{\alpha\beta}^\varepsilon$$

and let this chain be irreducible and aperiodic and have stationary probabilities $\{\nu_i^\varepsilon\}$, $i = 1,\ldots,m(\varepsilon)$.

Suppose that for some $a > 0$

$$(11) \quad \inf_{\varepsilon > 0} \min_{\alpha,i}\{\min_\alpha q_\alpha^\varepsilon(i), \min_i \nu_i^\varepsilon\} \geq a.$$

Let $\{\theta_n\}$ – the successive times of classes changing for $\{\xi_t^\varepsilon\}$, i.e.

$\rfloor \xi^{\varepsilon}_{\theta_n} \lfloor \neq \rfloor \xi^{\varepsilon}_{\theta_n - 1} \lfloor$ and let $\zeta_n = \rfloor \xi^{\varepsilon}_{\theta_n} \lfloor$. The process $\{\zeta_n\}$ is non-Markovian in general.

Theorem 5 (Kalashnikov (1981b))

If the relations (10) and (11) are fulfilled then

$$\sup_n \ d(\nu^{\varepsilon}_n, \zeta^{\varepsilon}_n) \xrightarrow[\varepsilon \to 0]{} 0$$

It is possible to estimate the convergence rate in this theorem by continuity estimates in terms of $P^{\varepsilon}(i)$ and (u^{ε}_{ij}), see Anichkin, Kalashnikov (1981), Kalashnikov (1981b).

5. Approximation of queueing systems

In order to solve different queueing problems one often approximates some governing laws by other ones which are simpler. Of course we can use for this case the previous results. For example the well-known relation $w_{n+1} = (w_n + s_n - e_n)^+$ for waiting times w_n in GI/GI/I/∞ system can be treated as defining the Markov chain $\{w_n\}$ an then use for its approximation one of the above-mentioned methods. But in this way we can break the structure of the system – the new (finite) Markov chain may have no physical sense. So it is necessary to consider some other approximations which preserve this sense, e.g. we can replace the arbitrary governing distribution functions in the initial model by Hyper-Erlang ones.

We are not ready to give here some general results. But we can give an example which illustrates the situation and uses general methods of continuity analysis, see Kalashnikov (1978a, 1981a, 1983a,b).

Let us consider the queueing system SM/M/c/r, see Anichkin (1983). It means that the system consists of c servers and has r waiting seats, service times have exponential distribution with parameter μ and customers arrive at jumps-times of some semi-Markov process $y(\cdot)$ with finite state space I. The process

$y(\cdot)$ is defined by transition matrix $P = (p_{ij})$, $i, j \in I$, and a matrix of distribution functions $F(x) = (F_{ij}(x))$, $i, j \in I$, where p_{ij} is a probability for $y(\cdot)$ to jump from state i to state j, $F_{ij}(x)$ is a conditional distribution function of holding time in state i given that the next state is j.

Our aim is to evaluate the difference between output flows in two systems of the above-mentioned types with "close" matrices $F(x)$ and $F'(x)$. A similar problem arises, for example, in the case when the real distributions $F(x)$ are changed into hyper-Erlang ones $F'(x)$ and hence the analysis of the system is reduced to the analysis of some Markov process.

Let $\{t_n\}$, $n \geqslant 0$, be the departure times from the system, $\alpha_n = (x_n, y_n, \theta_n)$ where x_n is the number of customers which the n-th customer leaves in the system after its departure, $y_n = y(t_n)$, θ_n – the residual time from t_n to the next jump of the process $y(\cdot)$. It is clear that the sequence $\{\alpha_n\}$ completely defines the output flow of the system.

Suppose that

$$\varepsilon = \sup_{i,j} \pi(F_{ij}, F'_{ij})$$

and matrix P is irreducible and aperiodic. Then from the results on continuity of regenerative processes (see Kalashnikov (1978a, 1981a, 1983a,b)) follows that for any $T = 0,1,2,\ldots$ the following inequality holds

$$(12) \quad \sup_n \pi(\alpha_n, \alpha'_n) \leqslant c_1 e^{-xT} + c_2 \varepsilon \sum_{k=1}^{T} (c\mu + 1)^k.$$

In inequality (12) the positive constants c_1, c_2 and x are completely defined by P and $F(x)$. Relation (12) implies that $\sup_n \pi(\alpha_n, \alpha'_n) \to 0$ if $\varepsilon \to 0$. Because hyper-Erlang distribution functions are π-dense in the class of all distribution functions on $(0,\infty)$ we can choose hyper-Erlang laws as F'_{ij}.

6. Exponential estimates for occurrence time of seldom event in regeneration process

Consider a problem which we often come across in reliability theory, queueing theory and other applied disciplines, see Soloviev (1983). The next results have the following peculiarity. Namely, though we receive in fact some limit theorems, the presence of quantitative estimates in it gives the possibility to use their assertions as approximating ones – prelimit laws are approximated by limit ones.

Let us have a regenerative process $X(t,\omega)$ which traditionally is sewed together from i.i.d. cycles $\{X^c(\cdot,\omega_i),\xi(\omega_i)\}$ where ω_i is an elementary event from triplet $(\Omega,\mathcal{F},P)$ and hence the sequence $\omega = (\omega_1,\omega_2,\ldots)$ is defined on denumerable product of similar triplets. Let $A \in \mathcal{F}$, $P(A) = q$ and let η_A be a random variable on $(\Omega,\mathcal{F},P)$ such that $\eta_A(\omega_1) \leqslant \xi(\omega_1)$ if $\omega_1 \in A$. We call this r.v. η_A to be occurrence-time of A on the cycle. Define the occurrence time of A in the regenerative process

$$(13) \quad \tau = \tau(\omega) = \sum_{i<k} \xi(\omega_i) + \eta_A(\omega_k), \quad \text{if} \quad k = \min\{j : \omega_j \in A\}.$$

It is easy to see that if $q > 0$ then equality (13) defines the r.v. $\tau(\omega)$ almost everywhere.

Naturally, in nonpathological **cases**, the distribution function of $q\tau$ tends to exponential one when $q \to 0$. Below we show quantitative variants of this assertion.

Introduce the necessary notations.

Let X, Y – scalar random variables;

$$\zeta_s(X,Y) = \zeta_s(P_X,P_Y)$$

$$= \sup\{|Ef(X) - Ef(Y)| : |f'(x) - f'(y)| \leqslant |x-y|^{s-1}, \forall x,y \in R^1\},$$

$$1 < s \leqslant 2,$$

an ideal metric of order s, see Zolotarev (1983);

$$\bar{A} = \Omega \setminus A; \quad R(x) = P(\xi \leqslant x/\bar{A}); \quad m_s = \int_0^\infty x^s dR(x), \quad s \geqslant 1;$$

$$\varepsilon_s(q) = 2q^{s-1} \frac{1 - q^{2-s}}{2 - s} \zeta_s(R(m_1 x),\mathrm{Exp}_1(x)) + q\zeta_1(R(m_1 x),\mathrm{Exp}_1(x)),$$

$$1 < s < 2,$$

$$\varepsilon_2(q) = (- 2q\,\ell n\,q)\zeta_2(R(m_1 x),\mathrm{Exp}_1(x)) + q\zeta_1(R(m_1 x),\mathrm{Exp}_1(x)).$$

<u>Theorem 6</u> (Kalashnikov, Vsehsviatski (1985))[*]

If $m_s < \infty$, $1 < s \leqslant 2$, then

$$(14)\quad \zeta_1(P_{q\tau/m_1}(\cdot),\mathrm{Exp}_1(\cdot)) \leqslant \varepsilon_s(q) + q + \frac{qE\eta_A}{m_1}$$

This inequality gives in fact the conditions the right to tend to 0 when $q \to 0$.

We need to note that Theorem 6 is of "uniform character", i.e. together with changing q all the process $X(\cdot,\omega)$ can change too — the assertion about the convergence will stay in power.

Similar assertions take place and for other metrics (e.g. uniform one) and for other normed multipliers — see Kalashnikov, Vsehsviatski (1984). The condition $s > 1$ in Theorem 6 can also be relaxed — see Kalashinikov (1983c).

The comparison with Soloviev (1983) shows that our estimates are better.

The proof of Theorem 6 and similar assertions uses both the properties of probability metrics (such as homogeneousity and regularity) and the "structure of the problem", namely, the representation of time τ as geometric sum of independent terms.

References

Anichkin, S.A. (1983). Hyper-Erlang approximation of probability distributions on $(0,\infty)$ and its application. Stability problems for stochastic models. Lect.Notes in Math, 1982, 1-16.

---

[*] New and better estimates are in the article of Kalashnikov and Vsehsviatski published in "Lect.Notes in Math., 1155, 1985, Springer-Verlag".

Anichkin, S.A. and V.V.Kalashnikov (1981). Continuity of random sequences and approximation of Markov chains. Adv.Appl.Prob., 13, 402-414.

Kalashnikov, V.V. (1978a). Qualitative analysis of the behaviour of complex systems by trial functions (in Russian), Nauka, Moscow.

Kalashnikov, V.V. (1978b). Solution of the approximation problem for a denumerable Markov chain. Engineering Cybernetics, v.16, N 3.

Kalashnikov, V.V. (1981a). Estimations of convergence rate and stability for regenerative and renovative processes. Colloquia Math.Soc.J.Bolyai. 24. Point Processes and Queueing Problems, Keszthely (Hungary), 1978, 163-180. North-Holland, Amsterdam.

Kalashnikov, V.V. (1981b). Methods of continuity estimates for states enlarging problem. Stability problems for stochastic models, 48-53, (in Russian) VNIISI, Moscow.

Kalashnikov, V.V. (1983a). Continuity estimates for queueing models and related problems. The 4-th International summer school on probability theory and math.statistics, Varna, 1982, 120-165 (in Russian). Publ.House of the Bulgarian Acad.Sci., Sofia.

Kalashnikov, V.V. (1983b). The analysis of continuity of queueing systems. Probability Theory and Math.Stat., Lec.Notes in Math., 1021, 268-278. Springer-Verlag, Berlin, Heidelberg, New York, Tokyo.

Kalashnikov, V.V. (1983c). An estimate of convergence rate in Renyi's theorem. Stability problems for stochastic models, 48-57 (in Russian), VNIISI, Moscow.

Kalashnikov, V.V. and S.Yu.Vsehsviatski (1985). Estimates of occurrence times of seldom events in regenerative processes, Engineering Cybernetics, v.23, N 2, 214-216.

Soloviev, A.D. (1983). Analytical methods of reliability computation and estimations. In B.V.Gnedenko (Ed.) "Problems of mathematical reliability theory" (in Russian), Radio i Sviaz, Moscow.

Zolotarev, V.M. (1976). Metric distances in spaces of random variables and their distributions, Math.USSR Sbornik, v.30, N 3.

Zolotarev, V.M. (1977). General problems of the stability of math.
models. Proc. 41-st session Intern.Stat.Institute, New Delhi,
XLVII (2), 382-401.
Zolotarev, V.M. (1983). Probability metrics. Theory Prob.Appl.,
v.XXVIII, N 2.

THE METHOD OF RENOVATING EVENTS AND ITS APPLICATIONS IN QUEUEING
THEORY

S.G.Foss

Novosibirsk State University

630090 Novosibirsk, USSR

The purpose of this talk is to present main results (some of
them are unpublished) obtained in queueing theory by so called method
of renovating events. This method was developed by A.A.Borovkov in
1978 and was published in his book "Asymptotic Methods in Queueing
Theory", 1980 - in Russian, 1984 - in English.

The method of renovating events (or, in other words, the renovating method, the renewal method) is some generalization of ideas
which are basic and traditional in the theory of semi-Markov random
processes. It allows us to find rather advanced results, related
to ergodic and stability theorems for various types of random processes in queueing theory. The main condition of the method is connected with the existence of random time intervals, on which the
process partially "forgets" its past. The very existence of the
renovating events is a condition close to necessity for ergodicity
of service processes. The renovating method is of certain interest
itself, since it is useful for solving some problems out of the
range of queueing theory.

Before representing the method of renovating events, let us
notice that the idea of this method is close to Deeblin's "coupling
method" which has become classical. However, the method of renova-

ting events has an advantage : it allows us to obtain both ergodic
and stability theorems for a broad class of stationary sequences in
the same way.

Assume the following setting of the problem.

Let $\{t_j; -\infty < j < \infty\}$ be a vector-valued, stationary, metri-
cally transitive sequence. Consider the vector-valued sequence
$\{w_n\}$ (the dimensions of w_n and t_j, in general, may be different),
which is defined by initial value w_o and recurrence relations

$$(1) \quad w_{n+1} = f(w_n, t_n) ; \quad n \geqslant 0$$

Notice that the basic types of service systems have the state
characteristics described by equations of the type (1); for example,
queueing systems and loss systems.

The renovating method allows us to clear up the conditions,
under which the sequence $\{w_{n+k}; k \geqslant 0\}$ converges as $n \to \infty$ under
to some stationary sequence $\{w^k; k \geqslant 0\}$; the stability conditions
of this latter sequence (or of the sequence $\{w_n\}$) under small
changes of the controlling sequence $\{t_j\}$; to estimate the rate of
convergence in this limit theorems.

If $w_o = 0$ and the function f is left-continuous and monotone in its first argument, then the ergodicity (as it was shown by
Loynes, 1962) always holds. But the given assumptions on w_o and
f are rather restrictive, and the method suggested by Loynes is
not successful in obtaining a stability theorem.

Let us introduce the following notation. Let $\mathcal{F}_{n,\ell}$ be a σ-
algebra generated by $t_n, \ldots, t_\ell$. We denote by T the one-to-one
measure preserving shift transformations on sets of σ-algebra
$\mathcal{F} = \mathcal{F}_{-\infty,\infty}$ that

$$T\{\omega : t_j \in B_j; j = 1,\ldots,k\} = \{\omega : t_{j+1} \in B_j; j = 1,\ldots,k\}$$

for any choice of Borel sets B_j in the domain of value t_j. The
corresponding transformations over $\mathcal{F}$-measurable functions is denoted
by U, so that $t_{j+1}(\omega) = Ut_j(\omega)$. The transformations T^{-1}, U^{-1}

and by those T^n, U^n for every integer n are defined similarly.

For the sake of simplicity we shall assume that the initial value $\mathfrak{w}_0$ is fixed.

Definition 1

The events $A_n \in \mathcal{F}_{-\infty, n+L}$; $n = 1,2,\ldots$ are called renovating on the intervals $[n, n+L]$, if random vectors $\mathfrak{w}_{n+k} = \mathfrak{w}_{n+k}(\omega)$, while $k > L$, $\omega \in A_n$ can be expressed in the form

$$(2) \quad \mathfrak{w}_{n+k} = \varphi(\mathfrak{k}_n, \ldots, \mathfrak{k}_{n+k-1}),$$

where φ depends only on the number of arguments and is defined by the choice of sequence A_n.

Definition 2

We shall say that the sequence of random vectors $\{\mathfrak{w}_n\}$ converges strongly to a random vector $\mathfrak{w}^o$ if

$$(3) \quad P\{U^{-k}\mathfrak{w}_k = \mathfrak{w}^o; \; k = n, n+1, \ldots\} \to 1$$

as $n \to \infty$.

Definition 3

The renovating events $\{A_n\}$ are called stationary if $A_n = T^n A_o$; $A_o \in \mathcal{F}_{-\infty, L}$.

The following statement is valid.

Theorem 1

Let there exists such a sequence of renovating events $\{A_j\}$ that

$$(4) \quad P\left\{ \bigcap_{\ell=\ell_o}^{\infty} \bigcup_{j=1}^{n} A_j T^{-\ell} A_{j+\ell} \right\} \to 1$$

as $n \to \infty$ for some $\ell_o > 0$. Then the sequence $\{\mathfrak{w}_n\}$ strongly converges to some random vector $\mathfrak{w}^o$; the stationary sequence $\{\mathfrak{w}^n \equiv U^n \mathfrak{w}^o\}$ satisfies the equations (1). If the sequence $\{A_n\}$ is stationary, then the inequality $P\{A_o\} > 0$ is necessary and sufficient for satisfying (4). Conversely, let the sequence of random vectors $\{w_n\}$ defined by (1) strongly converge to some random vector w^o. Then there exists a stationary sequence of renovating events $\{A_n\}$, $P\{A_o\} > 0$.

Let us make some additional remarks.

Remark 1

All the discussion above remains valid, if we consider a more general problem assuming that values of w_n and t_j lie in arbitrary metric spaces.

Remark 2

We shall assume below that the stationary events do exist. This assumption is not restrictive.

Remark 3

Theorem 1 permits us to obtain the estimates of the rate of convergence in ergodic theorems. It means that if

$$\mu \equiv \mu(\omega) = \min\{n \geqslant 1 : \omega \in A_n\} \quad \text{and} \quad \rho_n \equiv P\{w_n \neq w^n\}$$

then $\rho_n \leqslant P\{\mu > n - L\}$. We may propose another equivalent form of this statement. Let $\theta = \theta(\omega) = \min\{n \geqslant 1 : \omega \in A_n/A_o\}$. Then as one can clearly see $P\{\mu > n\} = E\{\theta; \theta > n\}$ and consequently $\rho_n \leqslant E\{\theta; \theta > n - L\}$. It means that we may obtain estimates of the rate of convergence of such type.

Let $G : \{1,2,...\} \to (0,\infty)$ be a monotonely non-reducing function and $\lim(G(n)/n) = \infty$. Therefore if $E\{G(\theta)\} < \infty$ then $\rho_n = 0(n/G(n))$, where function $0(x)$ may be rewritten in explicit form depending on $\{A_n\}$ and G.

Let us go over stability theorems.

We introduce the conditions under which small changes of controlling sequence $\{t_j\}$ lead to small changes in prestationary $\{w_n\}$ and stationary $\{w^n\}$ sequences.

More specifically, consider two controlling sequences $\{t_j^{(r)}\}$, $r = 1,2$ and corresponding sequences $\{w_n^{(r)}; n \geqslant 0\}$; $\{w^{(r)n}; n \geqslant 0\}$. Assume that both the sequences have stationary renovating events $\{A_n^{(r)}\}$; $P\{A_o^{(r)}\} > 0$; $r = 1,2$. Without any loss of generality we may consider that they are renovating on intervals of equal length L.

Let $F_o = \{f : \sup|f(x)| \leqslant 1\}$; $F \subseteq F_o$ be some subset of F_o; $d(\xi,\eta) = \sup_{f \in F}|Ef(\xi) - Ef(\eta)|$. This distance depends only on marginal distributions of random variables. It is possible to consider a more general case when the distance depends on the joint distributions of the processes.

Introduce the following notation :

$$g(n) = d(\varphi(t_1^{(1)},\ldots,t_n^{(1)}),\varphi(t_1^{(2)},\ldots,t_n^{(2)}))$$

and

$$\mu_r = \mu_r(\omega) = \min\{n \geqslant 1 : \omega \in A_n^{(r)}\}; \quad r = 1,2.$$

Then the following statement holds :

Theorem 2

For any $N < n - L$

$$(5) \quad d(w_n^{(1)},w_n^{(2)}) \leqslant g(N + L) + P\{\mu_1 > N\} + P\{\mu_2 > N\}$$

Corollary 1

Notice that the right side of the inequality does not depend on n. Therefore we may rewrite the left side as $\sup\limits_{n > N+L} (\ldots)$, so

$$\sup_{n > 1} d(w_n^{(1)},w_n^{(2)}) \leqslant \sup_{1 \leqslant n \leqslant N+L} d(w_n^{(1)},w_n^{(2)}) + \text{r.s. of (5)}.$$

As the left side of (5) does not depend on N, so

$$(6) \quad \sup_{n \geqslant 1} d(w_n^{(1)},w_n^{(2)}) \leqslant \inf_{N \geqslant 1} (\ldots)$$

Corollary 2

Inequality (5) also holds for stationary distributions. But the left side of the inequality does not depend on n in this case, therefore

$$d(w^{(1)1}, w^{(2)1}) \leqslant \inf_{N \geqslant 1} (g(N + L) + P\{\mu_1 > N\} + P\{\mu_2 > N\}).$$

Remark

The latter inequality such as (6) allows us to approximate one stationary distribution by another.

Let then $\{t_j^{(r)}\}$; $r = 0,1,\ldots$ be a family of controlling sequences depending on parameter r; $\{w^{(r)k}; k \geqslant 0\}$ and $\{w_n^{(r)}; n \geqslant 1\}$; $r = 0,1,\ldots$ be the corresponding stationary and pre-stationary sequences.

Assume that following conditions are satisfied :

(A) For any r there exists a sequence of stationary events $\{A_n^{(r)}\}$ which are renovating on intervals $[n,n+L]$ of equal length; random numbers $\mu_r(\omega) = \min\{n \geqslant 1 : \omega \in A_n^{(r)}\}$ are uniformly bounded, i.e. $\sup_r P\{\mu_r > k\} \leqslant p(k)$, where $p(k) \to 0$ as $k \to \infty$.

(B) The term $g(n,r) = d(\varphi(t_1^{(r)},\ldots,t_n^{(r)}),\varphi(t_1,\ldots,t_n))$ converges to 0 as $r \to \infty$ for any fixed n.

Theorem 3

If (A) and (B) are satisfied then

$$\sup_{n \geqslant n_o} d(w_n^{(r)}, w^o) \to 0 \quad \text{as} \quad n_o \to \infty, r \to \infty.$$

More than that, we may obtain estimates of the rate of convergence in terms of $g(n,r)$ and $p(k)$.

Remark

The condition (A) is always satisfied if

(A_1) The events A_k, $P\{A_k\} > 0$ admit a representation as open sets $A_k = \{\gamma_{k,1} > 0;\ldots;\gamma_{k,s} > 0\}$ for some $s \geqslant 1$, where $\gamma_{k,j}$ is measurable relative to $\mathcal{F}_{-\infty,k+L}$ $\cup \gamma_{k,j} = \gamma_{k+1,j}$ and the distribution of $(\gamma_{0,1}^{(r)}, \ldots, \gamma_{0,s}^{(r)})$ converge weakly as $s \to \infty$ to the distribution of $(\gamma_{0,1}, \ldots, \gamma_{0,s})$.

Consider the next example. Let the class of functions F, involved in the definition of the distance d, consist of only one function - indicator $I\{A\}$, where A is a set of continuity of all random variables $\varphi(\mathfrak{t}_1, \ldots, \mathfrak{t}_n)$; $n = 1, 2, \ldots$. Then (B) is justified if we assume that

(B$_1$) The finite-dimensional distributions of $\{\mathfrak{t}_j^{(r)}\}$ converge as $r \to \infty$ to the distributions of $\{\mathfrak{t}_j\}$;

(B$_2$) Functions $\varphi(y_1, \ldots, y_n)$ are continuous almost surely relative to the distribution of $(\mathfrak{t}_1, \ldots, \mathfrak{t}_n)$.

Theorem 4.

If (A$_1$), (B$_1$) and (B$_2$) are satisfied, then the finite-dimensional distributions of $\{\mathfrak{w}_{n+k}^{(r)}; k \geqslant 0\}$ converge weakly as $r \to \infty$, $n \to \infty$ to the finite-dimensional distributions of $\{\mathfrak{w}^k; k \geqslant 0\}$.

Remark

Verifying condition (B$_2$) there can arise the problem of finding a function φ relative to f and the events A_k. In applications the function φ usually may be put in the form $\varphi(y_1, \ldots, y_k) = f_k(0, y_1, \ldots, y_k)$, where $f_1 \equiv f$, $f_{k+1}(0, y_1, \ldots, y_{k+1}) = f(f_k(0, y_1, \ldots, y_k), y_{k+1})$.

Let us return to theorem 2. Assume that all random variables are defined on the same probability space. Note, that the definition of renovating events implies that $B_n^{(r)} = \overset{\ell}{\underset{k=0}{\cup}} A_{n+k}^{(r)}$ is the renovating event on interval $[n, n+\ell+L]$ and consequently the event $\mathfrak{D}_n = B_n^{(1)} \cap B_n^{(2)}$ is renovating for the both sequences simultaneously. Choose $\ell \geqslant 1$ so that inequality $P\{\mathfrak{D}_n\} > 0$ is justified. Let $\nu(\omega) = \min\{n \geqslant 1 : \omega \in \mathfrak{D}_n\}$. Then

$$d(\mathfrak{w}_n^{(1)}, \mathfrak{w}_n^{(2)}) \leqslant g(N + \ell + L) + P\{\nu > N\}$$

holds. The inequalities of such type may be obtained in a more

general case, when, firstly, sequences $\{w_n^{(r)}\}$ might not justify
(1) and then, the distance d might be of some more general type.

Ergodic theorems and stability theorems for many-server systems.

For simplicity here we only consider queueing systems $\langle G/G/m \rangle$
with service discipline FCFS and loss systems $\langle G/G/m \rangle_R$.

One of the fundamental characteristics of considered systems
is the vector of waiting times $w_n = (w_{n,1}, w_{n,2}, \ldots, w_{n,m})$, where
$w_{n,i}$ is the time from the arrival epoch of the n-th customer to the
liberation of i servers of the first $(n-1)$ customers.

For loss systems the following recurrence relation holds :

$$w_{n+1} = R(w_n + s_n \cdot e \cdot I\{w_{n,1} = 0\} - \tau_n \cdot i)^+$$

and for queueing systems

$$w_{n+1} = R(w_n + s_n e - \tau_n i)^+,$$

where τ_n are the interarrival times and s_n are the service times,
$x^+ = \max(x,0)$ for number x, $x^+ = (x_1^+, \ldots, x_m^+)$ for vector
$x = (x_1, \ldots, x_m)$. Symbol $R(x)$ denotes the vector, obtained from x
by ordering its coordinates by increasing; $e = (1,0,\ldots,0)$ and
$i = (1,1,\ldots,1)$; $I\{A\}$ – indicator of the set A.

The adduced equations are, obviously equations of the type (1)
while $t_n = (\tau_n, s_n)$.

Introduce some examples of renovating events. The simplest
renovation is of the type $\{w_{n,m} = 0\}$. However such event has probability 0 while $P\{s_n - \tau_n > 0\} = 1$.

Consider two types of renovating events :

a)

$$(9) \quad \{w_{n,1} = 0; \ w_{n,j} < \sum_{k=0}^{j-2} \tau_{n+k}; \ j = 2, \ldots, m\}$$

One can easily check up that this event is renovating on the interval $[n, n+m-2]$. It happens as, for example, for some $x > 0$

(10) $\{w_{n-1,m} < x\} \cap \{\tau_{n-1} > x;\ s_{n-1} < \tau_n + \ldots + \tau_{n+m-2}\}$

holds.

b)

(11) $\{w_{n,m} < x\} \cap \{s_{n+j} < a;\ \tau_{n+j} > b;\ j = 0,1,\ldots,L\}$,

where $a < mb$ and integer $L \geqslant 1$ is chosen according to the numbers $x > 0$ and $a,\ b > 0$ so that this event is renovating on the interval $[n,n+L]$.

One may find other examples of renovations in the book of A.A. Borovkov.

Usually the distribution of the sequence $\{w_n\}$ is unknown, so checking up the conditions of renovation $(P\{A_n\} > 0)$ may appear rather complicated. Nevertheless, taking some stationary sequence $\{v_n\}$, majorating $\{w_{n,m}\}$ (i.e. $v_n \geqslant w_{n,m}$ a.s.) and substituting v_n instead of $w_{n,m}$ into formulae (10)-(11) we obtain renovating events

(12) $\{v_{n-1} < x\} \cap \{\tau_{n-1} > x;\ s_{n-1} < \tau_n + \ldots + \tau_{n+m-2}\}$;

(13) $\{v_n < x\} \cap \{s_{n+j} < a;\ \tau_{n+j} > b;\ j = 0,\ldots,L\}$

being also stationary.

For queueing systems and loss systems such majorants may be constructed. For example, for queueing systems

$$w_{n,m} \leqslant w_m^r + \max_{i \leqslant j \leqslant m} w_{o,j},$$

where $\{w^n = (w_1^n,\ldots,w_m^n)\}$ is a stationary sequence of waiting times, and for loss systems -

$$w_{n,m} \leqslant w_k^n + \max_{1 \leqslant j \leqslant m} w_{o,j},$$

where $\{w^n = (w_1^n,\ldots,w_k^n)\}$ is a stationary sequence of the waiting times in k-server queueing system (for any $k \geqslant 1$).

The other examples of constructing majorants may be found in the above-mentioned book of A.A.Borovkov.

For many-server systems with arbitrary initial conditions and arbitrary stationary controlling sequences $\{(\tau_j, s_j)\}$ ergodic theorems and stability theorems are not obtained yet in such wholesome form as for dimension one.

In the multidimensional case the renovating method leads us to some rather restricting assertions, connected with the existence of some specific type of renovation (for example, (12) or (13)). For renovation of the type (12) condition $P\{A_n\} > 0$ becomes roughly speaking, the condition of unboundedness of random variables τ_j, and for renovation (13) − the condition of sufficiently rapid mixing.

The following question remains open : whether there exist renovating events (may be not of the type (12) or (13)) with positive probability for any queueing or loss systems (under some natural assumption on (τ_j, s_j)) ? If it is so, the assertion of existence of a renovating event does not lead to the loss of generality.

The answer to this question is positive at least in the following cases :

1) for single-server systems;

2) for many-server systems, if (τ_j, s_j) are independent for different j;

3) for many-server queueing systems, if random variables s_j have arithmetic distribution or $w_o = 0$.

Let us introduce certain ergodic theorems, obtained by using renovations of the type (12) or (13).

Theorem 5

Let $Es_j < \infty$ hold for loss systems and $Es_j < mE\tau_j$ for queueing systems. Assume that $P\{A_n\} > 0$, where A_n is of type (12) or (13). Then the statement of theorem 1 holds.

The condition of renovation for event of the type (12) is satisfied if, for example, there exists a set $\Omega \in \mathcal{F}_{-\infty,0}$ of positive probability where for any $\omega \in \Omega$ and $x > 0$

$$P_{\mathcal{F}_{-\infty,0}} \{\tau_1 > x; \ \tau_1 + \ldots + \tau_{m-1} > s_1\} > 0.$$

One can formulate the analogous condition in the case (13).

Theorem 5 implies

Theorem 6

Let (τ_j, s_j) be independent for different j, $Es_j < mE\tau_j$ holds for queueing systems and a) $Es_j < \infty$, b) random function τ_1 is not bounded – for loss systems. Then theorem 1 is valid. For loss systems the condition of unboundedness for τ_1 may be replaced by $P\{s_1 < m\tau_1\} > 0$.

In the case of GI/GI/m queues this fact is the famous Kiefer-Wolfowicz theorem. Whitt (1982) suggested a new proof of Kiefer-Wolfowicz theorem. In this proof he took advantage of GI/GI/m queue with cyclic discipline as a majorant.

Let us go over stability theorems.

Theorem 7

a) $P\{A_n\} > 0$ (where A_n is of type (12) or (13)),

b) finite-dimensional distributions of $(\tau_j^{(r)}, s_j^{(r)})$ converge weakly as $r \to \infty$ to finite-dimensional distributions of (τ_j, s_j);

c) $Es_j^{(r)} \to Es_j < \infty$ as $r \to \infty$.

Then finite-dimensional distributions of $\{w_k^{(r)}\}$ converge weakly as $r \to \infty$ to the distributions of $\{w^k\}$ if

d) $Es_1 < mE\tau_1$ – for queueing systems, and

d_1) $P\{s_1 - \tau_1 - \ldots - \tau_k = 0\} = 0$ for any $k \geqslant 1$ – for loss systems.

The conditions d_1) for loss systems is necessary for a.s. continuity of φ (theorem 4).

The propositions, similar to theorem 7, may be obtained in the terms of introduced above metric d.

The last part of the talk is dedicated to some estimates of the rate of convergence, which may be obtained by means of the renovating method in some particular case.

We begin from some estimates in ergodic theorems.

Theorem 8
Let
a) (τ_j, s_j) be independent for different j;
b) $mE\tau_1 > Es_1$ – for queueing systems, and $P\{m\tau_1 > s_1\} > 0$ – for loss systems;
c) $E(s_1)^\alpha < \infty$ for some $\alpha > 1$.
Then

$$P\{w_n \neq w^n\} \leq Cn^{1-\alpha},$$

where the constant C does not depend on n.

If c) is replaced with the condition
d) $E(\exp(\mu s_1)) < \infty$ for some $\mu > 0$, then

$$P\{w_n \neq w^n\} \leq C \exp(-\gamma n),$$

where $\gamma > 0$ and C do not depend on n.

The estimates in theorem 8 are unimprovable.

Adduce some estimates in stability theorems for queueing systems. The similar result could be formulated for loss systems.

Let $\lambda(\xi, \eta)$ be Levy-Prokhorov's distance between the distributions of random variables ξ and η.

Theorem 9
Assume that for any system the controlling variables $(\tau_j^{(r)}, s_j^{(r)})$ are independent for different j and
a) $mE\tau_1 > Es_1$;
b) $\varepsilon_r = \max(\lambda(\tau_j^{(r)}, \tau_1); \lambda(s_1^{(r)}, s_1)) \to 0$ as $r \to \infty$;
c) $E(s_1)^\alpha < \infty$, $\sup_r E(s_j^{(r)})^\alpha < \infty$ for some $\alpha > 1$.
Then

$$\lambda(w^{(r)o}, w^o) \leq C\varepsilon_r^{1-1/\alpha}.$$

If we include instead of c) the condition

d) $E(\exp(\mu s_1)) < \infty$, $\sup_r E(\exp(\mu s_1^{(r)})) < \infty$ for some $\mu > 0$, then

$$\lambda(w^{(r)o}, w^o) \leq c\varepsilon_r |\ln \varepsilon_r|.$$

This statement remains true when we use some other distances. The estimates of theorem 9 are also unimprovable. The constants participating in theorems 8-9 may be written explicitly.

Note, that the estimates in the theorems 8-9 are obtained using renovations of the type (13). These estimates remains valid if the condition of independence of (τ_j, s_j) for different j is replaced with a weaker one - the condition of sufficiently rapid mixing of $\{(\tau_j, s_j)\}$.

One point more : as it is shown in the case of single-server systems, the condition $P\{m\tau_1 > s_1\} > 0$ for loss systems is not necessary and apparently may be weakened.

The questions of existence and uniqueness of the stationary decision of the equation (1) were considered by Lisek (1979, 1982) for the GI/GI loss systems and some of their generalizations.

The results of type theorems 5-9 can be formulated for the other service systems, too : for example, for the systems with finite waiting room or for the G/G/∞ systems.

References

Achmarov, I. (1981). The rate of convergence in ergodicity and continuity theorems for systems with an infinite number of servers. Sibirsk.Mat.Zh., 21, 16-21 (in Russian).

Borovkov, A.A. (1984). Asymptotic methods in queueing theory. Wiley, New York. Translated from Russian, Izdat.Nauka, Moscow (1980).

Borovkov, A.A. (1984). A note on the rate of convergence in stability theorems. Teor.Verojat. i Primemen., 29, 119-120 (in Russian).

Foss, S.G. (1983). On conditions of ergodicity for many-server queues. Sibirsk.Mat.Zh., 24, 168-175 (in Russian).

Kalashnikov, V.V. (1979). Estimates of stability for renovative processes. Izv.Acad.Nauk.USSR, Tech.Kybernetika, 5, 85-89 (in Russian).

Kalashnikov, V.V. and S.A.Anichkin (1981). Continuity of random sequences and approximation of Markov chains. Adv.Appl.Prob., 2, 402-414.

Lisek, B. (1982). A method for solving a class of recursive stochastic equations. Z.Wahrscheinlichkeitstheorie verw.Geb., 60, 151-162.

Whitt, W. (1982). Existence of limiting distributions in the GI/G/s queue. Math.Oper.Res., 7, 88-94.

ON NON-TIME-HOMOGENEITY

Hermann Thorisson

Department of Mathematics
Chalmers University of Technology and the University
of Göteborg. S-412 96 Göteborg - Sweden

1. Introduction

The assumption of time-homogeneity is implicit in many stochastic
models. This is considered necessary for some calculations and limit
results. However, in many applications the time-homogeneity condition
is far from realistic. Here we propose two ways of attacking the
problem of non-time-homogeneity.

1.1. Reduction to time-homogeneity

It is wellknown that non-time-homogeneity can be removed from a
non-time-homogeneous Markov process (Z_t) by adding actual time to the
state of the process : (Z_t, t). This is not of much use. It is true
that the process becomes time-homogeneous, but at the cost of being
totally transient and in no way easier to handle.   .

In certain special cases, however, the non-time-homogeneous
process under consideration can be imbedded into a time-homogeneous
structure preserving the recurrence properties. In Section 2 we show
how a non-time-homogeneous Markov process (Z_t), associated with the

k-server queue with non-stationary Poisson arrivals, may be subordin-
ated to a time-homogeneous regenerative process, i.e., for each t,
Z_t can be obtained in a deterministic manner from that process at
time t. This can be used to investigate the ergodic properties of
(Z_t). For further results, see H. Thorisson "On regenerative and
ergodic properties of the k-server queue with non-stationary Poisson
arrivals".

1.2. <u>Taking limits the wrong way</u> (terminology suggested by Erhan
Çinlar).

The above approach is only successful in very special cases.
However, in order to obtain ergodic results reduction to time-
homogeneity is not necessary. The traditional way to obtain such
results is to start a process at time 0, consider its distribution
in the time-interval $[t,\infty)$ and check whether it stabilizes as $t \to \infty$.

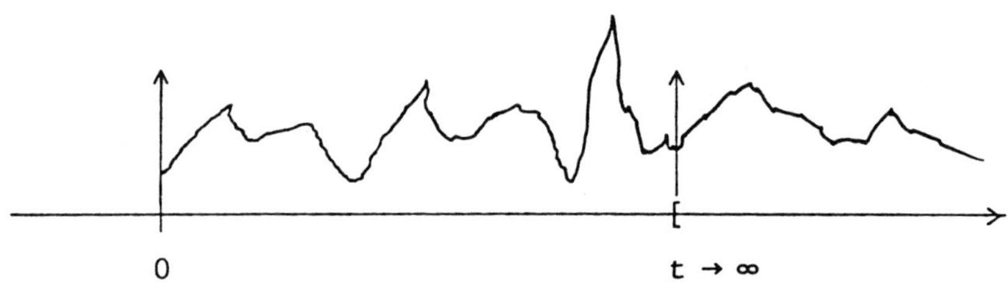

Figure 1

For this to work the mechanism governing the development of the
process must be time-homogeneous, or asymptotically so in some sense.

But why are we interested in limit distributions ? Probably not
so much because we wish to know how things will be in a far future if
they started now but rather because we wish to know how things are
now if they started long ago. Why then don't we start our process
at a time s and check what happens to its distribution in a fixed

time-interval $[t,\infty)$ as we send the starting point s backward to minus
infinity ?

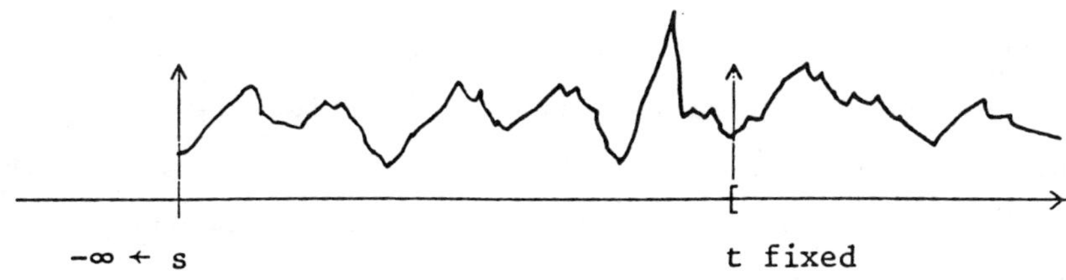

Figure 2

The answer is probably twofold. We are not used to time going
backward, and also in order to let it do so we have to work with a
family of processes, one process for each s, a far more complicated
procedure than in the traditional "forward limit" case where we only
have one process and consider it at different times t.

   The backward approach does _not_ require time-homogeneity, while
in the time-homogeneous case the backward and forward approaches are
 clearly equivalent.

    As far as I am aware, backward limits date back to a 1936 paper
by Kolmogorov "Zur Theorie der Markoffschen Ketten" dealing with
Markov chains on a finite state space. Kolmogorov's results were
elaborated by Blackwell in a 1945 paper "Finite non-homogeneous
chains", but otherwise the idea seems to have passed unnoticed.
However, during the last decade Cohn has extensively stidued Markov
chains on a countable state space drawing attention to the tail σ -
algebra. The results obtained in a forward setting in his 1974 paper
"On the tail events of a Markov chain" can easily be used to obtain
backward results. See his 1982 paper "On a class of non-homogeneous
Markov chains" for references, see also Geifeath's 1978 paper "Coupling methods for Markov processes". In Thorisson (1984) "Backward
limits of non-time-homogeneous regenerative processes" backward limits

are considered in the context of non-time-homogeneous regenerative processes introduced in Thorisson (1983) "The coupling of regenerative processes".

In Section 3 of this paper we prove a backward limit result for non-time-homogeneous Markov chains in discrete time and space under a time-uniform aperiodicity an non-null recurrence condition extending the wellknown conditions from the time-homogeneous case. The proof is based on coupling and the completeness of the space of probability measures on a given measurable space with respect to total variation. Also we show that if a family of random sequences has a backward limit in total variation, then coupling can be used to prove that result.

Let $(\Omega, F, \mathbb{P})$ be the probability space supporting all the random elements in this paper, let $\mathbb{E}$ denote expectation and let ε be small enough.

2. The queue $M_t/G/k$: from non-time-homogeneity to time-homogeneity

2.1.Preliminaries

Consider the $M_t/G/k$ queue and put

$$Z_t = (Q_t, V_{1t}, \ldots, V_{kt})$$

where Q_t = the number of customers present in the system at time t and for $i = 1, \ldots, k$

V_{it} = the residual service time of the customer being served by the i'th server at time t.

By $M_t/G/k$ we understand the k-server queueing system where customers arrive in the time-interval $(0,\infty)$ according to a non-stationary Poisson process $n(\cdot)$, $n(A)$ = the number of customers arriving in the Borel set A, and line up to be served under the "first come, first served" discipline, the service times $v_1, v_2, \ldots$ of the $(Q_0 - k)^+$ customers waiting for service at time 0 and the customers arriving in $(0,\infty)$ are i.i.d, and $n(\cdot)$, $(v_n)_1^\infty$ and Z_0 are independent.

Under these assumptions $(Z_t)_{[0,\infty)}$ clearly is a non-time-homogeneous Markov process, the non-time-homogeneity being induced by the non-stationarity of the Poisson arrival process $n(\cdot)$. Here we shall sketch how the random character of the non-time-homogeneity can be removed, under a natural stability condition, by subordinating $(Z_t)_{[0,\infty)}$ to a certain time-homogeneous process.

For convenience, let $n(\cdot)$ be defined on the whole real line $(-\infty,\infty)$ although it is only active on the positive half-line $(0,\infty)$. Let a be the intensity measure of $n(\cdot)$,

$$a(A) = \mathbf{E}[n(A)]$$

for all Borel subsets A of $(-\infty,\infty)$, and assume that a is atomless. Let θ_t be the shift transformation for measures on $(-\infty,\infty)$,

$$\theta_t a(A) = a(t + A) \text{ for all Borel subsets A of } (-\infty,\infty).$$

Let L_d be the non-negative lattice with span d,

$$L_d = \{0, d, 2d, \ldots\}.$$

2.2. The theorem

It is wellknown (see e.g. Cinlar, 1975, "Introduction to Stochastic Processes", Theorem 4.7.4) that

$$n(0, t] = \bar{n}(0, a(0, t]], \quad t \in (0,\infty),$$

where $\bar{n}(\cdot)$ is a <u>stationary</u> Poisson process with intesity 1. Our result is based on and related to this :

Theorem 1

Let d and c be positive and finite constants and let

$$c < k / E[v_1].\tag{1}$$

If

$$\frac{1}{d}\,\theta_t a\,(0,d] \leqslant c \quad \text{for all } t\tag{2}$$

then the copy of the $M_t/G/k$ system can be chosen in such a way that there exists a mapping $f(\cdot,\cdot)$, determined by d, such that

$$Z_t = f(\theta_t a,\ \bar{Z}_t), \quad \text{for all } t \in [0,\infty),\tag{3}$$

where $(\bar{Z}_t)_{[0,\infty)}$ is a stochastic process, regenerative in the traditional sense and satisfying the following properties :

(a) The regeneration times S_n, $n \geqslant 0$, of $(\bar{Z}_t)_{[0,\infty)}$ are L_d-valued, the distribution of $X_n = S_n - S_{n-1}$, $n \geqslant 1$, is supported by no sublattice of L_d and $E[X_1] < \infty$.

(b) The distribution of $(\bar{Z}_t)_{[0,\infty)}$ and $(S_n)_0^\infty$ is independent of a and the distribution of $(\bar{Z}_{S_0+t})_{t \in [0,\infty)}$ and $(S_n-S_0)_{n=0}^\infty$ is independent of the distribution of Z_0.

(c) $(\bar{Z}_t)_{[0,\infty)}$ is zero delayed, i.e. $\mathbb{P}(S_0=0) = 0$, if and only if $Q_0 = 0$ a.s.

Remark 1

    The existence of constants d and c such that (1) and (2) hold is clearly equivalent to the following condition :

$$\liminf_{t \to \infty} \frac{1}{t}\,\sup_s\,\theta_s a\,(0,t] < k / E[v_1].\tag{4}$$

Remark 2

    The time-homogeneous $\bar{Z}_t$ takes care of the randomness of Z_t while $\theta_t a$ takes care of the non-time-homogeneity. Observe that if $n(\cdot)$ is stationary then $\theta_t a = $ constant times Lebesgue measure on $(-\infty,\infty)$;

in particular, $\theta_t a$ does not depend on t and thus the non-time-homogeneity vanishes.

2.3. The proof

In order to obtain the desired copy of the $M_t/G/k$ system we divide $(0,\infty)$ into the sub-intervals $(r,r+d]$, $r \in L_d$, and attach to each interval a random element ξ_r determining the arrivals in the interval and the service times of those who arrive in it :

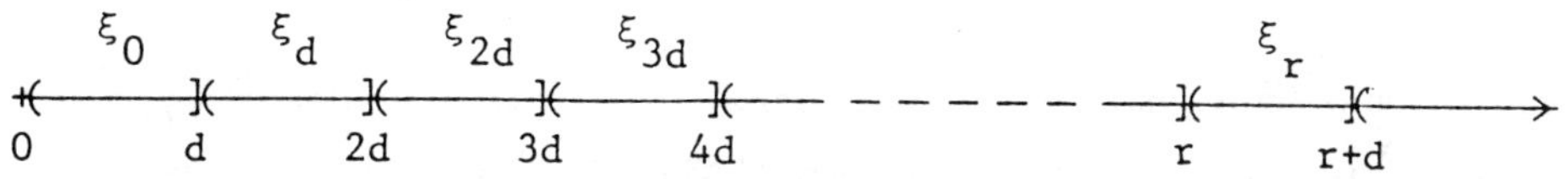

Figure 3

Let $n_r(\cdot)$, $r \in L_d$, be i.i.d. stationary Poisson processes with intensity 1 and independent of the i.i.d. random variables v_{ri}, $r \in L_d$, $i \geqslant 1$, governed by the service time distribution. Put

$$\xi_r = (n_r(\cdot), (v_{ri})_{i=1}^{\infty})$$

and let $(\xi_r)_{L_d}$ be independent of $(Z_0, v_1, \ldots, v_{(Q_0-k)^+})$. Clearly the ξ_r's are i.i.d. with distribution independent of a. Define $n(\cdot)$ on $(r,r+d]$ by

$$\theta_r n(0,x] = n_r(0, \theta_r a(0,x]], \quad x \in (0,d],$$

and let

$$v_{ri}, \quad 1 \leqslant i \leqslant n_r(0, \theta_r a(0,d]],$$

be the service times of the customers arriving in the time-interval $(r,r+d]$.

We now modify the $M_t/G/k$ system in a certain way ending up with
a time-homogeneous group-arrival system :

Let the $n_r(0, \theta_r a(0,d]]$ customers arriving in the time-interval
$(r, r+d]$ wait until time $r+d$ before their service can start, and add

$$n_r(0,cd] - n_r(0, \theta_r a(0,d]] \qquad\qquad (5)$$

customers arriving at time $r+d$ with service times

$$v_{ri}, \; n_r(0, \theta_r a(0,d]] < i \le n_r(0,cd] .$$

Observe that the number of customers in (5) is non-negative due to
the condition (2). The queue length in this <u>modified system</u> clearly
coincides in the time-set L_d with the queue length in a <u>k-server</u>
<u>system with groups</u> of $n_r(0,cd]$ customers arriving at times $r+d$,
$r \in L_d$, served under the "first come, first served" discipline with
service times

$$v_{ri}, \; 1 \le i \le n_r(0,cd] .$$

Observe that this <u>group-arrival system</u> does not depend on a.   The
groups are i.i.d. and independent of the i.i.d. service times.   Thus
this system is time-homogeneous.

Denote the queue length in the group-arrival system by $\tilde{Q}_t$.   It
is easily seen (see Thorisson,   "The queue GI/GI/k : Finite moments
of the cycle variables and uniform rates of convergence", Lemma 2) that
the above <u>modification</u> yields a queue length process denominating the
original $M_t/G/k$ queue length process and thus

$$\tilde{Q}_r = 0 \;\; \Rightarrow \;\; Q_r = 0, \qquad r \in L_d . \qquad\qquad (6)$$

We now define the regeneration times $(S_n)_0^\infty$:

$$S_n = \text{the } (n+1)\text{th} \quad r \in L_d \quad \text{such that} \quad \tilde{Q}_r = 0.$$

Clearly $(S_n)_0^\infty$ is an L_d-valued renewal process. The S_n's are a.s.-finite and $\mathbb{E}[X_1] < \infty$ since the traffic intensity of the group-arrival system is less than one, this follows from (1) :

$$\frac{\mathbb{E}[\sum_{i=1}^{n_r(0,cd]} v_{ri}]}{k\,d} = \frac{c\,d\,\mathbb{E}[v_1]}{k\,d} < 1.$$

Also $(S_n)_0^\infty$ is aperiodic since

$$\mathbb{P}(X_1 = d) \geqslant \mathbb{P}(n_r(0,cd] = 0) = e^{-cd} > 0.$$

Thus (a) holds, (b) is obvious from what follows and (c) holds since $Q_0 = 0 \Leftrightarrow \tilde{Q}_0 = 0 \Leftrightarrow S_0 = 0.$

The sequence of random times S_0, S_1, ... splits $(\xi_r)_{L_d}$ into i.i.d. cycles $(\xi_r : r \in L_d \cap [S_n, S_{n+1}))$, $n \geqslant 0$, that are independent of the delay-cycle $(\xi_r : r \in L_d \cap [0, S_0))$. Now $\tilde{Q}_{S_n} = 0$ and (6) yields

$$Q_{S_n} = 0.$$

Thus the $M_t/G/k$ system is idle at the random times S_0, S_1, ... and if $t \in [S_n, S_{n+1})$ then clearly Z_t is determined by $\theta_t a$ and

$$\bar{Z}_t = ((\xi_r : r \in L_d \cap [S_n, t)), t-S_n).$$

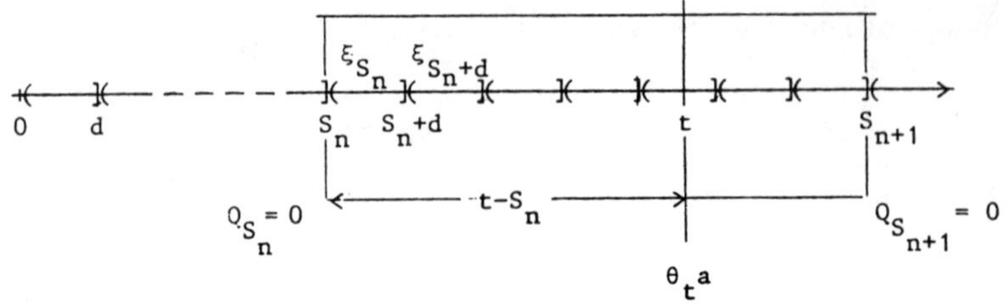

Figure 4

Further, Z_t is determined by $\theta_t a$ and $\bar{Z}_t$ in the same way for all t and thus there exists a mapping $f(\cdot,\cdot)$ such that (3) holds for $t \geqslant S_0$. It is easy to define $\bar{Z}_t$ for $t < S_0$ in a convenient way from the delay-cycle, and since $\bar{Z}_t$ is defined within a cycle it is clear that $(\bar{Z}_t)_{[0,\infty)}$ is a regenerative process.

2.4. On ergodicity

We shall sketch here how Theorem 1 can be applied to derive ergodic results. Let $\|\cdot\|$ denote the total variation norm. (The difference of two probability measures is a signed measure with total mass 0. For such signed measures ν it holds that $\|\nu\| = 2 \sup_A \nu(A)$.) Then, with sup$_a$ denoting supremum over a satisfying (2), the limit part of

$$\sup_a \| \mathbb{P}(f(\theta_t a, \bar{Z}_t) \in \cdot) - \mathbb{P}(f(\theta_t a, \vec{Z}_0^\star) \in \cdot) \|$$

$$\leqslant \| \mathbb{P}(\bar{Z}_t \in \cdot) - \mathbb{P}(\vec{Z}_0^\star \in \cdot) \| \to 0 \quad \text{as } t \to \infty, \ t \in L_d, \tag{7}$$

where $(\vec{Z}_t^\star)_{[0,\infty)}$ is the periodically stationary version of $(\bar{Z}_t)_{[0,\infty)}$, follows from Thorisson, 1983, "The coupling of regenerative processes", Corollary 2; the inequality is obvious. Thus it holds that

$$Z_t \overset{\text{tot}}{\sim} f(\theta_t a, \vec{Z}_0^\star) \tag{8}$$

uniformly in a satisfying (2) for large t, where $\overset{\text{tot}}{\sim}$ means that the

left hand and the right hand sides have, in the sense of total
variation, approximately the same distribution. In fact we have only
established (8) for $t \in L_d$, but that restriction can be removed, see
Thorisson, " On regenerative and ergodic properties of the k-server
queue with non-stationary Poisson arrivals".

Now consider an $M_t/G/k$ system starting at an arbitrary time
$s \in (-\infty,\infty)$. Let Z_t^s indicate this (thus $Z_t^0 = Z_t$). Let the distribution of Z_s^s be λ (let λ be fixed, i.e. independent of s) and let P_{st}
be the transition probabilities. Then Z_t^s has distribution λP_{st}.
Also if we replace a in Theorem 1 by $\theta_s a$ and let Z_0 have distribution λ then Z_{t-s} gets the same distribution as Z_t^s. Thus, if we in
(8) replace t by t-s and a by $\theta_s a$ we obtain (observe that (2) holds
if and only if it holds with a replaced by $\theta_s a$ and that $\theta_{t-s}\theta_s a = \theta_t a$)
the following <u>backward limit</u> result :

$$\lambda P_{st} \xrightarrow{\text{tot}} \pi_t \qquad\qquad\qquad\qquad (9)$$

uniformly in a satisfying (2) as $s \downarrow -\infty$, where π_t is the distribution
of $f(\theta_t a, \vec{Z}_0^*)$ and $\xrightarrow{\text{tot}}$ denotes convergence in total variation.

3. Backward limits

3.1. The theorem

Let (Z_n) be a non-time-homogeneous Markov chain on a countable
state space E. Let

$$p_{ij}(m,n) = \mathbb{P}(Z_n = j \mid Z_m = i), \qquad m \leqslant n, \qquad i, j \in E,$$

be the transition probability from state i at time m to state j at
time n. Let $p_{ij}(m,n)$ be defined for all integers m and n satisfying
$m \leqslant n$. Let $P(m,n)$ be the matrix with entries $p_{ij}(m,n)$.

Let

$$f_{ij}(m,k) = \mathbb{P}(Z_{m+1} \neq j, \ldots, Z_{m+k-1} \neq j, Z_{m+k} = j \mid Z_m = i), \quad k \geq 1,$$

be the probability that starting from state i at time m it takes k steps to hit state j. Put $f_{ij}(m,\infty) = \mathbb{P}(Z_n \neq j, \forall n > m \mid Z_m = i)$.

Write $p_{i\cdot}(m,n)$ and $f_{ij}(m,\cdot)$ to denote the probability measures defined by

$$p_{iA}(m,n) = \sum_{j \in A} p_{ij}(m,n), \quad A \subset E,$$

and

$$f_{ij}(m,B) = \sum_{k \in B} f_{ij}(m,k), \quad B \subset \{1,2,\ldots,\infty\},$$

respectively.

Theorem 2

If

$$f_{ii}(m,[1,\infty)) = 1, \quad -\infty < m < \infty,$$

$$\sum_{k=1}^{\infty} \sup_m f_{ii}(m,[k,\infty)) < \infty$$

(time-uniform non-null recurrence)

$$\gcd\{k : \inf_m f_{ii}(m,k) > 0\} = 1 \quad \text{(time-uniform aperiodicity)}$$

then for all $j \in E$ and all integers n

$$p_{ij}(m,n) \to \pi_{ij}(n) \quad \text{as } m \downarrow -\infty \quad \text{(backward limit)} \tag{10}$$

where $\pi_{i\cdot}(n)$ is a probability measure satisfying (regard $\pi_{i\cdot}(n)$ as a row vector)

$$\pi_i \cdot (n) = \pi_i \cdot (m) P(m,n), \qquad -\infty < m < n < \infty.$$

3.2. The Proof

Let $(Z_n)_m^\infty$ and $(Z_n')_{m'}^\infty$, where $m' \le m$, be independent Markov chains with transition probabilities $p_{ij}(m,n)$ and starting in state i at time m and m' respectively. Put

$$T_{mm'} = \inf \{ k \ge 0 : Z_{m+k} = Z_{m+k}' \}.$$

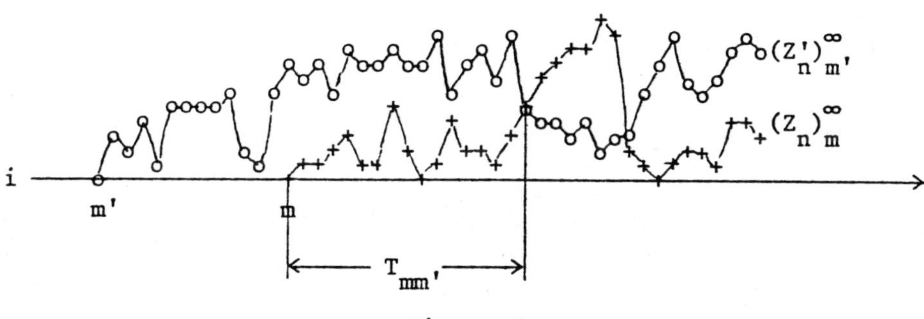

Figure 5

Then T_{mm}' is a coupling epoch for the chains (see Griffeath, "Coupling methods for Markov processes,", Thorisson, "The coupling of regenerative processes" and "On maximal and distributional coupling") and thus the coupling inequality holds :

$$\| p_i \cdot (m,n) - p_i \cdot (m',n) \| \le 2 \, \mathbb{P}(T_{mm'} > n-m). \qquad (11)$$

Here $\| \cdot \|$ denotes the total variation norm, see Section 2.4. Under the conditions of the theorem it can be proved (see Thorisson, "The coupling of regenerative processes", Proposition 1.2 and Lemma 2.6) that there exists a finite random variable T with distribution independent of m and m' and such that T dominates $T_{mm'}$ stochastically. Thus $T_{mm'}$ in (11) can be replaced by T without violating the inequality to obtain

$$\sup_{\substack{m' \\ m' \leqslant m}} \| p_{i\cdot}(m,n) - p_{i\cdot}(m',n) \| \leqslant 2\, \mathbb{P}(T > n-m) \to 0 \text{ as } m \downarrow -\infty.$$

$$(12)$$

Thus the sequence of probability measures $p_{i\cdot}(m,n)$, $m = n$, $n-1$, ...,
is Cauchy convergent. Since probability measures on a given measurable space form a complete metric space with respect to total variation, there exists a limit probability measure $\pi_{i\cdot}(n)$. The total
variation convergence easily renders the final statement of the
theorem and the proof is complete.

3.3. Some remarks

Remark 3

If the chain is <u>time-homogeneous</u>, i.e. $p_{ij}(m,n) = p_{ij}(0,n-m)$,
then the conditions of the theorem reduce to the familiar conditions
that i is non-null recurrent and aperiodic. Further, $\pi_{ij}(n)$ will not
depend on n and (10) may be rewritten as

$$p_{ij}(m,n) \to \pi_{ij} \text{ as } n \to \infty \qquad \text{(traditional forward limit)}.$$

Remark 4

The conditions of the theorem are only sufficient conditions and
far from being necessary. They extend the conditions from the time-
homogeneous case, but certainly it is possible to utilize the non-
time-homogeneity and prove (10) under conditions that exclude the
possibility of time-homogeneity. Natural questions are: What conditions
characterize (10) ?   When is $\pi_{ij}(n)$ independent of i ?

Remark 5

It follows from the final statement of Theorem 1 that there
exists a Markov chain $(Z_n^{\star})_{-\infty}^{\infty}$ governed by the transition probabilities

$p_{ij}(m,n)$ and developing through integer times from minus to plus infinity (doubly infinite non-homogeneous chain).

In his 1936 paper, "Zur Theorie der Markoffschen Ketten". Kolmogorov proved in the finite state space case, using the method of convergent sub-sequences, that there exists a probability vector solution to

$$\pi(n) = \pi(m)P(m,n), \qquad -\infty < m < n < \infty, \qquad (13)$$

and that the solution is unique if and only if

$$P(m,n) \to \pi(n) \qquad \text{as } m \downarrow -\infty.$$

Observe that in the time-homogeneous case (13) becomes the wellknown formula $\pi = \pi P$ where $P = P(0,1)$.

In his 1945 paper, "Finite non-homogeneous chains", Blackwell applies martingale convergence to prove in the finite state space case that there is an integer $h \geqslant 1$ and time-dependent partitions $C_{m1} \cup \ldots \cup C_{mh} \cup N_m = E$ such that for each $k = 1, \ldots, h$ and $i_m \in C_{mk}$ it holds that $p_{i_m j}(m,n) \to \pi_j^k(n)$ as $m \downarrow -\infty$ where $\pi^k(\cdot)$'s and if $(Z_n^{\star})_{-\infty}^{\infty}$ is a doubly infinite Markov chain governed by the $p_{ij}(m,n)$'s then $\mathbb{P}(Z_n^{\star} \in N_n \text{ i.o. for } n \leqslant 0) = 0$.

If we assume the existence of a probability vector solution to (13) then Cohn's results in "On the tail events of a Markov chain", can be used to obtained backward limits (consider the reversed chain $(Z_{-n}^{\star})_{-\infty}^{\infty}$). Observe that with with our method the existence of a solution to (13) comes as a consequence of the backward limit result.

Remark 6

It follows from the proof that (10) holds in the strong sense of total variation :

$$p_{i.}(n-k,n) \overset{tot}{\to} \pi_{i.}(n) \qquad \text{as } k \to \infty. \tag{14}$$

Further, in (12) we can drop the supremum and replace $p_{i.}(m',n)$ by the limit $\pi_{i.}(n)$ and m by n-k to obtain that (14) holds <u>uniformly</u> in n = ..., -1, 0, 1, ... (<u>time-uniform backward ergodicity</u>). Cf. " Thorisson " Backward limits of non-time homogeneous regenerative processes to see how results on <u>time-uniform rates of convergence</u> can be established and how to obtain a <u>non-time-homogeneous analogue of</u> <u>discrete renewal theory</u>.

3.4. Backward limits and coupling

Finally, we sketch a coupling criteria for backward convergence. For each integer m let $(Z_i^m)_{i=m}^{\infty}$ be a random sequence on a <u>general</u> state space $(E, \mathscr{E})$. Suppose for an integer n there exists for each pair of sequences $(Z_{n-k+i}^{n-k})_{i=0}^{\infty}$ and $(Z_{n-k+i}^{n-k'})_{i=0}^{\infty}$, k < k', a coupling (or more generally <u>distribution coupling</u>, see Thorisson, "The coupling of regenerative processes" and "On maximal and distributional coupling") with coupling epoch $T_{n-k,n-k'}$ such that

$$\sup_{k' \in \{k,k+1,...\}} \mathbb{P}(T_{n-k,n-k'} > k) \to 0 \quad \text{as } k \to \infty \tag{15}$$

<u>(backward successful at n)</u>.

Then

$$\left\| \mathbb{P}\left((Z_i^{n-k})_{i=n}^{\infty} \in \cdot\right) - \mathbb{P}\left((Z_i^{n-k'})_{i=n}^{\infty} \in \cdot\right) \right\| \leq 2\,\mathbb{P}(T_{n-k,n-k'} > k) \tag{16}$$

and applying (15) and the completeness of probability measures with respect to total variation yields

$$\mathbb{P}\left((Z_i^{n-k})_{i=n}^{\infty} \in \cdot\right) \xrightarrow{tot} \pi_n \quad \text{as } k \to \infty \tag{17}$$

<u>(backward ergodicity at n)</u>

where π is a probability measure on $(E^{\{n,n+1,...\}}, \mathscr{E}^{\{n,n+1,...\}})$.

Further, if (15) holds for all n (<u>backward successful</u> couplings) then the collection of probability measures π_n, n = ..., $-$ 1, 0, 1, ..., is consistent and extends uniquely to a probability measure π on $(E^{\{...,-1,\ 0,\ 1,\ ...\}}, \mathscr{E}^{\{...,\ -1,\ 0,\ 1,\ ...\}})$ and thus for all integers n

$$\mathbb{P}\left((Z_i^{n-k})_{i=n}^{\infty} \in \cdot\right) \xrightarrow{\text{tot}} \pi(E^{\{...,n-2,n-1\}} {}_x \cdot) \text{ as } k \to \infty \qquad (18)$$

(<u>backward ergodicity</u>).

Finally, suppose (15) holds uniformly for all integers n or equivalently

$$T_{mm'} \overset{D}{\leqslant} T, \quad -\infty < m' < m < \infty, \ \underline{\text{(uniformly successful)}} \qquad (19)$$

where T is a finite random variable with distribution independent of m and m' and $\overset{D}{\leqslant}$ denotes stochastic domination. Then we may replace $T_{n-k,n-k'}$ in (16) by T, take supremum in n, let $k \to \infty$ and obtain that (18) holds uniformly in n (<u>time-uniform backward ergodicity</u>).

Conversely, there always exists a maximal distributional coupling, i.e., a distributional coupling such that (16) holds with identity, see Thorisson "On maximal and distributional coupling". Since convergence implies Cauchy convergence we have that (17) implies (15). Further, if (18) holds uniformly in n then the Cauchy convergence is uniform in n and thus so is the convergence in (15) implying (19).

Our observations can be summarized as follows :

Backward ergodicity at n $\Leftrightarrow \exists$ couplings backward successful at n; Backward ergodicity $\Leftrightarrow \exists$ backward successful couplings.
     Time-uniform backward ergodicity $\Leftrightarrow$ uniformly successful couplings.

References

D. Blackwell (1945), Finite non-homogeneous chains. Ann. Math. 46, 594-599.

E. Çinlar (1975), Introduction to Stochastic Processes. Prentice-Hall Inc. Englewood Cliffs, New Jersey.

H. Cohn (1974), On the tail events of a Markov chain. Z. Wahrscheinlichkeitsth. 29, 65-72.

H. Cohn (1982), On a class of non-homogeneous Markov chains. Math. Proc. Comb. Phil. Soc. 92, 527-534.

D. Griffeath (1978), Coupling methods for Markov processes. Studies in Probability and Ergodic Theory. Adv. Math. Supplementary Studies 2.

A.N. Kolmogorov (1936), Zur Theorie der Markoffschen Ketten. Math. Ann. 112, 155-160.

H. Thorisson (1983), The coupling of regenerative processes. Adv. Appl. Prob. 15, 531-561.

H. Thorisson (1985), The queue GI/GI/k: Finite moments of the cycle variables and uniform rates convergence. Stochastic Models, 1(2), 221-238.

H. Thorisson, On regenerative and ergodic properties of the k-server queue with non-stationary Poisson arrivals. To appear in J. Appl. Prob.

H. Thorisson (1984), Backward limits of non-time-homogeneous regenerative processes. Preprint, Dept. Math. Göteborg.

H. Thorisson, On maximal and distributional coupling. To appear in Annals of Probability.

A SEMI-MARKOV QUEUE WITH EXPONENTIAL SERVICE TIMES

J.H.A. de Smit and G.J.K. Regterschot

Twente University of Technology

Enschede, The Netherlands

1. Introduction

In de Smit (1985) a general model for the single server semi-Markov queue was studied. Its solution was reduced to the Wiener-Hopf factorization of a matrix function. For the general case a formal factorization can be given in which the factors have probabilistic interpretations but cannot be calculated explicitly (see Arjas (1972)). Explicit factorizations can only be given in special cases. The present paper deals with the case in which customers arrive according to a Markov renewal process, while an arriving customer requires an exponential service time whose mean depends on the state of the Markov renewal process at his arrival. For this special case we shall obtain an explicit factorization. The model is formally described as follows.

Customers arrive at a single server at time epochs $T_1, T_2, \ldots;$ with $T_1 = 0$. The interarrival times are denoted by $A_n = T_n - T_{n-1}$, $n = 2, 3, \ldots;$ and the service time of the n-th customer by S_n, $n = 1, 2, \ldots$. The queue discipline is first-come-first-served. Let $(Y_n, n = 1, 2, \ldots)$ be an irreducible aperiodic Markov chain with finite state space $\{1, 2, \ldots, N\}$ and assume that for all choices of n, x, y and j

(1) $P(S_n \leq x, A_{n+1} \leq y, Y_{n+1} = j \mid Y_1,\ldots,Y_n,S_1,\ldots,S_{n-1},A_2,\ldots,A_n)$

$$= P(S_n \leq x \mid Y_n)P(A_{n+1} \leq y, Y_{n+1} = j \mid Y_n),$$

while the conditional probabilities in the right hand member of (1)
do not depend on n and

$$P(S_n \leq x \mid Y_n = i) = 1 - e^{-\lambda_i x}, \ x \geq 0, \ i = 1,2,\ldots,N,$$

with $\lambda_i > 0$.

Special cases of this model were studied by Çinlar (1967) and
Latouche (1983). References to other related papers can be found
in de Smit (1985).

We denote

$$A_{ij}(\phi) = E(\exp(-\phi A_{n+1})1(Y_{n+1} = j) \mid Y_n = i),$$

Re $\phi \geq 0$, $i = 1,2,\ldots,N$; $j = 1,2,\ldots,N$; where 1(B) is the indicator function of the event B, i.e. 1(B) = 1 if B occurs and
1(B) = 0 otherwise.

Let 1 be the N-dimensional column vector with all elements
equal to 1. $A(\phi)$ is the N×N-matrix with elements $A_{ij}(\phi)$. We
assume that there exists a $\delta > 0$ such that $A(\phi)$ can be continued
analytically to the region Re $\phi > -\delta$. P = A(0) is the transition
matrix of the Markov chain $(Y_n, n = 1,2,\ldots)$ and the row vector
π satisfying

$$\pi P = \pi$$

and

$$\pi \underline{1} = 1$$

is its stationary distribution. Moreover we introduce the N×N-
matrices

$$\Lambda = \mathrm{diag}(\lambda_1,\ldots,\lambda_N)$$

and

$$\bar{\alpha} = - A'(0).$$

The elements of $\bar{\alpha}$ are

$$\alpha_{ij} = - A'_{ij}(0) = E(A_{n+1} \, 1(Y_{n+1} = j) \mid Y_n = i),$$

$i = 1,\ldots,N; \quad j = 1,\ldots,N.$

We define the mean interarrival time

$$\alpha = \pi \bar{\alpha} \underline{1},$$

the mean service time

$$\beta = \pi \Lambda^{-1} \underline{1}$$

and the traffic intensity

$$\rho = \beta/\alpha.$$

In section 2 we consider a system of Wiener-Hopf-type equations which is solved using a factorization method as described in de Smit (1983). In section 3 we obtain explicit results for the distributions of the actual waiting time of the n-th customer W_n, the virtual waiting time at time t denoted by W_t^*, the number of customers in the system just before the arrival of the n-th customer denoted by C_n, the number of customers in the system at time t denoted by C_t^*, the number of customers during the first busy period M_1, the length of the first busy period P_1 and the length of the first busy cycle L_1. We shall denote $x^+ = \max(0,x)$, $x^- = \min(0,x)$, while δ_{ij} is Kronecker's symbol, i.e. $\delta_{ii} = 1$, and $\delta_{ij} = 0$ for $i \neq j$. I is the N×N-identity matrix.

2. A system of Wiener-Hopf-type equations

Let $U_n = S_1 + \ldots + S_n$, $n = 1,2,\ldots$; $U_0 = 0$; be the cumulative service time of the first n customers and assume that the system is initially empty, i.e. $W_1 = 0$. For $i = 1,2,\ldots,N$; $j = 1,2,\ldots,N$; $\text{Re } \phi \geqslant 0$, and $\text{Re } \eta \geqslant 0$, $\text{Re } \theta \geqslant 0$, $|r| < 1$, or $\text{Re } \eta \geqslant 0$, $\text{Re } \theta > 0$, $|r| \leqslant 1$, or $\text{Re } \eta > 0$, $\text{Re } \theta \geqslant 0$, $|r| \leqslant 1$, we define

$$Z_{ij}(r,\phi,\eta,\theta) = \sum_{n=1}^{\infty} r^n E(\exp(-\phi W_n - \eta T_n - \theta U_{n-1}) 1(Y_n = j) \mid Y_1 = i)$$

and

$$V_{ij}(r,\phi,\eta,\theta) = \sum_{n=1}^{\infty} r^{n+1} E((1 - \exp(\phi [W_n + S_n - A_{n+1}]^-))$$

$$\times \exp(-\eta T_{n+1} - \theta U_n) 1(Y_{n+1} = j) \mid Y_1 = i).$$

Let $Z(r,\phi,\eta,\theta)$ and $V(r,\phi,\eta,\theta)$ be the N×N-matrices with elements $Z_{ij}(r,\phi,\eta,\theta)$ and $V_{ij}(r,\phi,\eta,\theta)$, respectively.

It is easily shown (see de Smit, 1985) that Z and V satisfy the following system of Wiener-Hopf-type equations for $\mathrm{Re}\ \phi = 0$, and $\mathrm{Re}\ \eta \geqslant 0$, $\mathrm{Re}\ \theta \geqslant 0$, $|r| < 1$, or $\mathrm{Re}\ \eta \geqslant 0$, $\mathrm{Re}\ \theta > 0$, $|r| \leqslant 1$, or $\mathrm{Re}\ \eta > 0$, $\mathrm{Re}\ \theta \geqslant 0$, $|r| \leqslant 1$,

(2) $Z(r,\phi,\eta,\theta)H(r,\phi,\eta,\theta) = rI + V(r,-\phi,\eta,\theta),$

where

$$H(r,\phi,\eta,\theta) = ((\phi + \theta)I + \Lambda)^{-1} \Lambda M(r,\phi,\eta,\theta)$$

with

$$M(r,\phi,\eta,\theta) = (\phi + \theta)\Lambda^{-1} + I - rA(\eta - \phi).$$

In what follows we shall obtain from (2) an explicit expression for Z using a factorization method.   To this end we first find a factorization of H, i.e. we determine two matrices H^+ and H^- such that for $\mathrm{Re}\ \phi = 0$,

(3) $H(r,\phi,\eta,\theta) = H^+(r,\phi,\eta,\theta)H^-(r,\phi,\eta,\theta)$

and

(4)   for $\mathrm{Re}\ \phi > 0$ $H^+(r,\phi,\eta,\theta)$ is analytic and for $\mathrm{Re}\ \phi \geqslant 0$ it is bounded and continuous and its inverse exists and is also bounded;

(5)   for $\mathrm{Re}\ \phi < 0$ $H^-(r,\phi,\eta,\theta)$ is analytic and for $\mathrm{Re}\ \phi \leqslant 0$ it is bounded and continuous and its inverse exists and is also bounded.

If we can determine H^+ we have the following solution for Z (see de Smit, 1985).

<u>Theorem 1</u>

If a factorization according to (3)-(5) exists, then we have for $\text{Re } \phi \geqslant 0$ and $\text{Re } \eta \geqslant 0$, $\text{Re } \theta \geqslant 0$, $|r| < 1$, or $\text{Re } \eta \geqslant 0$, $\text{Re } \theta > 0$, $|r| \leqslant 1$, or $\text{Re } \eta > 0$, $\text{Re } \theta \geqslant 0$, $|r| \leqslant 1$,

$$(6) \quad Z(r,\phi,\eta,\theta) = r\,[H(r,0,\eta,\theta)]^{-1}H^+(r,0,\eta,\theta)\,[H^+(r,\phi,\eta,\theta)]^{-1}.$$

In order to find the factorization we prove an auxiliary lemma.

<u>Lemma 1</u>

For   (i) $|r| < 1$, $\text{Re } \eta \geqslant 0$, $\text{Re } \theta \geqslant 0$,

or   (ii) $|r| \leqslant 1$, $\text{Re } \eta > 0$, $\text{Re } \theta \geqslant 0$,

or   (iii) $|r| \leqslant 1$, $\text{Re } \eta \geqslant 0$, $\text{Re } \theta > 0$,

or   (iv) $r = 1$, $\eta = 0$, $\theta = 0$ and $\rho < 1$,

$\det M(r,\phi,\eta,\theta)$ (considered as a function of ϕ) has exactly N zeros in the left half plane $\text{Re } \phi < 0$;   in case (iv) $\det M(1,\phi,0,0)$ equals zero at $\phi = 0$ and it has no zeros elsewhere on the imaginary axis.

<u>Proof</u>

For $d > 0$ let C_d be the closed contour consisting of the part of the imaginary axis running from $-\,id$ to id and the half circle with centre at the origin and radius d running anti-clockwise from id to $-\,id$. Let $d_0 = 2 \max\limits_i \lambda_i$ then in the cases (i)-(iii) we have on C_d for $d > d_0$

$$|(\phi + \theta)\lambda_i^{-1} + 1| > |r| \sum_{j=1}^{N} |A_{ij}(\eta - \phi)|,$$

so that from Theorem 1 of Appendix 2 of de Smit (1983) it follows that the total order of the zeros of $\det M(r,\phi,\eta,\theta)$ inside C_d is the same as the total order of the zeros of $\det(\phi + \theta)\Lambda^{-1}$

inside C_d. (Note that these determinants have no poles inside C_d). In case (iv) the above inequality still holds on C_d except at $\phi = 0$; while det $M(1,0,0,0) = 0$. This implies the last statement of the lemma (see Lemma 1 of Appendix 2 of de Smit, 1983). It remains to prove that in case (iv) det $M(r,\phi,0,0)$ has exactly N zeros in the left half plane Re $\phi < 0$. The matrix $M^*(r,\phi)$ is obtained from $M(r,\phi,0,0)$ by adding all columns of $M(r,\phi,0,0)$ to its first column, i.e.

$$M^*_{i1}(r,\phi) = \phi\lambda_i^{-1} + 1 - r \sum_{j=1}^{N} A_{ij}(-\phi), \quad i = 1,2,\ldots,N;$$

and

$$M^*_{ij}(r,\phi) = M_{ij}(r,\phi,0,0), \quad i = 1,2,\ldots,N; \quad j = 2,\ldots,N.$$

Let $\mu(r)$ be the function implicitly defined by

$$(7) \quad \det M(r,\mu(r),0,0) = \det M^*(r,\mu(r)) = 0,$$

and $\mu(1) = 0$. With the implicit function theorem we shall prove that there is a neighbourhood of 1 in which the function $\mu(r)$ is uniquely defined and continuously differentiable. Since we have assumed that there exists a $\delta > 0$ such that $A(\phi)$ can be continued analytically to the region Re $\phi > -\delta$, there exists in $\mathbb{R} \times \mathbb{C}$ a neighbourhood of $(1,0)$ in which det $M^*(r,\phi)$ is continuously differentiable with respect to both r and ϕ. Moreover, det $M^*(1,0) = 0$ and

$$M^*_{i1}(1,\phi) = \phi(\lambda_i^{-1} - \sum_{j=1}^{N} \alpha_{ij}) + o(\phi), \quad \phi \to 0,$$

so that

$$(8) \quad \det M^*(1,\phi) = \phi \det M^* + o(\phi), \quad \phi \to 0,$$

where

$$M^*_{i1} = \lambda_i^{-1} - \sum_{j=1}^{N} \alpha_{ij}, \quad i = 1,2,\ldots,N;$$

and

$$M^*_{ij} = \delta_{ij} - P_{ij}, \quad i = 1,2,\ldots,N; \quad j = 2,\ldots,N.$$

Since $\pi(I - P) = 0$, we have

$$(9) \quad \pi M^* = (\beta - \alpha)e_1,$$

where $e_1 = (1,0,\ldots,0)$. Since $I - P$ has rank $N - 1$ and $\beta < \alpha$, π is the only solution of (9) and hence $\det M^* \neq 0$. With (8) it follows that

$$\frac{\partial}{\partial \phi} \det M^*(1,\phi) = \det M^* \neq 0,$$

so that all conditions of the implicit function theorem are satisfied and consequently in a neighbourhood of 1 the function $\mu(r)$ is uniquely defined and continuously differentiable. For r close to 1 we thus have

$$\mu(r) = - (1 - r)\mu'(1) + o(1 - r), \quad r \to 1,$$

and

$$M_{i1}^*(r,\mu(r)) = - (1 - r)\left[\mu'(1)(\lambda_i^{-1} - \sum_{j=1}^{N} \alpha_{ij}) - 1\right] + o(1 - r),$$
$$r \to 1.$$

Since $\det M^*(r,\mu(r)) = 0$, we have

$$0 = \frac{d}{dr} \det M^*(r,\mu(r))\Big|_{r=1} = \det M^\circ$$

where

$$(10) \quad M_{i1}^\circ = \mu'(1)\left(\lambda_i^{-1} - \sum_{j=1}^{N} \alpha_{ij}\right) - 1$$

and

$$M_{ij}^\circ = \delta_{ij} - P_{ij}, \quad i = 1,2,\ldots,N; \quad j = 2,\ldots,N.$$

Since $\det M^\circ = 0$ we must have

$$\pi M^\circ = 0$$

and in particular

$$\sum_{i=1}^{N} \pi_i M_{i1}^\circ = 0$$

or with (10)

$$\mu'(1) = \frac{1}{\beta - \alpha} = - \frac{1}{\alpha(1 - \rho)} \, ,$$

so that

(11) $\quad \mu(r) = (1 - r)/(\alpha(1 - \rho)) + o(1 - r), \ r \to 1.$

The inequalities

$$\left| \phi \lambda_i^{-1} + 1 \right| > |r| \sum_{j=1}^{N} \left| A_{ij}(- \phi) \right|,$$

$i = 1, 2, \ldots, N;$ hold for $|r| < 1$, Re $\phi = 0$, and for $r = 1$, Re $\phi = 0$, $\phi \neq 0$. This implies (see Lemma 1 of Appendix 2 of de Smit, 1983) that for $|r| < 1$ det $M(r, \phi, 0, 0)$ has no zeros on the imaginary axis except at 0.

For $r < 1$ (case (i)) there are exactly N zeros in the left half plane Re $\phi < 0$. These zeros are continuous functions of r. From (11) we see that for $\rho < 1$ and $r < 1$ and sufficiently close to 1 $\mu(r)$ lies in the right half plane. Since det $M(1, \phi, 0, 0)$ has no zeros on the imaginary axis except at 0, we see that for $r \uparrow 1$ the zero $\mu(r)$ is the only zero which tends to the imaginary axis. As a consequence the N zeros of det $M(r, \phi, 0, 0)$, which for $r < 1$ lie in the left half plane Re $\phi < 0$, will remain there for $r = 1$. This completes the proof. $\square$

We denote the zeros in the left half plane Re $\phi < 0$ of det $M(r, \phi, \eta, \theta)$ by $\mu_1(r, \eta, \theta), \ldots, \mu_N(r, \eta, \theta)$ and assume that the following condition is satisfied.

Condition 1

$\mu_1(r, \eta, \theta), \ldots, \mu_N(r, \eta, \theta)$ are all distinct and distinct from $- (\theta + \lambda_i)$, $i = 1, \ldots, N$.

As in de Smit (1983) we see that this condition will always be satisfied if $\lambda_1, \ldots, \lambda_N$ are distinct except possibly for countably many values of one of the variables r, η and θ if the two other ones are fixed. From the definition of H it follows that

$\mu_1(r,\eta,\theta),\ldots,\mu_N(r,\eta,\theta)$ are the zeros in the left half plane of det $H(r,\phi,\eta,\theta)$. For $k = 1,2,\ldots,N$; let B_k be a (non-unique) non-zero row vector satisfying

$$B_k H(r,\mu_k(r,\eta,\theta),\eta,\theta) = 0,$$

and let B be the $N\times N$-matrix the k-th row of which is B_k. Moreover let

$$D = \text{diag}(\mu_1(r,\eta,\theta),\ldots,\mu_N(r,\eta,\theta)).$$

The $N\times N$-matrix L is defined by

$$L_{ij} = \frac{B_{ij}}{\theta + \lambda_j + \mu_i(r,\eta,\theta)} \ , \quad i = 1,\ldots,N; \quad j = 1,\ldots,N;$$

and we impose the following condition.

Condition 2
―――――――――

   Det $L \neq 0$.

The remark made after Condition 1 also applies here.
Define

$$C = L^{-1},$$

and

$$K(r,\phi,\eta,\theta) = I + C(\phi I - D)^{-1}B.$$

Note that the matrices B, D, L and C depend on r, η and θ. For convenience we have omitted these arguments.

   We next give the following factorization theorem.

Theorem 2
―――――――――

If Conditions 1 and 2 hold then

(i) $\det K(r,\phi,\eta,\theta) = \prod_{i=1}^{N}\left[\dfrac{\theta + \lambda_i + \theta}{\phi - \mu_i(r,\eta,\theta)}\right],$

(ii) for Re $\phi = 0$,

$$H(r,\phi,\eta,\theta) = H^{+}(r,\phi,\eta,\theta)H^{-}(r,\phi,\eta,\theta),$$

where

(a) $H^+(r,\phi,\eta,\theta) = K^{-1}(r,\phi,\eta,\theta)$ satisfies (4),

and

(b) $H^-(r,\phi,\eta,\theta) = K(r,\phi,\eta,\theta)H(r,\phi,\eta,\theta)$ satisfies (5).

Proof

We easily see (cf. the proof of Theorem 3.1 in Regterschot and de Smit, 1985) that

$$(12) \quad \det K(r,\phi,\eta,\theta) = g(r,\phi,\eta,\theta)/\prod_{i=1}^{N} (\phi - \mu_i(r,\eta,\theta))$$

where $g(r,\phi,\eta,\theta)$ is a polynomial of ϕ of degree N, and

$$(13) \quad \det H(r,\phi,\eta,\theta) = f(r,\phi,\eta,\theta) \prod_{i=1}^{N} \left[\frac{\phi - \mu_i(r,\eta,\theta)}{\phi + \lambda_i + \theta} \right],$$

where $f(r,\phi,\eta,\theta)$ is bounded and bounded away from 0 for $\text{Re } \phi \leqslant 0$. Since

$$\lim_{\phi \to -\lambda_k - \theta} (\phi + \lambda_k + \theta)H_{ij}(r,\phi,\eta,\theta) = - r\delta_{ik}\lambda_k A_{kj}(\theta + \lambda_k + \eta)$$

we have

$$\lim_{\phi \to -\lambda_k - \theta} (\phi + \lambda_k + \theta)H^-_{ij}(r,\phi,\eta,\theta)$$

$$= - r\lambda_k A_{kj}(\theta + \lambda_k + \eta) \left[\delta_{ik} - \sum_{\ell=1}^{N} C_{i\ell} \frac{B_{\ell k}}{\lambda_k + \theta + \mu_\ell(r,\eta,\theta)} \right] = 0,$$

if $\lambda_1,\ldots,\lambda_N$ are distinct. We see that the conclusion still holds if some of the λ_k coincide. Hence $H^-(r,\phi,\eta,\theta)$ has no poles at $\phi = -\lambda_k - \theta$, $k = 1,\ldots,N$. From (12) and (13) we have

$$\det H^-(r,\phi,\eta,\theta) = g(r,\phi,\eta,\theta)f(r,\phi,\eta,\theta)/\prod_{k=1}^{N} (\phi + \lambda_k + \theta)$$

and consequently $g(r,\phi,\eta,\theta) = \prod_{k=1}^{N} (\phi + \lambda_k + \theta)$, which proves part (i) and (ii) (a) of the theorem. In order to prove part (ii) (b) it remains to check that $H^-(r,\phi,\eta,\theta)$ has no poles at $\mu_k(r,\eta,\theta)$, $k = 1,2,\ldots,N$. Since

$$H^-_{ij}(r,\phi,\eta,\theta) = H_{ij}(r,\phi,\eta,\theta)$$
$$+ \sum_{\ell=1}^{N} \frac{1}{\phi - \mu_\ell(r,\eta,\theta)} C_{i\ell} \sum_{m=1}^{N} B_{\ell m} H_{mj}(r,\phi,\eta,\theta)$$

we have

$$\lim_{\phi \to \mu_k(r,\eta,\theta)} (\phi - \mu_k(r,\eta,\theta)) H^-_{ij}(r,\phi,\eta,\theta)$$

$$= C_{ik} \sum_{m=1}^{N} B_{km} H_{mj}(r,\mu_k(r,\eta,\theta),\eta,\theta) = 0.$$

This completes the proof. $\square$

From Theorems 1 and 2 we now have the following solution of (2).

Theorem 3

For $\mathrm{Re}\ \phi \geqslant 0$ and $\mathrm{Re}\ \eta \geqslant 0$, $\mathrm{Re}\ \theta \geqslant 0$, $|r| < 1$, or $\mathrm{Re}\ \eta \geqslant 0$, $\mathrm{Re}\ \theta > 0$, $|r| \leqslant 1$, or $\mathrm{Re}\ \eta > 0$, $\mathrm{Re}\ \theta \geqslant 0$, $|r| \leqslant 1$, we have

$$(14)\quad Z(r,\phi,\eta,\theta) = r\,[H(r,0,\eta,\theta)]^{-1}\,[K(r,0,\eta,\theta)]^{-1} K(r,\phi,\eta,\theta),$$

provided Conditions 1 and 2 are satisfied.

3. Waiting time, queue length and busy period

Iff $\rho < 1$ the pair (W_n,Y_n) converges for $n \to \infty$ weakly to a random vector (W,Y) and the pair (C_n,Y_n) converges to a random vector (C,Y). If moreover the marginal distribution of A_n is non-arithmetic then W^*_t converges for $t \to \infty$ to a random variable W^* and C^*_t converges to a random variable C^* (see de Smit, 1985). By inserting (14) into the general results in de Smit (1985) we obtain explicit expressions for the distributions of (W,Y), (C,Y), W^* and C^*. The calculations are straightforward and the results are given below. We denote $\mu_k = \mu_k(1,0,0)$,

$k = 1,\ldots,N$; while I_k is the $N{\times}N$-matrix in which the k-th diagonal element equals 1 while all other elements equal 0.

Actual waiting time

Assume $\rho < 1$ and denote

$$Z_j(\phi) = E(\exp(-\phi W)1(Y = j)).$$

while $Z(\phi) = (Z_1(\phi),\ldots,Z_N(\phi))$, then for $\mathrm{Re}\ \phi \geqslant 0$,

$$(15)\quad Z(\phi) = \pi K^{-1}(1,0,0,0)K(1,\phi,0,0).$$

Virtual waiting time

Assume $\rho < 1$ and A_n non-arithmetic then for $\mathrm{Re}\ \phi \geqslant 0$,

$$(16)\quad E(\exp(-\phi W^*)) = 1 - \rho + \frac{1}{\alpha} Z(\phi)(\phi I + \Lambda)^{-1}\underline{1}.$$

Number of customers in the system at arrival epochs

Let $\rho < 1$ and define

$$R_j(s) = E(s^C 1(Y = j))$$

and $R(s) = (R_1(s),\ldots,R_N(s))$, then for $|s| < 1$,

$$(17)\quad R(s) = \pi + \pi K^{-1}(1,0,0,0)(1 - s)$$

$$\cdot \sum_{k=1}^{N} \left[\frac{1}{\mu_k} CI_k B(\mu_k I + \Lambda)^{-1}\Lambda A(-\mu_k)\{I - sA(-\mu_k)\}^{-1} \right.$$

$$\left. - K(1,-\lambda_k,0,0)I_k A(\lambda_k)\{I - sA(\lambda_k)\}^{-1} \right].$$

Number of customers in the system in continuous time

For $\rho < 1$ and A_n non-arithmetic we have for $|s| < 1$,

$$(18)\quad E\!\left(s^{C^*}\right) = 1 + \pi K^{-1}(1,0,0,0)(1 - s)$$

$$\cdot \frac{1}{\alpha}\left[\sum_{k=1}^{N} \frac{1}{\mu_k^2} CI_k B(\mu_k I + \Lambda)^{-1}\Lambda\{A(-\mu_k) - I\}\{I - sA(-\mu_k)\}^{-1} \right.$$

$$\left. + \frac{1}{\lambda_k} K(1,-\lambda_k,0,0)I_k\{A(\lambda_k) - I\}\{I - sA(\lambda_k)\}^{-1} \right]\underline{1}.$$

<u>Busy period</u>

Define for $i = 1,\ldots,N;$ $j = 1,\ldots,N;$

$$D_{ij}(r,\eta,\theta) = E(r^{M_1}\exp(-\eta L_1 - \theta P_1)1(Y_{M_1+1} = j) \mid Y_1 = i),$$

and let $D(r,\eta,\theta)$ be the N×N-matrix with elements $D_{ij}(r,\eta,\theta)$.
Then for $|r| < 1$, Re $\eta \geqslant 0$, Re $\theta \geqslant 0$, or $|r| \leqslant 1$, Re $\eta > 0$,
Re $\theta \geqslant 0$, or $|r| \leqslant 1$, Re $\eta \geqslant 0$, Re $\theta > 0$,

(19) $D(r,\eta,\theta) = I - K(r,0,\eta,\theta)H(r,0,\eta,\theta).$

Expressions (15)-(19) generalize well-known results for the ordinary
GI$|$M$|$1 queue. We see that the stationary waiting time distributions
are mixtures of exponentials and the stationary queue length distributions are mixtures of geometric distributions. As in de Smit
(1983) and Regterschot and de Smit (1985), the derivations imply
a numerical algorithm for calculating the parameters of these stationary distributions. It should be noted that the above results
are valid only if Conditions 1 and 2 are satisfied. In particular
they should hold for $r = 1$, $\eta = 0$ and $\theta = 0$, in order that
expressions (15)-(18) be valid. We have studied some particular
cases numerically and it turned out that in those cases the conditions were satisfied. These results will be reported elsewhere.
We conjecture that the conditions are always satisfied, although
we have not been able to prove it.

<u>References</u>

Arjas, E. (1972). On the use of a fundamental identity in the
     theory of semi-Markov queues. Adv.Appl.Prob. <u>4</u>, 271-284.
Çinlar, E. (1967). Queues with semi-Markov arrivals. J.Appl.Prob.
     <u>4</u>, 365-379.
de Smit, J.H.A. (1983). The queue GI$|$M$|$s with customers of dif-
     ferent types or the queue GI$|$H$_m|$s. Adv.Appl.Prob. <u>15</u>, 392-419.

de Smit, J.H.A. (1985). The single server semi-Markov queue.
    To appear in Stochastic Process.Appl.

Latouche, G. (1983). An exponential semi-Markov process with appli-
    cations to queueing theory. Report Laboratoire d'Informatique
    Théorique, Université Libre de Bruxelles.

Regterschot, G.J.K. and J.H.A. de Smit (1985). The queue $M|G|1$
    with Markov modulated arrivals and services. To appear in
    Math.Oper.Res.

SECTION VIII. BRANCHING
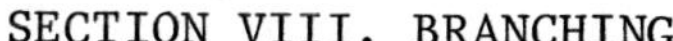

ESTIMATION THEORY FOR MULTITYPE BRANCHING PROCESSES

Søren Asmussen

Institute of Mathematical Statistics

University of Copenhagen

1. Introduction

We consider a p-type Galton-Watson process $\{Z_n\}_{n\in\mathbb{N}}$, i.e.
$Z_n = (Z_n(1)\ldots Z_n(p))$,

$$Z_{n+1} = \sum_{i=1}^{p} \sum_{k=1}^{Z_n(i)} Z_{n,k}^{(i)}$$

where the $Z_{n,k}^{(i)} \in \mathbb{N}^p$ are independent for all n, i, k and with
the same distribution for any fixed i. The interpretation of
$Z_{n,k}^{(i)}$ is as the offspring vector produced by the kth individual
of type i in the nth generation.

This model may be argued to have a prominent rôle among the
many generalizations and modifications of the usual (one-type)
Galton-Watson process : not only are the relevant examples abundant,
but also (e.g. Asmussen and Hering, 1983, Ch.I) a suitable interpretation of the types most often makes it possible to use the model
as approximation to a more general branching process. Detailed investigations into the multitype Galton-Watson process are therefore highly
useful for assessing conjectures for more general processes, and

moreover is the setting particular well behaved from the mathematical point of view. This is due to the fact that the behaviour of the mean, a fundamental quantity in branching processes, can be completely described by using standard matrix algebra and Perron-Frobenius theory.

We let P_i, E_i etc. refer to the case of an initial ancestor or type i and denote by $M = (m(i,j))$ the offspring mean matrix,

$$m(i,j) = E_i Z_1(j) = EZ_{n,k}^{(i)}(j).$$

We assume M to be positively regular with Perron-Frobenius root $\rho > 1$ and let u, v be the corresponding right and left eigenvectors[1] normalized by

$$vu' = \sum_{i=1}^{p} u(i)v(i) = 1, \quad |v| = v1' = \sum_{i=1}^{p} v(i) = 1$$

Also all second moments $E_i Z_1(j)^2$ are assumed to be finite.

As is well-known, $\{W_N\} = \{\rho^{-N} Z_N u'\}$ is then a non-negative martingale with $W = \lim W_N$ satisfying $E_i W = u(i)$. For this reason, $u(i)$ is often in demographic terms said to represent the reproductive value of type i. Also $\rho^{-n} Z_n \to Wv$ so that v is the limiting stable typevector of the population. For simplicity, we assume throughout that the extinction probability is zero. Then $0 < W < \infty$ a.s.

In the present note, we shall survey the problems concerning the estimation of the growth rate ρ (and to some extent also other relevant parameters) given that our observations are based upon generations $0,\dots,N$. We first briefly recall some classical theory for the one-type case, proceed to give some main features of the difficulties connected with the case $p > 1$, and conclude with a discussion incorporating what the author sees as the main open problems in the area.

---

1) all vectors being row vectors, this means $Mu' = \rho u'$, $vM = \rho v$. The symbol 1 is used both for the number one and the vector $(1\dots1)$.

2. Estimating the offspring mean of a Galton-Watson process

For $p = 1$, all vectors (e.g. Z_n) reduce to real numbers and we may suppress i, j in the notation (e.g. $Z_{n,k}$ is the number of children of the kth individual in the nth generation).

From the limit result $\rho^{-n}Z_n \to W$ a.s. it is straightforward to give estimators $\hat{\rho}^{1)}$ for ρ which are based on observations from the N first generations only and are strongly consistent as $N \to \infty$, $\hat{\rho} \to \rho$ a.s. An apparent possibility is the so-called <u>Lotka-Nagaev estimator</u> $\hat{\rho}_{LN} = Z_N/Z_{N-1}$. However, some further reflection indicates that since our concern is to estimate the mean ρ of the $Z_{n,k}$, it would be natural to consider the empirical mean. Now exactly the $Z_{n,k}$ with $n \leqslant N - 1$ depend on the N first generations only. Their total number is $\Sigma_0^{N-1} Z_n$ and the empirical mean is thus

$$
(1) \quad \frac{\displaystyle\sum_{n=0}^{N-1} \sum_{k=1}^{Z_n} Z_{n,k}}{\displaystyle\sum_{n=0}^{N-1} Z_n} = \frac{Z_1 + \ldots + Z_N}{Z_0 + \ldots + Z_{N-1}}
$$

Though the setting is not that of traditional parametric statistics with fixed sample size, it may in fact be proved that the estimation (1) is in some appropriate sense the maximum likelihood estimator, based upon the sampling scheme where the observations are $\{Z_{n,k} : n \leqslant N - 1, k \leqslant Z_n\}$ and also upon the more restricted one where we observe only $Z_0,\ldots,Z_n$. See Dion and Keiding (1978) and also Keiding and Lauritzen (1978). For this reason, we denote the estimator (1) by $\hat{\rho}_{ML}$. What regards asymptotic properties, it is

---

1) The notation $\hat{\rho}$ is used for <u>any</u> estimator for ρ, not only as frequently in the statistical literature for the maximum likelihood estimator $\hat{\rho}_{ML}$.

immediate from the definitions that $\hat{\rho}_{LN}$ and $\hat{\rho}_{ML}$ are both strongly consistent. They are also asymptotically normal in the sense
that in the limit $N \to \infty$ the distributions of

$$(W\rho^N)^{\frac{1}{2}}(\hat{\rho}_{LN} - \rho), \quad (W\rho^N)^{\frac{1}{2}}(\hat{\rho}_{ML} - \rho)$$

become normal (the limiting variances σ^2_{LN}, σ^2_{ML}, say, are easily
computed in terms of $\sigma^2 = \mathrm{Var}\ Z_{n,k}$, but need not concern us here;
as expected, $\sigma^2_{LN} > \sigma^2_{ML}$).

3. Estimators and sampling schemes for the multitype case

The interpretation of (1) as an empirical mean immediately
generalizes to the case $p > 1$ and suggests that

$$(2) \quad \hat{m}_{ML}(i,j) = \frac{\displaystyle\sum_{n=0}^{N-1} \sum_{k=1}^{Z_n(i)} z_{n,k}^{(i)}(j)}{\displaystyle\sum_{n=0}^{N-1} Z_n(i)}$$

is the maximum likelihood estimator of $m(i,j)$ and $\hat{M}_{ML} = (\hat{m}_{ML}(i,j))$
that of M. Indeed this is so. Therefore ρ, being a functional
of M, can be maximum likelihood estimated as the same functional
of $\hat{M}_{ML}$. I.e. $\hat{\rho}_{ML}$ is the Person-Frobenius root of $\hat{M}_{ML}$ (which
exist for n large enough). Also the asymptotic properties carry
over from the case $p = 1$: one has $\hat{\rho}_{ML} \to \rho$ a.s. and also
$(W\rho^N) (\hat{\rho}_{ML} - \rho)$ is asymptotically normal as $N \to \infty$, see Asmussen
and Keiding (1978).

This may look like a complete solution of the multitype case.
However, the difficulty is that the sampling scheme seems unrealis-
tically detailed in many cases and that therefore $\hat{M}_{ML}$ and $\hat{\rho}_{ML}$
are not observable. Therefore alternatives need to be discussed,
and we shall consider amongst other the following possibilities :

Full information sampling

$\hat{M}_{ML}$ and $\hat{\rho}_{ML}$ are observable. Sufficient conditions are that either
all $Z_{n,k}^{(i)}(j)$ with $n \leqslant N - 1$ are observable, or that the numbers

$$Z_n(i)$$
$$\sum_{k=1}^{Z_n(i)} Z_{n,k}^{(i)}(j)$$

of type j children produces by type i parents in generations
$n = 0,\ldots,N - 1$ are so.

Type vector sampling

The observations are the vectors $Z_0,\ldots,Z_n$, i.e. the $Z_n(i)$ with
$i = 1,\ldots,p,\ n = 0,\ldots,N$.

Sampling based on total generation sizes only

The observations are $|Z_0|,\ldots,|Z_N|$.

It was first argued by Becker (1977) that the last possibility
is of relevance, and he suggested the estimator

$$\hat{\rho}_B = \frac{|Z_1| + \ldots + |Z_N|}{|Z_0| + \ldots + |Z_{N-1}|}$$

which reduces to $\hat{\rho}_{ML}$ for $p = 1$. As noted in Becker (1977), it
is trivial from $|Z_N| \cong \rho^N W$ that $\hat{\rho}_B \to \rho$ a.s. as $N \to \infty$. However,
more refined asymptotic properties are not straigthformard to derive.
From the limiting behaviour of $\hat{\rho}_{ML}$ one hopes for the efficiency
of an estimator $\hat{\rho}$ to be comparable to that of $\hat{\rho}_{ML}$ in the sense
that

$$\text{(CE)} \quad (W\rho^N)^{\frac{1}{2}}(\hat{\rho} - \rho) \xrightarrow{D} N(0,\sigma^2) \quad \text{for some} \quad \sigma^2 = \sigma^2(\hat{\rho}),$$

However, it was found in Asmussen and Keiding (1978) that $\hat{\rho}_B$ as
well as other relevant estimators do not in general possess property
(CE). As will next be explained, these phenomena are closely related to the more refined limit theory of the multitype Galton-Watson
process, (Kesten and Stigum, 1966; Athreya, 1969 and 1971;
Asmussen, 1977).

4. The asymptotic properties of $\hat{\rho}_B$ and $\hat{\rho}_{LN}$

It will now for a while be convenient to specialize to the case $p = 2$. This is due to the fact that the spectral properties of M, though in principle determined by the Jordan canonical form also for $p > 2$, become much simpler for $p = 2$ where M can be given diagonal form

$$M = \rho u \otimes v + \rho_1 u_1 \otimes v_1 = \rho u'v + \rho_1 u_1' v_1$$

with ρ_1 real with $|\rho_1| < \rho$ and $Mu_1' = \rho_1 u_1'$, $v_1 M = \rho_1 v_1$, $vu_1' = v_1 u' = 0$, $v_1 u_1' = 1$.

Now any vector z can in a unique way be written as $z = \alpha v + \alpha_1 v_1$ where $\alpha = zu'$, $\alpha_1 = zu_1'$. Taking $z = Z_N$, we have $\alpha \simeq \rho^N W$ and it is seen that the rate of convergence to the stable type vector v is given by the magnitude of $\alpha_1 = Z_N u_1'$. The behaviour of this linear functional is motivated by the observation that $\{\rho_1^{-1} Z_N u_1'\}$ is a martingale with

$$(3a) \quad \text{Var } \rho_1^{-N} Z_N u_1' \simeq \left(\frac{\rho}{\rho_1^2}\right)^N \tau^2 \to \infty, \quad \rho_1^2 < \rho,$$

$$(3b) \quad \text{Var } \rho_1^{-N} Z_N u_1' \simeq N\tau^2 \to \infty, \qquad \rho_1^2 = \rho,$$

$$(3c) \quad \text{Var } \rho_1^{-N} Z_N u_1' \to \tau^2, \qquad\qquad \rho_1^2 > \rho,$$

for a suitable constant τ^2. Thus martingale theory suggests a.s. convergence for $\rho_1^2 > \rho$ and a CLT (possibly with a random normalization) for $\rho_1^2 \leq \rho$, and in fact

$$(4a) \quad \frac{Z_N u_1'}{(W\rho^N)^{\frac{1}{2}}} \xrightarrow{D} N(0, \tau^2), \qquad \rho_1^2 < \rho,$$

$$(4b) \quad \frac{Z_N u_1'}{(WN\rho^N)^{\frac{1}{2}}} \xrightarrow{D} N(0, \tau^2), \qquad \rho_1^2 = \rho,$$

$$(4c) \quad \frac{Z_N u_1'}{\rho_1^N} \xrightarrow{a.s.} W_1, \qquad \rho_1^2 > \rho.$$

Here (4c) is immediate from L^2-theory for martingales, whereas for $\rho_1^2 \leqslant \rho$ there are some difficulties in a direct application of the martingale CLT. A suitable tool is provided by Th.2.2 of Asmussen and Keiding (1978), which gives conditions on a sequence $\{a_n\}$ of vectors which are sufficient (and close to necessary) for the asymptotic normality of

$$(5) \quad \sum_{n=0}^{N-1} \{Z_{n+1} - Z_n M\} a'_{N-n}$$

For (4a-b) one then just expands $Z_n u'_1$ in a telecosping series corresponding to $a_n = \rho_2^{n-1} u_1$.

To see that these results are highly relevant for determining the asymptotic properties of estimators of ρ, we shall consider $\hat{\rho}_{LN} = |Z_N|/|Z_{N-1}|$. Assuming that (as will typically be the case), u is not proportional to 1, we may normalize by $u + u_1 = 1$ and get

$$\begin{aligned}
|Z_{N-1}|(\hat{\rho}_{LN} - \rho) &= |Z_N| - \rho|Z_{N-1}| = Z_N 1' - \rho Z_{N-1} 1' \\[2mm]
&= (Z_N - Z_{N-1}M)1' + Z_{N-1}(M1' - 1') \\[2mm]
&= (Z_N - Z_{N-1}M)1' + (\rho_1 - \rho)Z_{N-1}u' \\[2mm]
&= S_N + T_N \quad \text{(say)}.
\end{aligned}$$

Here S_N is of the form (5) with $a_1 = 1'$, $a_n = 0$ $n > 1$, and Asmussen and Keiding (1978) yield the CLT for $S_N/(W\rho^N)^{\frac{1}{2}}$. Thus comparing with (4) we see that T_N dominates S_N if $\rho_1^2 \geqslant \rho$, and from $|Z_{N-1}| \cong \rho^{N-1}W$ we get

$$(6a) \quad \left(\frac{W\rho^N}{N}\right)^{\frac{1}{2}} (\hat{\rho}_{LN} - \rho) \overset{D}{\to} N(0, \rho(\rho_1 - \rho)^2 \tau^2), \quad \rho_1^2 = \rho,$$

$$(6b) \quad \left(\frac{\rho}{\rho_1}\right)^N (\hat{\rho}_{LN} - \rho) \overset{a.s.}{\to} \frac{\rho(\rho_1 - \rho)W_1}{\rho_1 W}$$

For $\rho_1^2 < \rho$, S_N and T_N are of the same magnitude. However, $S_N + T_N$ can (up to a constant term) be expanded on the form (5) and Asmussen and Keiding (1978) yield

(6c) $(W\rho^N)^{\frac{1}{2}}(\hat{\rho}_{LN} - \rho) \xrightarrow{D} N(0,\omega^2)$ for somme ω^2.

I.e., the desired conclusion (CE) holds only if $\rho_1^2 < \rho$, whereas
otherwise the rate of convergence, as given by (6a-b), is slower.
A not too difficult variant of the analysis yields a trichotomy of
just the same form for $\hat{\rho}_B$, see Asmussen and Keiding (1978).

5. Further work and open problems

$1°$ Type-vector sampling seems in many cases to be the natural
possibility. Of course, estimates based on total generation sizes
can still be used but better ones should exist. No results in this
direction are known (the author does not follow the analysis of
Khan and Rehman, 1980).

$2°$ A similar trichotomy as for the limiting behaviour of $\hat{\rho}_B$,
$\hat{\rho}_{LN}$ was found in Heyde (1981). However, the model there can be
reduced to the multitype Galton-Watson process, see Asmussen (1982).

$3°$ For sampling based on total generations sizes only, the
estimates $\hat{\rho}_B$, $\hat{\rho}_{LN}$ appear only to be crude first guesses, and an
apparent drawback is their poor asymptotic properties for $\rho_1^2 \geqslant \rho$.
It seems reasonable to ask whether better estimates with the property (CE) can be found. A partial result in this direction was
given in Asmussen (1982) for the case $p = 2$, $\rho_1^2 > \rho$. The idea is
simply first to estimate the variation caused by $Z_n u_1'$ and next to
eliminate it for the estimation of ρ. The resulting estimator is
given as the largest solution of the quadratic

(7) $(|z_{N-2}|^2 - |z_{N-3}||z_{N-1}|)\eta^2 + (|z_{N-3}||z_N| - |z_{N-2}||z_{N-1}|)\eta$

$\qquad + (|z_{N-1}|^2 - |z_{N-2}||z_N|) = 0$

and has property (CE). A main deficit is, however, the lack of robustness properties. It would seem reasonable from that point of view to require that any estimator of ρ should yield reasonable results for any sequence of observations which grows asymptotically exponentially. This is not the case here since when $|Z_N| = W\rho^N$ exactly, the coefficients in (7) vanish. Therefore the results of Asmussen (1982) are presumably not more than a first opening in this direction.

$4°$ What can be said about estimation of other functionals of M ? For full information sampling, M can be maximum likelihood estimated and standard transformation techniques apply just as for ρ, but in the case of less information it would not seem conceivable that all functionals are estimable. Particular important cases are :

$4a°$ The second largest eigenvalue ρ_1. For example, we need to know ρ_1 (or more precisely whether $\rho_1^2 \gtrless \rho$ in order to use the estimate (6) or to use the asymptotic results (6a-c) for $\hat{\rho}_B$, $\hat{\rho}_{LN}$;

$4b°$ The reproductive value vector u, important not only because of its intrinsic interest but also since if u is known or can be estimated, the estimator

$$(8) \quad \hat{\rho}_u = \frac{Z_1 u' + \ldots + Z_N u'}{Z_0 u' + \ldots + Z_{N-1} u'}$$

behaves just as good as $\hat{\rho}_{ML}$. See Asmussen and Keiding (1978), Dion and Nanthi (1982) and Carvalho (1983).

$4c°$ The stable type vector v. For type vector sampling, an obvious estimator is $Z_N/|Z_N|$, but for say sampling based on total generation sizes only, it would seem reasonable to conjecture that consistent estimates do not always exist.

In Lockhart (1982), the class of estimable parameters is identified for p = 1 and the approach may be potentially useful also in some of the cases considered here.

5° A sample scheme intermediate between full information sampling and sampling based on total generation sizes only has been
considered in Dion and Nanthi (1982), the results and proofs of
which, however, seem to require some amendments.

6° Asymptotic normality of the estimators is most often argued
to be of relevance for giving confidence intervals. To this end,
one therefore needs to replace $W\rho^N$ by a function (say $|Z_N|$) of
the observations of the same magnitude, and also to estimate the
variances. For the last point, see Nanthi (1982), Dion and Nanthi
(1982) and Carvalho (1983).

References

Asmussen, S. (1977). Almost sure behaviour of linear functionals of
    supercritical branching processes. Trans.Amer.Math.Soc. $\underline{231}$,
    233-248.

Asmussen, S. (1982). On the rôle of a certain eigenvalue in estima-
    ting the growth rate of a branching process. Austr.J.Statist.
    $\underline{24}$, 151-159.

Asmussen, S. and H.Hering (1983). Branching processes. Birkhäuser,
    Basel Boston Stuttgart.

Asmussen, S. and N.Keiding (1978). Martingale central limit theorems
    and asymptotic estimation theory for multitype branching pro-
    cesses. Adv.Appl.Probab. 10, 109-129.

Athreya, K.B. (1969). Limit theorems for multitype continuous time
    Markov branching processes. Z.Wahrscheinlichkeitsth.verw.Geb.
    $\underline{12}$, 320-332; ibid. $\underline{13}$, 204-214.

Athreya, K.B. (1971). Some refinements in the limit theory of super-
    critical multitype Markov branching processes. Z.Wahrschein-
    lichkeitsth.verw.Geb. $\underline{20}$, 47-57.

Becker, N. (1977). Estimation for discrete time branching processes
    with applications to epidemics. Biometrics $\underline{33}$, 515-522.

Carvalho, M.L.S. (1983). Etude asymptotique de quelques estimateurs
    du processus de ramification multitype. Doctoral dissertation,
    l'Université Pierre et Marie Curie, Paris.

Dion, J.-P. and N.Keiding (1978). Statistical inference in branching
    processes. In : Advances in Probability 5 (Joffe-Ney ed.).
    Marcel Dekker, New York.

Dion, J.-P. and K.Nanthi (1982). Estimation in multitype branching
    processes. Technical report, Université de Montréal.

Heyde, C.C. (1981). On Fibonacci (or lagged Bienaymé-Galton-Watson)
    processes. J.Appl.Probab. 17, 1079-1082.

Keiding, N. and S.L.Lauritzen (1978). Marginal maximum likelihood
    estimates and estimation of the offspring mean in a branching
    process. Scand.J.Statist. 5, 106-110.

Kesten, H. and B.P.Stigum (1966). Additional limit theorems for
    indecomposable multidimensional Galton-Watson processes. Ann.
    Math.Statist. 37, 1211-1223.

Khan, I.A. and S.Rehman (1980). On estimation of the matrix of the
    first moments of a two-dimensional branching process. Pure
    Appl.Math.Sci. 12, 41-48.

Lockhart, R. (1982). On the non-existence of consistent estimates
    in Galton-Watson processes. J.Appl.Probab. 19, 842-846.

Nanthi, K. (1982). Estimation of the variance for the multitype
    Galton-Watson process. J.Appl.Probab. 19, 421-426.

EXACT DISTRIBUTION OF KIN NUMBER IN MULTITYPE GALTON-WATSON PROCESS

A. Joffe[1] and W.A.O'N. Waugh [2]

[1]Université de Montréal, C.P.6128, Quebec H3C 3J7, Canada

[2]The University of Toronto, Ontario M5S 1A1, Canada

1. Introduction

For more background on this subject we refer the reader to
Waugh (1981), Joffe and Waugh (1982) and (1985). Here we borrow
freely from those articles.

The "kin number problem" concerns the distribution of the number
of relatives of a randomly chosen individual in a population, for
example the number of the individual's siblings, or children, or
aunts. It has numerous applications such as to social policy
(housing needs, etc.) to demography, to anthropology (where family
relationships may affect a person's status in a tribe) and to cancer
research (where an inquiry into the health of a cancer victim's
relatives depends on their number and affinity).

A simple and important aspect of the kin number problem is the
relationship between the distribution of the number of siblings of
a randomly chosen individual and the distribution of the number of
children of a randomly chosen parent. An early reference to this
aspect of the problem and to a method of dealing with it by using
additional information, or bith-order, is given by Burks (1933).
The point is easily stated : suppose we sample, at random, an indi-

vidual, from the population. Adopting a convenient term from anthropology let us call this individual Ego. Then a n-sized sibling group is n times more likely to be the source of Ego than a one-child group, and the sibling group to which Ego belongs will be larger, in expectation, than the group of offspring of a randomly chosen mother.

More recent work in which the kin number problem is treated by way of the life-table and maternity function is given by Goodman, Keyfitz and Pullum, (1974), (1975) and Le Bras (1973). A difficulty of this approach is that it does not allow for the gestation period nor for postpartum infertility, which may be highly relevant to the small sample phenomena of numbers within Ego's immediate family. A large-scale survey of kin numbers of 2075 "Egos" has been carried out in France at the Institut d'Etudes Démographiques (Gokalp, 1979) which includes a great deal of detailed information such as the distance between the permanent residences of a parent and (adult) child. A situation in anthropology where the joint distribution of generation sizes is relevant is described by Adams (1981). In the Amazonian Carib tribe which she investigated, before the advent of the mining industry, groups of brothers formed cooperating groups so that there was an advantage to Ego in belonging to a large sibling group, but a cost in giving birth to a large number of offspring.

Pedigree studies in cancer research such as those of McKusick (1965) and Lynch et al. (1977a,b) clearly depend on the size of the family net, and the latter state "when interpreting the genetic significance of the spectrum of tumors in a given family, one must realize that the number and ages of persons at risk for cancer, as well as other causes or morbidity and mortality, will influence considerably the probability of cancer occurrence". The pedigrees of Lynch et al. are in fact mainly four-generation family nets. Family size is also referred to in the discussion of industrial carcinogenesis by Morgan (1979).

Galton-Watson branching processes are the simplest model for the evolution of a population in which the kin number problem may be studied in details; for large populations this has been done in Waugh (1981), Jagers (1982) and exact distributions were derived in Joffe and Waugh (1982).

The large population approximation can be made rigorous for a much more general class of processes namely the Grump-Mode-Jagers branching processes. This has been achieved in Nerman and Jagers (1984) and also the method of Neveu (1985) should apply here.

In pedigree studies of the incidence of a disease in the family of a patient, it is natural to consider more than one type. In a simple case we may consider those afflicted with the disease and those who are not as the two types. Various states of the disease, or of propensity to contract it, lead to more than two types. For this and other applications, and for its theoretical interest, we consider the problem in the framework of a multitype Galton-Watson process. The large population has been studied in Joffe and Waugh (1985) and also in Nerman (1984).

In the monotype case, an individual, Ego, is sampled at random from the whole population in a certain generation. In each preceding generation, Ego has a uniquely defined ancestor. Let A be Ego's ancestor in some fixed generation, which can be taken as generation # 0. Let $\mathcal{T}(A)$ be the sub-family tree stemming from A, that is, the set of all individuals descending from it who are realized. The main problem studied is to find the joint distribution of the sizes of the generations in $\mathcal{T}(A)$.

In the multitype case it is necessary to take account of type of Ego and of her ancestor A, also the sizes of the generations must be considered in terms of the different types. We shall write $\mathcal{T}(A) = a$, where $a = 1,2,\ldots$, to indicate that A is of·type a, and similarly $\mathcal{T}(E) = e$ to indicate Ego's type. We shall refer to the groups consisting of all members of Ego's generation in $\mathcal{T}(A)$ as her "cousinage" ("sorority" if Ego is in generation # 1). The sub-groups of Ego's cousinage who are of the same type as Ego will be called her "type-cousinage".

The multitype case involves various possible procedures for sampling Ego, and hence A and $\mathcal{C}(A)$. For example it may be decided in advance to sample Ego at random from all members of her generation who belong to some fixed type. This would happen if an inheritable disease was being studied, one type was defined to be victims of the disease, and sampling was among victims. Another possibility would be to sample Ego from all members of her generation who belong to some fixed type, and whose ancestors also belonged to a fixed type. Here, for example, Ego's type might be that of victims, their ancestors also being victims – a special case where the two fixed types are the same. Yet another possibility would be to sample at random from the entire population in a fixed generation just as in the monotype case. These three possibilities are studied in Joffe and Waugh (1985).

In the next section we give some new results on the kin number problem for a multitype Galton-Watson branching process. They give exact distribution which should allow to develop the formula related to different sampling procedures of Ego in a small population. By taking their limits one should be able to recover the results of Joffe and Waugh (1985) for large populations.

2. Multitype Galton-Watson processes

Let N^{+d} be the set of d-dimensional vectors whose coordinates are non-negative integers and for each i, $i = 1,\ldots,d$ let $p_i(\underset{\sim}{j}) = p_i(j_1,\ldots,j_d)$ be a probability distribution in N^{+d}. It represents the probability law of the children produced by a parent of type i.

A multitype Galton-Watson process is a Markov chain $\underset{\sim}{Z} = (Z_{n,1},\ldots,Z_{n,d})$ whose state space is N^{+d} and its transition probability is given by

$$(1) \quad \underline{P}(\underset{\sim}{i},\underset{\sim}{j}) = p_1^{*i_1} * p_2^{*i_2} \cdots p_d^{*i_d}(\underset{\sim}{j}), \ \underset{\sim}{i} = (i_1,\ldots,i_d), \ \underset{\sim}{j} = (j_1,\ldots,j_d)$$

where $*$ denotes the product of convolution.

For $\underset{\sim}{s} = (s_1,\ldots,s_d) \in R^d$ with $0 \leqslant s_i \leqslant 1$ we introduce the probability generating function of P_i :

$$(2) \quad f_i(\underset{\sim}{s}) = \sum_{j \in N^{+d}} P_i(j) s_1^{j_1} \ldots s_d^{j_d} \,.$$

Let

$$(3) \quad \underset{\sim}{f}(\underset{\sim}{s}) = f(f_1(\underset{\sim}{s}),\ldots,f_d(\underset{\sim}{s})).$$

It defines a map of the unit cube of R^d into itself. Then it is well-known that if $\underset{\sim}{Z_0} = \underset{\sim}{i}$ the generating function of $\underset{\sim}{Z_n}$ is given by

$$(4) \quad E_{\underset{\sim}{i}} s_1^{Z_{n,1}} \ldots s_d^{Z_{n,d}} = f_{n,1}^{i_1}(\underset{\sim}{s}) \cdot f_{n,2}^{i_2}(\underset{\sim}{s}) \ldots f_{n,d}^{i_d}(\underset{\sim}{s})$$

where

$$(5) \quad \underset{\sim}{f}_n = (f_{n,1},\ldots,f_{n,d}) = \underset{\sim}{f} \circ \ldots \circ \underset{\sim}{f}, \text{ the } n\text{-th iterate of (3).}$$

For more details on multitype Galton-Watson processes, see Athreya and Ney (1972).

Each individual in the n-th generation has a probability $1/|Z_n|$ of being chosen as Ego $(|Z_n| = Z_{n,1} + \ldots + Z_{n,d}$ denotes the total number of individuals in the n-th generation). Then Ego has, for each $k \leqslant n$ a uniquely k-th ancestor A who belongs to generation $\# n - k$. We are interested in the stochastic structure of $\mathcal{T}(A)$, and in particular in the joint distribution of generation sizes and types in it. Those generations will be denotes by $\underset{\sim}{W_0}$, $\underset{\sim}{W_1},\ldots,\underset{\sim}{W_k}$ (here $\underset{\sim}{W_0} = A$, Ego's ancestor and $\underset{\sim}{W_k}$ is Ego's cousinage in the n-th generation of the original Galton-Watson process).

To simplify the notation we assume that we start with one individual of type i. Then we have the following theorems.

Theorem 1

$$(6) \quad P(\text{ego of type } e, A \text{ of type } a, W_k = i \mid Z_{n-k})$$

$$= Z_{n-k,a} \; j_e \; P(\xi_1^a = j) \int_0^1 s^{|j|-1}$$

$$\prod_{\ell=1}^{d} f_{k,\ell}^{Z_{n-k,\ell}}(s,\ldots,s)/f_{k,a}(s,\ldots,s)ds$$

where ξ_1^a is a random vector whose distribution is given by the generating function $f_{k,a}(s_1,\ldots,s_d)$.

Theorem 2

$$(7) \quad P(\text{ego of type } e, A \text{ of type } a, W_k = j \mid Z_0 = i)$$

$$= j_e \; P(\xi_1^a = j) \int_0^1 s^{|j|-1}$$

$$\frac{\partial}{\partial S_a} f_{n-k,i}(f_{k,1}(s,\ldots,s),\ldots,f_{k,d}(s,\ldots,s))ds$$

To prove Theorem 1 one introduces the random vectors ξ_j^i, $j = 1,\ldots,Z_{n-k,i}$, $i = 1,\ldots,d$, representing the descendance in the n-th generation of each individual of the n-k-th generation. Those random vectors are independent and their generating functions are given by $f_{k,i}(s_1,\ldots,s_d)$.

By the very definition of the sampling procedure one has

$$(8) \quad P(\text{ego of type } e, A \text{ of type } a, W_k = j \mid Z_{n-k};$$

$$\xi_j^i, j = 1,\ldots,Z_{n-k,i}, i = 1,\ldots,d)$$

$$= \frac{\displaystyle\sum_{\ell=1}^{Z_{n-k,a}} \xi_{\ell,e}^a \; V_j(\xi_\ell^a)}{\displaystyle\sum_{j=1}^{d} \sum_{k=1}^{Z_{n-k,j}} |\xi_k^j|}$$

$$\text{where} \quad V_j(\ell) = \begin{cases} 1 & \text{if } \ell = j \\ 0 & \text{if } \ell \neq j \end{cases}.$$

Taking expectation with respecto to the $\underset{\sim}{\xi}$ gives (6) (using independence, symmetry and the formula $E \dfrac{1}{j + X} = \displaystyle\int_0^1 s^{j-1} Es^X ds$). (7) is obtained from (6) by taking expectation with respect to $\underset{\sim}{Z}_{n-k}$. The joint distribution of $W_0, \ldots, W_k$ is easily derived from Theorem 2 by the use of the following formula

$$\underline{P}_{Z_0} (\underset{\sim}{W}_0 = a, \underset{\sim}{W}_1 = j_1, \ldots, \underset{\sim}{W}_k = j_k)$$

$$= \underline{P}(\underset{\sim}{Z}_1 = \underset{\sim}{j}_1, \ldots, \underset{\sim}{Z}_k = \underset{\sim}{j}_k) = \frac{P_{Z_0}(\underset{\sim}{W}_k = \underset{\sim}{j}_k, W_0 = a)}{P(\underset{\sim}{Z}_k = j_k)}$$

Remarks

1. In the supercritical case one should be able to obtain the limit theorem as n goes to infinity.

2. To use branching process techniques, the model adopted is of one-sex type, either the female line is followed, or the male line. It is a great challenge to develop a mathematical model which could give even approximate results without the one-sex restriction.

References

Adams, K.J. (1981). The rôle of children in the changing socio-
    economic strategies of the Guyanese Caribs. Canad.J.Anthropol.
Athreya, K.B. and P.Ney (1972). Branching processes. Springer-Verlag.
Burks, B.S. (1933). A statistical method for estimating the distri-
    bution of sizes of completed fraternities in a population re-
    presented by a random sampling of individuals. J.Amer.Statist.
    Assoc. 28, 388-394.
Gokalp, C. (1979) Le réseau familial. Population 6, 1077-1094.

Goodman, L.A., N.Keyfitz and T.W.Pullum (1974). Family formation and the frequency of various kinship relationships. Theoret. Popn Biol. 5, 1-27.

Goodman, L.A., N.Keyfitz and T.W.Pullum (1975). Addendum to "Family formation and the frequency of various kinship relation-ships". Theoret.Popn Biol. 8, 376-381.

Jagers, P. (1982). How probable is it to be firstborn ? and other branching process applications to kinship problems. Math. Biosci. 59, 1-15.

Joffe, A. and W.A.O'N.Waugh (1982). Exact distributions of kin numbers in a Galton-Watson process. J.Appl.Prob. 19, 767-775.

Joffe, A. and W.A.O'N.Waugh (1985). The kin number problem in a multitype Galton-Watson process. J.Appl.Prob.

Le Bras, H. (1973). Parents, grands-parents, bisaïeuls. Population 1, 9-38.

Lynch, H.T., H.D.Guirgis, P.M.Lynch, J.F.Lynch and R.D.Harris (1977a). Familial cancer syndromes, a survey. Cancer, 39, 1867-18881.

Lynch, H.T., H.D.Guirgis, P.M.Lynch, J.F.Lynch and R.D.Harris (1977b). Rôle of heredity in multiple primary cancer. Cancer 40, 1849-1854.

McKusick, V.A. (1965). Some computer applications to problems in human genetics. Methods of Information in Medicine IV 4, 183-189.

Morgan, W.K.C. (1979). Industrial carinogens : the extent of the risk. Thorax 34, 431-433.

Nerman, O. (1984). The growth and composition of supercritical branching populations on general typespaces. Preliminary manuscript, Department of Mathematics, Chalmers University of Technology and the University of Göteborg.

Nerman, O. and P.Jagers (1984). The stable doubly infinite pedigree process of supercritical branching populations. Z.Wahrschein-lichkeitstheorie verw.Gebiete 65, 445-460.

Neveu, J. (1985). Recent results on branching processes. Those proceedings.

Waugh, W.A.O'N. (1981). Application of the Galton-Watson process to the kin number problem. Adv.Appl.Prob. 13, 631-649.

ARBRES DE GALTON-WATSON

J.Neveu

Laboratoire de Probabilités,

Tour 56 - 4 Place Jussieu, 75230 Paris Cedex 05, France

La notion d'arbre est intéressante à formaliser pour étudier les processus de branchement. En effet l'étude de ces processus me paraît trop s'appuyer sur des outils purement analytiques alors qu'une utilisation plus intensive de la propriété intuitive du branchement permet des raisonnements plus probabilistes.

Un arbre ω peut être défini comme un sous-ensemble non vide de l'espace U des suites finies $u = j_1,\ldots,j_n$ d'entiers strictement positifs tel que

a) $uj \in \omega$ si $j \leqslant \nu_u(\omega)$ pour un entier $\nu_u(\omega) \geqslant 0$,

b) $u \in \omega$ si $uj \in \omega$ pour un $j \in N^*$ $(u \in U)$.

L'espace Ω des arbres est muni de la tribu $\mathcal{F}$ engendrée par les $\Omega_u = \{\omega : \omega \ni u\}$ et de la filtration $(\mathcal{F}_n, n \in N)$ engendrée par les applications "$n^{\text{ème}}$ génération" $z_n(\omega) = \omega \cap N^{*n}$ de Ω dans l'espace des parties finies de $N^{*n} \subset U$. Les v.a. $\nu_u : \Omega_u \to N^*$ sont alors $\mathcal{F}_n$-mesurables si u est de longueur $< n$; on définit aussi naturellement des translations $T_u : \Omega_u \to \Omega$ sur l'espace des arbres.

A toute loi $q = (q(j), j \in N^*)$ sur N^* est associée une probabilité unique P sur $(\Omega,\mathcal{F})$ telle que $P(\nu = j) = q(j)$ $(j \in N^*)$ qui possède la propriété de branchement suivante : Pour tout $n \in N$, conditionnellement par rapport à $\mathcal{F}_n$ les translations T_u ($u \in z_n$) sont indépendantes entre elles et de même loi P. Cela signifie de manière précise que

$$E^{\mathcal{F}_n}[\prod_{u \in z_n} f_u \circ T_u] = \prod_{u \in z_n} \int_\Omega f_u dP$$

quelles que soient les fonctions boréliennes $f_u : \Omega \to R_+$ ($u \in U$).

Bien entendu les v.a. entières $Z_n = \text{Card}(z_n)$ forment alors lorsque n parcourt N un processus de Galton-Watson de loi de reproduction q mais le modèle précédent permet aussi de considérer les processus des descendants ($Z_n \circ T_u$, $n \in N$) d'un individu u particulier, ainsi que de nombreuses autres variables aléatoires intéressantes qui ne sont pas des fonctions des Z_n seulement.

Le formalisme précédent peut être étendu à des arbres marqués de manière à considérer des processus de branchement à plusieurs types, spatiaux ou/et avec âge à temps continu. Le détail de la formalisation et des exemples d'application paraîtront prochainement (Neveu et Chauvin).

Références

Chauvin, B. Arbres et processus de Bellman-Harris. A paraître.
Joffe, A. 1978). Remarks on the structure of trees. Adv.Probab.,
       Vol.5, Ed. A.Joffe et P.Ney.
Neveu, J. Arbres et processus de Galton-Watson. A paraître.

SECTION IX. APPLICATIONS IN MEDICINE

SOME REMARKS ON SEMI-MARKOV PROCESSES IN MEDICAL STATISTICS

D.R. Cox

Department of Mathematics
Imperial College of Science and Technology
London SW7 2BZ, England

1. Introduction

The object of this paper is to give a brief account of potential applications of semi-Markov processes in medical statistics, especially in connection with studies involving "quality of life".

Over recent years there has been some interest in the analysis of survival data particularly from the following point of view. For each of n independent individuals we observe the smaller of a random variable T_i, representing failure time, and a constant c_i, representing potential censoring time. Also we observe an indicator random variable specifying which type of outcome has arisen. That is, we observe

$$X_i = \min(T_i, c_i), X_i' = \begin{cases} 1 & (T_i \leq c_i), \\ 0 & (T_i > c_i), \end{cases} \tag{1}$$

for i=1,...,n. Also for each individual there is a vector z_i of explanatory variables (treatment indicators, etc...), often constant but sometimes a function of time. A primary objective is to study how the distribution of T_i depends on z_i.

To define T_i, a time origin is required for each individual and appropriate choice of this is important. Often in clinical trials, the time origin is the instant of entry into the study. Throughout this paper an assigned value of time usually refers to different points in real time for different individuals.

One of several ways of studying the above system is via a generalization of the hazard, namely

$$h(t;z) = \lim_{\Delta \to 0+} \frac{P(t \leqslant T < t + \Delta \mid t \leqslant T, z)}{\Delta} \; . \qquad (2)$$

If z is a function of time it is normally assumed that only its value at t is relevant in (2). The form may be specified fully parametrically or at least partly nonparametrically. A special form is

$$h(t;z) = h_o(t) \exp(\beta^T z) \; , \qquad (3)$$

where $h_o(t)$ is baseline hazard corresponding to an individual with z=0 and β is an unknown parameter vector. The function $h_o(t)$ may be specifically parametrically, e.g. as corresponding to a Weibull distribution, or left unspecified. There is an extensive literature reviewed by Cox and Oakes (1984) on such ideas, application not being restricted to medical investigations.

2. Quality of life

A severe limitation on studies that concentrate solely on failure time is that no explicit account is taken of the patient's state during treatment, although to some extent this can be covered by a further analysis considering, for example, end of a period of remission of symptoms as failure.

The rather unfortunate term "quality of life" is often used for assessments of patients status. It is a very welcome and important

development in clinical trials that increasing emphasis is being
placed on such matters; for a good review, see Fayers and Jones
(1983).

Observations obtained can be of a number of different types :
simple indicators of status (e.g. in hospital or not); physical and
biochemical measurements (e.g. blood pressure, ECG classification);
measures of physical performance (e.g. treadmill work rate); adverse
reactions to treatments, which may be physician assessed or self-
assessed; judgements of well being based on physicians 'assessment
or patients' self-assessment questionnaire. The investigation of
the interrelation of such measures may raise major conceptual and
statistical problems outside the scope of the present discussion.

Note , however, that if we are to represent the "state" of a
patient mathematically in an at all realistic way a quite rich state
space is required. In the present paper, however, we confirm attention to countable and indeed finite state spaces. Also, while in
practice observations of state are sometimes available at only quite
widely spaced time points, we shall suppose this information to be
available in continuous time.

3. A simple representation

Suppose then that $Y(t)$ denotes the state of an individual at
time t. The states are labelled $\{0,1,2,\ldots\}$, with 0 denoting the
absorbing state (death). Let

$$I_r(t) = \begin{cases} 1 & Y(t)=r \ , \\ \\ 0 & Y(t) \neq r \ . \end{cases} \qquad (4)$$

Our objectives in representing the system by a formal stochastic
process may be

(a) to study possible relations with a biological model,

(b) to obtain guidance on data reduction, especially for studying
     the effect of explanatory variables,

(c) to study the distribution of random variables such as

$$T_w = \int_o^\infty \sum_{r=o}^\infty w_r I_r(t)dt, \tag{5}$$

where $w_o = 0$ and w_r is the utility per unit time of residence in
state r. The state corresponding to full health can be taken as
defini ng   the unit of utility,

(d) to examine the effect of censoring on the estimation of various
     properties, including the distribution of failure time, the spe-
     cial case of (5) with $w_r = 1$ $(r \neq 0)$,

(e) to develop methods for predicting the future of an individual
     with a given "history".

The model has affinity with some considered in reliability
theory for relating time to system failure and component properties
(Gertsbakh, 1984), although the objectives and emphases here are
different.

The choice of weights in (5) raises formidable problem in practice. One approach is via a sensitivity analysis. For example, in
comparing two treatments it may be possible to delimit regions in the
space of the $\{w_r\}$ where the superiority of one of the treatments is
reasonably clearly established. Another approach, particularly appropriate when the states are ordered, is to take

$$w_r = \begin{cases} 1 & (r \geq k), \\ 0 & (r < k, \end{cases}$$

so that the value of (5), $T^{(k)}$, say, is the total time spent in state
k or better. The marginal distributions of $T^{(1)}$, $T^{(2)}$,...., then provide one fairly easily interpreted summary of the properties of the
system : see also Temkin (1978).

4. Some special processes

The simplest model attached to the state space of Section 3 is, of course, a time-homogeneous Markov process in which, in particular, the individual residence times in states are independently exponentially distributed (Fix and Neyman, 1951). While often too simple to be at all realistic, it is nevertheless very useful for preliminary work. There are special cases in which

(a) there is a "birth-death" process in which only transitions between adjacent states are allowed,

(b) all transitions are downwards,

(c) if the transition from r to s has positive probability, that from s to r has zero probability.

Estimation of parameters in such models is immediate, for censored or uncensored data, and calculation of theoretical properties is straightforward, at least in principle.

One broad generalization of such simple Markov models is to make the defining parameters functions of time; recall the comments in Section 1 on the definition of time. If the dependence on explanatory variables is of a product form generalizing (3), a quite general discussion of nonparametric estimation is available (Johansen, 1983; Andersen et al., 1982), in particular generalizing the product limit estimate (Kaplan and Meier, 1958). Full parametric estimation is, of course, possible too. See also Cox (1972).

The next possibility is to consider a semi-Markov model (Weiss and Zelen, 1965). In terms of the complete transition intensity

$$h_{rs}(t;H_t,z) = \lim_{\Delta \to 0+} \frac{P\{Y(t+\Delta)=s \mid Y(t)=r, H_t,z\}}{\Delta} \tag{6}$$

where H_t is the history of the process up to t, the dependence on

H_t is restricted to u_t, the current residence time in the state at t, i.e.

$$h_{rs}(t;H_t,z) = \pi_{rs}(u_t;z) \quad ; \qquad (7)$$

of course time heterogenity could be introduced as well.

The special cases (a) - (c) above arise, (c) being considered by Voelkel and Crowley (1984). In addition there is the possibility that

$$\pi_{rs}(u_t;z) = \rho_r(u_t;z)\,\sigma_{rs}(z) \quad , \qquad (8)$$

which may be called a process of proportional intensities (Cox and Isham, 1980, pp. 54-58). The dependency on z might be confined to one of the factors in (8). Under (8), the individual residence times depend only on the current state and not on the state to which exit occurs. If plausible inherently and consistent with data, (8) offers one route to a quite major reduction in the number of parameters to be fitted.

Methods of estimation associated with such processes, subject to censoring, were considered by Lagakos et al (1978) and made rigorous by Gill (1980) and, in the special case (c), by Voelkel and Crowley (1984).

Forms of dependence on H_t in (b) other than on the current residence time u_t and state are certainly possible and some are worth considering, for example in order to examine sensitivity of conclusions to model structure. One possibility is to introduce a dependence on the number of times the current state (or some controlling state) has been entered. Another is to introduce a dependence on total discounted sojourn time.

Suppose then that there is a set, C, of states which are called rate-controlling and that for each such state we consider

$$V_m(t) = \int_0^t e^{-\kappa_m u} I_m(t-u) du \qquad (m \in C), \qquad (9)$$

where $\kappa_m \geq 0$ are discounting constants. A class of processes can then be considered in which the complete transition intensities (6) are functions of r, s and of the history H_t only via the observed value $v_m(t)$ $(m \in C)$. For example, C might consist of a single critical state with all the transition rates determined by the total (discounted) time spent in that state. If in (9) $\kappa_m = 0$, then $V_m(t)$ is the total sojourn time in state m in $(0,t)$. Whether such processes are regarded as semi-Markov processes is a matter of definition; in fact they are all, of course, Markov processes with a suitably extended state space.

As the simplest nontrivial special case we may suppose there to be three states $\{0,1,2\}$ of which 0 is absorbing and 1 is rate-controlling.

Write

$$V(t) = \int_0^t e^{-\kappa u} I_1(t-u) du. \qquad (10)$$

Then, ignoring the dependence on z, we may define the process via the transition intensities (6), by specifying $\pi_{10}(v)$, $\pi_{12}(v)$, $\pi_{21}(v)$, $\pi_{20}(v)$. Here $\pi_{rs}(v)$ is the value of $h_{rs}(t;H_t)$ when the history H_t implies that $V(t) = v$. If the joint distribution of $Y(t)$ and $V(t)$ is defined by

$$p_r(v;t) = \lim_{\Delta \to 0+} \frac{P\{Y(t)=r,\ v < V(t) < v+\Delta\}}{\Delta}$$

for $r=1,2$, then

$$- \kappa \frac{\partial (vp_2)}{\partial v} + \frac{\partial p_2}{\partial t} = -p_2 \pi_{2.} + p_1 \pi_{12} \quad ,$$

$$\frac{\partial p_1}{\partial v} - \kappa \frac{\partial (vp_1)}{\partial v} + \frac{\partial p_1}{\partial t} = -p_1 \pi_{1.} + p_2 \pi_{21} \quad , \tag{11}$$

where $\pi_{2.} = \pi_{21} + \pi_{20}$, $\pi_{1.} = \pi_{12} + \pi_{10}$.

Detailed investigation will not be made here; it is clear that by taking Laplace transforms and elimination of say p_1, reduction to a second-order ordinary differential equation is achieved, with further reduction and explicit solution if $\kappa = 0$.

Fitting such a model to empirical data would, however, not require explicit solution of the probabilistic properties of the model.

5. Analysis in the presence of censoring

Finally we outline some unsolved problems when semi-Markov models are used to aid the analysis of censored data, role (d) of Section 3.

The general point is that even if interest is focused on total failure time (first passage time to state zero), use of the product limit estimate in the presence of censoring would ignore information contained in the state occupied at censoring. Under some circumstances this information could be appreciable. Lagakos (1976, 1977) has investigated a very special case via a three-state Markov model with no "reverse" transitions; see, also, Dinse and Lagakos (1982).

To illustrate a more general treatment, but still via a special case, suppose again that there are just three states, that a semi-Markov model of proportional intensity form (8) applies and that transitions occur only between adjacent states. Then, with 0 an absorbing state, the process is specified by two probability densities $f_1(.)$, $f_2(.)$, concerning residence times in states 1, 2, respectively , and the probability ϑ that exit from state 1 is to state 0.

Assume, again for simplicity, that initially all individuals have just entered state 1. It is easy to show that the Laplace transform of the density of failure time, that is, the first passage time $T^{(1)}$ to state zero is

$$f^\star_{T^{(1)}}(s) = E\{\exp(-sT^{(1)})\}$$
$$= \frac{\vartheta f^\star_1(s)}{1-(1-\vartheta)f^\star_1(s)f^\star_2(s)} \quad , \tag{12}$$

where $f^\star_r(s)$ is the Laplace transform of $f_r(.)$ (r=1,2). If, in line with the discussion of Section 3, there is interest in $T^{(2)}$, the total time spent in state 2, this has Laplace transform

$$f^\star_{T^{(2)}}(s) = \frac{\vartheta}{1-(1-\vartheta)f^\star_2(s)} \quad , \tag{13}$$

having an atom ϑ of probability at zero.

For statistical analysis, either to analyze a single homogeneous set of data or to assess the effect of explanatory variables, there are for uncensored data three broad possibilities :
(i)   to estimate the distribution of $T^{(1)}$, or $T^{(2)}$, directly, this not requiring assumptions about the structure of the process ;
(ii)   on the basis of some special model, e.g. the one above, to estimate $f^\star_1$, $f^\star_2$ and $f^\star_{T^{(1)}}$, finally inverting the Laplace transform in some numerical fashion ;
(iii) to follow the procedure of (ii), with, however, a parametric form for f_1 and f_2, Erlangian being particularly convenient although not necessarily appropriate!

Three unsolved issues are first how best to implement (ii), secondly how much efficiency is gained by using (ii) or (iii) rather than (i), and thirdly how sensitive methods (ii) and (iii) are to the special assumptions made.

When the data are subject to censoring, method (i) is replaced by the product limit estimate for the distribution of $T^{(1)}$, but there is no analogue for $T^{(2)}$, unless it can be assumed that the state 2 would not have been reentered a further time between censoring and failure. The assumption that censoring is uninformative usually has to be viewed critically.

The issues raised in this Section typify a rather broad class of problems, frequently complicated further in practice by observation being possible only at discrete time points. Clearly there is much scope for further work.

I am grateful to the Science and Engineering Research Council for their support.

References

P.K. Andersen, O. Borgan, R. Gill and N. Keiding (1982), Linear non-parametric tests for comparison of counting processes, with applications to censored survival data (with discussion). Int. Statist. Rev. $\underline{50}$, 219-258.

D.R. Cox (1972), The statistical analysis of dependencies in point processes. In stochastic point processes, ed. P.A.W. Lewis, 55-66, New York:Wiley.

D.R. Cox and V. Isham (1980), Point processes. London:Chapman & Hall.

D.R. Cox and D. Oakes (1984), Analysis of survival data. London: Chapman & Hall.

G.E. Dinse & S.W. Lagakos (1982), Nonparametric estimation of life-time and disease onset distributions from incomplete observa-tions. Biometrics $\underline{38}$, 921-932.

P.M. Fayers and D.R. Jones (1983), Measuring and analysing quality
     of life in cancer clinical trials : a review. Statistics in
     Medicine 2, 429-446.

E. Fix and J. Neyman (1951), A simple stochastic model of recovery,
     relapse and loss of patients. Human Biology 23, 205-241.

I.B. Gertsbakh (1984), Asymptotic methods in reliability theory :
     a review. Adv. Appl. Prob. 16, 147-175.

R.D. Gill (1980), Nonparametric estimates based on censored observa-
     tions of a Markov renewal process. Z. Wahr. 53, 97-116.

S. Johansen (1983), An extension of Cox's regression model. Int.
     Statist. Rev. 51, 165-174.

E.L. Kaplan and P. Meier (1958), Nonparametric estimation from in-
     complete observations. J. Am. Statist. Assoc. 53, 457-481.

S.W. Lagakos (1976), A stochastic model for censored survival time
     in the presence of an auxiliary variable. Biometrics 32, 551-
     560.

S.W. Lagakos (1977), Using auxiliary variables for improved estima-
     tes of survival time. Biometrics 33, 399-404.

S.W. Lagakos, C.J. Sommer and M. Zelen (1978), Semi-Markov models
     for partially censored data. Biometrika 65, 311-317.

N. Temkin (1978), An analysis for transient states with application
     to tumor shrinkage. Biometrics 34, 571-580.

J.G. Voelkel and J. Crowley (1984), Nonparametric inference for a
     class of semi-Markov processes with censored observations. Ann.
     Statist. 12, 142-160.

G.H. Weiss and M. Zelen (1965), A semi-Markov model for clinical
     trials, J. Appl. Prob. 2, 269-285.

SEMI-MARKOV AND NON-HOMOGENEOUS MARKOV MODELS IN MEDICINE

Daniel Commenges

Université de Bordeaux II

146 rue Leo Saignat, 33076 Bordeaux Cedex, France

1. Introduction

Longitudinal studies form the most interesting part of medical studies. All therapeutical trials and the majority of epidemiological studies can be considered as longitudinal studies but most often this aspect is ignored or grossly simplified when the data are analysed. For instance a very simple therapeutical trial can be described as follows. Patients are included in the trial, each at a certain moment of their history. When included they are assigned randomly to one of two treatment groups. A particular variable R, called here the response variable, can be observed till the end of the study. A simple method of analysis consists in comparing the values of R in the two groups observed after a time T_e of treatment. A Pearson's X^2 or a Student's t test for instance can be used according to the statistical nature of the variable.

Such procedures have drawbacks which can be summarized by saying that they lose some information by ignoring the dynamical aspect of the data. Three particular ways in which information is lost are :
- the comparison is made at a particular value of T_e and it may be that there exist larger differences at other values of T_e;

- patients observed for less than T_e are lost;
- only one observation of R is used whereas using several observations could reduce the influence of rapid random variation or
  of mesurement errors.

    This has two consequences :
- loss of power of the test procedure;
- insufficient description of the action of the drug.

    Methods able to take account of the dynamical dimension of the
data have been well developed for survival data. This development
started with actuarial analysis and more recently has involved three
main steps :
- estimation of the survival function (Kaplan and Meier, 1958);
- comparison of k groups : logrank test (Mantel, 1966; Peto and
  Peto, 1972);
- multiple regression (Cox, 1972).
Most recently these problems have been reexamined in the light of
the theory of counting processes (Andersen, Borgan, Gill and
Keiding, 1982).

2. The response process

    Survival data however are of a very particular type and many
real situations have to be oversimplified to be put in this framework. In the survival process, R takes the value 1 until the time
of death and then takes the value 0. Thus an observation can be
summarized by the time of death. In the general case we are interested in all the values that R takes over a certain period of
time, that is, the stochastic process $R(t)$, $t \leqslant T$, and we have
several censored observations of this process. Measurement errors
may also be present.

    As for survival data there are three main problems :
- description of the data;
- comparison of k groups;

- management of concomitant variables, either by non-parametric
  adjustment of by some regression model.

     According to the nature of R the response process can have a
discrete or a continuous state-space. We will consider here only
the discrete case. In a particular application the first task is
to define the states of interest and to draw a flow-chart which
specifies the possible paths between them. In the general case
any graph is possible. Particular cases are :

- survival data

There are two states, life and death, one being absorbing.

- competitive risks

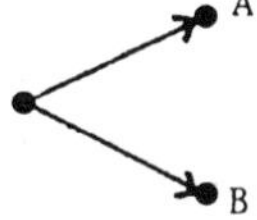

all states are absorbing except one. These processes are trivial
Markov processes.

- successive stages

- irreversible evolution.

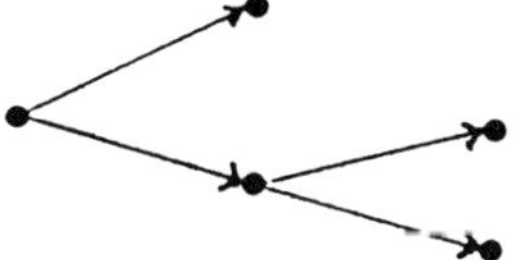

We give some examples of applications.

Hyperthyroidic patients may be treated by radio-active iodine which destroys part of their thyroid. For many patients a normal function is restored but some of them may become hypothyroidic. This evolution however can be reversible. Thus there are three states and all paths are possible.

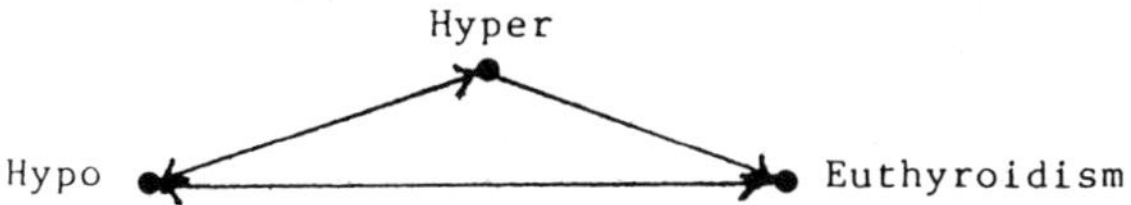

Typical absence epilepsy is a form of childhood epilepsy which after some time may be complicated by the appearance of grand-mal epilepsy. From both forms of the disease remissions can be obtained, which we consider as two different states. Thus we have the preceding flow-chart.

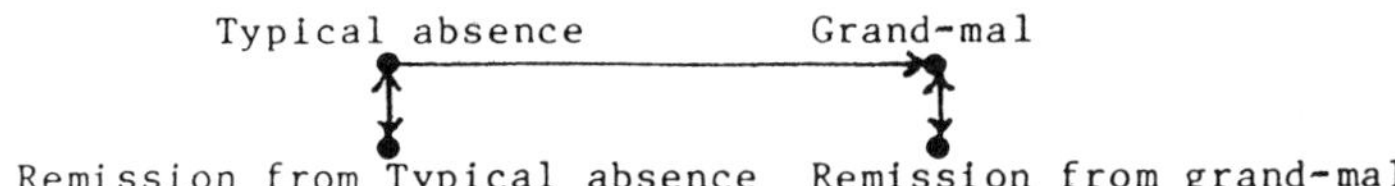

Dementia is a degenerative disease of the brain in old people, partly due to a bad cerebral circulation. Six states have been defined going from health, state 1, to death, state 6. Most patients patients have a monotone degradation of their health status but a certain number of patients experience transitory improvements. The flow-chart is as above.

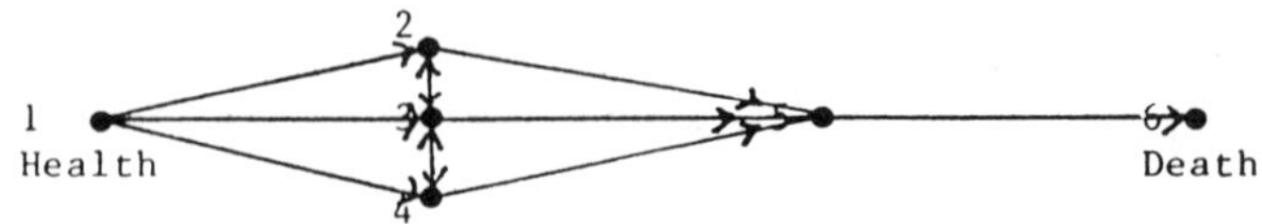

3. What hypotheses must be made ?

We have to make some hypotheses if we wish to make some inference from our data. Those of a Markov type are natural in this
context.

Homogeneous Markov (HM) models have been used for a long time.
Aaling (1958) applied such a model to pulmonary tuberculosis.
Iosifescu and Tautu (1973) give a good account of many biomedical
applications. In a HM model, sojourn time distributions are exponential and this constraint seems to be too rigid. In particular
for survival processes there are many cases where an exponential
survival function is not adequate and more general models are often
preferred.

Semi-Markov (SM) models have developed more recently : Weiss
and Zelen (1965); Lagakos, Sommer and Zelen (1978); Voelkel and
Crowley (1984). In these models state changes form a homogeneous
Markov chain and the sojourn time distributions are not restricted
to be exponential but are arbitrary.

Non-homogeneous Markov (NHM) models have been studied even more
recently : Flemming (1978); Aalen and Johansen (1978). Very few
applications of NHM models seem to have been undertaken. In these
models the transition probabilities depend on time.

Thus in both SM and NHM models the future depends on a certain
time : in SM models this is the time since the last entry in the
present state while in NHM models this an "absolute" time (relative
however to an origin of time in general different for each individual).

Both models can be viewed as particular cases of what can be
called the non-homogeneous semi-Markov model. However this model
is probably too general for application to medical studies where
the number of observations is not very large. Thus, if we exclude
the too rigid HM model, the choice lies between SM and NHM models.

SM models seem natural in many instances. They are particular-

ly adapted when patients can be stabilized in a certain state.
Hyperthyroidic patients for instance may become hypothyroidic after
treatments. If they have stayed for a certain time in this state
this indicates that a large part of their thyroid has been destroyed
and that their chance to revert to another state is low.

NHM models are natural if a certain event has a particular
importance in the history of the patient and thus the transition
probabilities are likely to depend on the time elapsed since this
event. The onset of a disease or of a treatment, the birth of the
individual (which leads to consider age-dependent probabilities)
are such events.

Although in a particular application we may be oriented towards
a SM or a NHM model by qualitative considerations as above, it is
interesting to examine the validity of each model at the light of
the data. In the uncensored case Anderson and Goodman (1957) gave
methods for testing the homogeneity and the order of a Markov chain.
For SM processes one may test separately that state changes form a
homogeneous Markov chain and that sojourn times do not depend on
previous sojourn times nor on an absolute time. This could be done
by use of Cox' model in which these times would appear as explanatory variables (as suggested by Gail, Santner and Brown, 1980).
There is still the problem of finding a global statistic for the
model. In conclusion it seems that these test-of-fit problems have
not yet received definite answers.

4. Inference problems

We focuss on non-parametric models which are often used in
medical applications and which have the advantage that no further
hypotheses are needed in addition to those mentioned in the preceding section.

4.1. Estimation problems

Lagakos, Sommer and Zelen (1978) gave estimators of the distribution functions which describe the probabilistic behaviour of a SM process. These were further studied by Gill (1980).

Temkin (1978) developed for her particular application an estimator of the "probability of being in response" that is, the probability of being in a certain state as a function of time. This is certainly the most important estimation problem from a practical point of view.

This problem finds a simple solution for NHM processes in a way which is a straightforward generalization of the method of Kaplan and Meier. The estimator, separately proposed by Flemming (1978) and Aalen and Johansen (1978) is as follows. We define a partition of time $\{t_m\}$ so fine that in each interval at most one jump is observed. The estimated transition matrix $\hat{P}(t_m, t_{m+1})$ is the identity if no jump occurs or a stochastic matrix with only one off diagonal element positive and equal to $1/R_i(u-)$ where $R_i(u-)$ is the number of individuals at risk, at position (i,j) if the jump occured at time u.

4.2. Two-sample test problem

One of the most important problems in medical studies is to compare the evolution of two or more samples. We have presented (Commenges et al., 1984) a simple method for testing the hypothesis that two samples follow the same NHM chain. This approach can be extended without modification to NHM processes.

Suppose that we observe $n_{ij}(k)$ jumps from state i to state j, $j = 1,\ldots,m$, at times u_k^i, $k = 1,\ldots K_i$. We denote by $R_i(t)$ the number of individuals who are in state i at time t. The variation of $R_i(t)$ is due to jumps from state i to other states, jumps from other states to state i and censoring in state i.

We suppose that we can observe $R_i(u_k^i-)$, that is, the number of individuals in state i just before time u_k^i (for survival processes this number is called "number at risk"). We denote by $n_{ij}^g(k)$ and $R_i^g(u_k^i-)$ the numbers relative to treatment groups g, $g = 1,2$, and we can arrange these numbers in a contingency table (Table 1). Thus, for each departure state i we have K_i contingency tables and for all states we have a total of ΣK_i contingency tables. Conditionally on the marginal totals the tables are independent and we may use exactly the same argument as Mantel (1966) who applied the Mantel-Haenszel (MH) procedure to contingency tables constructed from a survival process : he thus obtained the so-called Mantel-Haenszel test, approximately equivalent to the logrank test.

We may apply directly the MH procedure to all the contingency tables. Consider the vector $n_i^1(k)$ formed by the elements $n_{ij}^1(k)$, $j = 1,\ldots m$, of the first row of the contingency table ik. Conditionally on the marginal totals of the table, the expectation and the covariance matrix $V_i(k)$ of $n_i^1(k)$ can be computed following

Table 1. Contingency table countaining the numbers of patients going from state i to j, $j = 1,\ldots,m$ at time u_k^i in the two groups to be compared.

arrival state	1	2	.....	m	number at risk
group 1	$n_{i1}^1(k)$	$n_{i2}^1(k)$	...	$n_{im}^1(k)$	$R_i^1(u_k^i-)$
group 2	$n_{i1}^2(k)$	$n_{i2}^2(k)$	...	$n_{im}^2(k)$	$R_i^2(u_k^i-)$
Total	$n_{i1}(k)$	$n_{i2}(k)$	...	$n_{im}(k)$	$R_i(u_k^i-)$

Mantel and Haenszel (1959) and we have :

$$E\{n_{ij}^{1}(k)\} = n_{ij}(k) \cdot R_{i}^{1}(k)/R_{i}(k)$$

$$\mathrm{cov}\{n_{ir}^{1}(k), n_{is}^{1}(k)\} = \frac{R_{i}^{1}(k)R_{i}^{2}(k)n_{ir}(k)\{\delta_{rs}R_{i}(k) - n_{is}(k)\}}{\{R_{i}(k)\}^{2}\{R_{i}(k) - 1\}}$$

where $R_{i}(k)$ is an abbreviation for $R_{i}(u_{k}^{i}-)$.

Consider now the vector $x = \sum_{i,k} n_{i}^{1}(k)$. Conditionally on the marginal totals of all the tables, its expectation and covariance matrix are :

$$E(x) = \sum E\{n_{i}^{1}(k)\}$$

and

$$V = \sum V_{i}(k)$$

Let $v = x - E(x)$. Then the variable :

$$X_{1}^{2} = v'V^{-1}v$$

has approximately a χ^{2} distribution. The number of degrees of
freedom (d.f.) is $m - 1$ since the m terms sum to a fixed marginal total. Thus V^{-1} must be a pseudo-inverse or, more simply,
X_{1}^{2} can be computed from $m - 1$ components of v with the corresponding covariance matrix. This procedure is powerful if the
difference between the two groups go in the same direction, (i)
whatever the time, (ii) whatever the departure state. Such a
situation would occur if we were to compare two protocols for
hyperthyroidic patients (see section 2) and one protocol would give
a better control of the thyroid function so that the state "euthyroidism" would be favoured whatever the departure state whereas
"hyper" and "hypo" would be unfavoured. On the contrary, if the
two treatments differed by the fact that treatment 1 would be more
aggressive on the thyroid function than treatment 2 and if only few
patients could go directly from "hyper" to "hypo", then at the
departure of "hyper", "euthyroidism" would be favoured in treatment

1 whereas on the same treatment it would be more unlikely to go back
from "hypo" to "euthyroidism". However "hypo" would be always
favoured and "hyper" unfavoured. An objection to the use of this
procedure is that in many flow-charts, all paths are not possible
so that different states do not have the same set of directly attainable states and thus, the contingency tables do not have the same
number of columns. Such a situation occurs for instance, in the
example of epilepsy in section 2.

A more standard alternative hypothesis is that the differences
may depend on the departure state. We can apply the procedure
separately for each departure state. The basic statistic is then
$x(i)$ such that :

$$x_j(i) = \sum_k n^1_{ij}(k)$$

and for each departure state we compute $X^2(i)$ which has approximately a χ^2 distribution with $m_i - 1$ d.f., where m_i is the
number of states directly attainable from i. A global statistic
is :

$$X^2 = \sum_i X^2(i)$$

which has a χ^2 distribution with $\sum(m_i - 1)$ d.f. .

If this statistic indicates a difference between the two groups
the examination of the $X^2(i)$ may give further information : for
instance the difference may be concentrated at the departure of a
given state.

We may use a compromise between the two procedures, obtained
by grouping only a subset of the departure states.

A great advantage of these procedures is that it is possible
to adjust for concomitant factors. If we have C homogeneous
classes we just apply the same procedures to the $C.\sum K_i$ contingency tables.

We can extend the procedure to the comparison of two samples
from a SM process. We observe jumps from state i at sojourn

times $u_1^i, \ldots, u_{Ki}^i$. We can construct for time u_k^i exactly the same contingency table as above and apply exactly the same procedures. The only difference is that the times u_k^i do not have the same meaning : they are "absolute" times for NHM processes, sojourn times for SM processes.

Thus, it appears that the MH procedure is a very flexible tool in this context and we can use it even in a freer way by reverting to its original spirit. The MH procedure allows us to adjust for factors which may have an influence on the transition probabilities. Among them are the sojourn time and the time from first entrance in the study. We can make a partition of both time scales and constitute contingency tables in which the jumps occur for approximately homogeneous sojourn times and first entrance times.

We could also consider (as suggested by the referee) a complementary test obtained by constituting contingency tables relative to arrival states instead of departure states. The interpretation of such a test however would not be so easy and we do not treat it here.

Finally if we consider the last example of section 2 about dementia we see that the states are clearly ordered and the MH procedure cannot use this information. In the same spirit however we can use a procedure based on a Kruskal-Wallis statistic such as proposed by Hsieh, Crowley and Tormey (1983).

5. Conclusion

Both NHM and SM processes can be considered as generalizations of the survival process. The inference methods which can be used for these processes are also generalizations of those used for the survival process in the sense that both the Lagakos-Sommer-Zelen estimator for SM processes and the Aalen-Johansen estimator for NHM processes reduce to the Kaplan-Meier estimator for the survival process. The two-sample tests based on the MH procedure reduce to the Mantel-Haenszel test. Probably the application of the MH procedure needs some theoretical justification. Such a justification can be given either by the use of partial likelihood (Cox, 1975) or by using the results of the theory of counting processes (Andersen et al., 1982).

It is possible to choose between NHM and SM models on qualitative grounds but it is necessary to examine this choice in the light of the data. A problem of particular importance - and not yet perfectly resolved - is to test the basic Markov hypotheses because these hypotheses tend to multiply the number of observed independent events. If they are false the actual sizes of the tests based on them will be much larger than the nominal ones and this can lead to erroneous conclusions. Also it is a first important result of the analysis to know if the states which have been defined in a preliminary stage of the study are the states of a Markov process.

We have focussed here on non-parametric approaches but we must recognize that an advantage of semi-parametric models such as Cox' model (see Kay, 1984) over non-parametric ones is that dependencies from the past can be easily expressed by covariables so that the Markov properties can be tested.

References

Aalen, O.O. and S.Johansen (1978). An empirical transition matrix for non-homogeneous Markov chains based on censored observations. Scand.J.Statist.5, 141-150.

Aaling, D.W. (1958). The after-history of pulmonary tuberculosis : a stochastic model. Biometrics 14, 527-547.

Andersen, P.K., O.Borgan, R.Gill and N.Keiding (1982). Linear non-parametric tests for comparison of counting processes, with applications to censored data. Int.Statist.Rev. 50, 219-258.

Anderson, T.W. and L.A.Goodman (1957). Statistical inference about Markov chains. Ann.Math.Statist. 28, 89-110.

Commenges, D., P.Barberger-Gateau, J.F.Dartigues, P.Loiseau and

R.Salamon (1984). A non-homogeneous Markov chain model for
follow-up studies with application to epilepsy. Meth.Inform.
Med. 23, 109-114.

Cox, D.R. (1972). Regression models and life tables. J.Roy.Stat.
Soc. B 34, 187-220.

Cox, D.R. (1975). Partial likelihood. Biometrika 62, 269-276.

Flemming, T.R. (1978). Nonparametric estimation for non-homogeneous
Markov processes in the problem of competing risks. Ann.
Statist. 6, 1057-1070.

Gail, M.H., T.J.Santner and C.C.Brown (1980). An analysis of comparative carcinogenesis experiments based on multiple times to
tumor. Biometrics 36, 255-266.

Gill, R.D. (1980). Nonparametric estimation based on censored
observations of Markov renewal process. Z.Wahr.Verw.Geb. 53,
97-116.

Iosifescu, M. and P.Tautu (1973). Stochastic processes and applications in biology and medicine. Springer-Verlag.

Hsieh, F.Y., J.Crowley and D.C.Tormey (1983). Some test statistics
for use in multistate survival analysis. Biometrika 70, 111-119.

Kaplan, E.L. and P.Meier (1958). Nonparametric estimation from
incomplete observations. J.Amer.Stat.Assoc. 53, 457-481.

Kay, R. (1984). Multistate survival analysis : an application to
breast cancer. Meth.Infor.Med. 23, 157-162.

Lagakos, S.W., C.J.Sommer and M.Zelen (1978). Semi-Markov models
for partially censored data. Biometrika 65, 311-317.

Mantel, N. (1966). Evaluation of survival data and two rank-order
statistics arising in its consideration. Cancer Chemotherapy
Report 50, 163-170.

Mantel, N. and W.Haenszel (1959). Statistical aspects of the analysis
of data from retrospective studies of disease. J.Nat.Cancer
Inst. 22, 719-740.

Peto, R. and J.Peto (1972). Asymptotically efficient rank invariant
test procedures. J.Roy.Statist.Soc. A 135, 185-207.

Temkin, N. (1978). An analysis for transient states with application
    to tumor skrinkage. Biometrics 34, 571-580.
Voelkel, J. and J.Crowley (1984). Nonparametric inference for a class
    of semi-Markov processes with censored observations. Ann.Stat.
    12, 142-160.
Weiss, G.H. and M.Zelen (1965). A semi-Markov model for clinical
    trials. J.Appl.Prob. 2, 269-285.

APPLICATION OF SEMI-MARKOV PROCESSES TO HEALTH CARE PLANNING :
A MODEL TO STUDY THE FLOW OF PATIENTS FOLLOWING THE DIALYSIS-
TRANSPLANTATION PLAN

R.Polesel [1] and G.Romanin-Jacur [2]

[1] Seminario Matematico - Università di Padova

[2] Istituto di Elettrotecnica e di Elettronica
Università di Padova and LADSEB-CNR

1. Introduction and problem statement

Among the applications of Operations Research in medicine and
health care, those aiming at a good utilization of available facilities are particularly important. In this context a major problem
is in forecasting future care-needs for a given pathology in a given
geographical area, in order to compare the present resources with
those which are expected to be required in the future, and to act
accordingly (see, e.g., Boldy, 1981).

In the present paper we consider forecasting the demand for
apparatus and medical assistance related to patients with End Stage
Renal Disease (E.S.R.D.) being treated at the Padova University
Children's Hospital (P.U.C.H.).

E.S.R.D. in children is an infrequent disease which needs very
peculiar medical care. On one hand, the low incidence of E.S.R.D.
leads to the necessity of considering a large (and heterogeneous)
population supplying patients to one care delivery centre, as it
is not economically convenient to employ several expensive resources
at a very low utilization level. On the other hand these patients
are very similar as concerns their medical treatment : at the consi-

dered hospital the so called "Dialysis-Transplantation Plan" is adopted; this treatment is followed by all children suffering from E.S.R.D.

At the beginning of this care plan, the E.S.R.D. patient is established on hemodialysis until a matching kidney is available for transplantation; in the post-transplant period, in order to oppose the rejection, which almost always takes place, the patient needs strict medical assistance (periodical checks and other kinds of treatment). If a rejection takes place, he/she will come back to dialysis, waiting for successive transplantation, when possible. This cycle may wholly repeat once or twice again. As will be shown in the following, this plan seems very suitable for description by a semi-Markov model. Provided that the forecasting period is five years, a stationary model like this seems to be appropriate, that is, possible future changes in the treatment of patients would not considerably affect the model parameters.

In recent literature several semi-Markov models have been adopted to describe the flow of patients through various care units (hospital wards, home care, day-hospital, care plans, etc.). In particular we refer to Kao (1973 and 1974), Hershey et al. (1981) and Weiss et al. (1982), to note that a model of this kind has great theoretical capabilities to represent health care systems.

Kao (1973) suggests a previous partitioning of the population into various patient groups, so that, in each group, patients have the same transition probabilities and identically distributed waiting times, thus assuming that the state space is common to all patient groups. Weiss et al. (1982) give a methodology to specify and apply a semi-Markov model (giving an example for the case of an obstetric unit), whereby they suggest widening the state space (possibly by redefining it) until it includes all care units where any patient may stay in during his/her clinical path.

In our opinion, a general model concerning a whole group of pathologies, like renal disease, may be usefully decomposed into different models, where the state spaces and/or the population do not necessarily coincide; a restricted model, avoiding excessive generalizations, clearly can be more accurate and precise. In the specific case, the dialysis-transplantation plan can be thought as a subset of the whole health-care system for pediatric renal diseases. In Polesel (1983) we give an example of a semi-Markov model inserted in another wider semi-Markov model, which avoids both complex enlargements of the state space and further population subdivisions.

In the following sections we propose a semi-Markov model describing the recovery progress of patients following the Dialysis-Transplantation Plan; more precisely, in section 2, the mathematical model is presented in detail, while in section 3 we describe its implementation on a computer, and we briefly discuss the results obtained.

2. The mathematical model

2.1. Disease dynamics

As described in the previous section, the clinical history of patients following the "Dialysis-Transplantation Plan" may be described by the digraph in figure 1, where nodes represent care units and arcs represent allowed movements.

Here we point out that :

(i) From an analysis of the disease evolution, we can say that the behaviour of every patient in each clinical state depends only on the time elapsed since the patient entered that state;

(ii) the distinction between states 1 and 3 is necessary as a different behaviour has been noted between the patients under dialysis for the first time and those dialyzed after rejection; the same holds for the patients wearing a transplanted kidney for

node	clinical status	care unit
1	ESRD (waiting for transplant)	hemodialysis
2	Post-transplant (1st time)	periodical checks
3	ESRD after graft rejection	hemodialysis
4	Post-transplant (2nd, 3rd)	periodical checks
5	Death	----

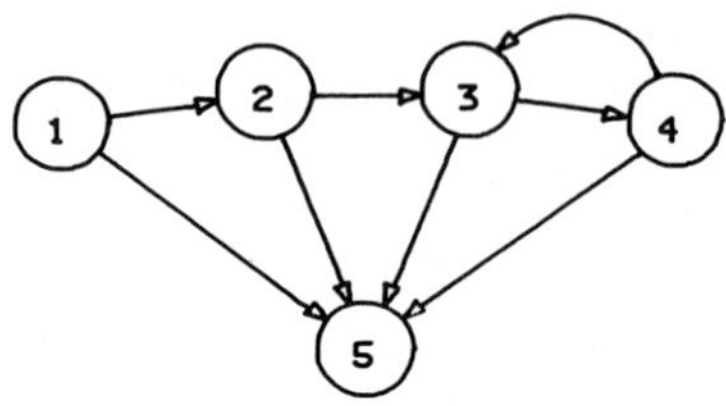

Fig.1. Clinical history of the patient flow related to End State
       Renal Disease.

the first time and those already transplanted twice or more times
(states 2 and 4).

We thereby build up a model of such a clinical history, based on
the following main assumptions :

(A) the clinical history of each patient with ESRD can be thought
     as the evolution of a semi-Markov discrete-time process;

(B) the clinical history of each patient is independent on those
     of other patients;

(C) the clinical history of each patient has finite length, i.e.,
     the semi-Markov process has an absorbing state;

(D) the number of new patients in each month is a random variable,
     and we can estimate its mean and variance.

By developing hypothesis (A) we will assume that, at every time
unit of his/her history, each patient lies in one of the states of
set $E = \{1,2,3,4,5\}$ (see figure 1) corresponding to the set of
care-units; each patient makes a transition from state i to state
j with probability p_{ij} (for any state k previous to i), after
waiting in state i the time τ_{ij} where τ_{ij} is a random variable

having $h_{ij}(\cdot)$ as discrete time probability density function and $H_{ij}(\cdot)$ as probability distribution function.

Let $\widetilde{w}_i(t)$ be the probability that the sojourn time in state i is more than t time units. Then we have :

$$\widetilde{w}_i(t) = \sum_{j\in E} \sum_{m\geqslant t+1} p_{ij}h_{ij}(m) = \sum_{j\in E} \sum_{m\geqslant t+1} c_{ij}(m)$$

where $c_{ij}(\cdot)$ is called the "core probability function" (see Howard, 1971).

Let then $\phi_{ij}(t)$ the probability that the patient is in state j in time t, given that he entered state i at time zero (multistep transition probability); then :

$$\phi_{ij}(t) = \sum_{k\in E} \sum_{m=1}^{t} c_{ik}(m)\phi_{kj}(t - m) + \delta_{ij}\widetilde{w}_i(t)$$

$$\phi_{ij}(0) = \delta_{ij}$$

where δ_{ij} is the Kronecker delta.

The following results are standard and easily derived (see, e.g., Howard, 1971) :

$$\begin{cases} \pi_i(t) = \sum_{j\in E} \pi_j(0)\phi_{ji}(t) \\[2em] \pi_i(0) \quad \text{known} \end{cases}$$

where $\pi_i(t)$ represents the state probability at time t;

$$\gamma_{ki}(s \mid r) = \frac{c_{ki}(s + r)}{\widetilde{w}_k(r)}$$

is the probability that the patient will make the next transition to state i in exactly s time units, given that he is in state k and has been there for exactly r time units

$$\Psi_{ki}(t \mid r) = \sum_{j\in E} \sum_{s=1}^{t} \gamma_{kj}(s \mid r)\phi_{ji}(t - s) +$$

$$+ \delta_{ki} \sum_{j\in E} \sum_{s\geqslant t+1} \gamma_{kj}(s \mid r) \qquad (s > 0)$$

is the probability that the patient <u>will be in</u> state i at time t,
given that he presently is in state k and has been there for exactly
r time units.

We can observe there that the estimation of p_{ij} and $h_{ij}(\cdot)$
and the knowledge of $\pi_i(0)$, $\forall$ i, j $\in$ E will give us the probabilistic knowledge of the state of one patient at time t, given his
present state and the time has already been in it.

2.2. Semi-Markov population model

To the semi-Markov process describing the disease dynamics, the
following "population model" is naturally associated (see Howard,
1971) : Let us consider the set {0,1,2,3,4,5} including the five
states previously described and the new state 0 (which will be associated with the external population) and let us think of these
states as infinite capacity "compartments". Let us then consider a
population (the patients) where each individual moves among compartments 1,...,5, according to the semi-Markov law defined in section
2.1; let the movements of all individuals be independent on one
another (hypothesis (B)). Let us further suppose (hypothesis (D))
that the individuals' flow from compartment 0 to one of the compartments in {1,2,3,4}, at every time t, is described by a r.v. X_t
with mean μ_t and variance σ_t^2. The situation may be represented
by the block-diagram in figure 2.

If we refer to the application to the real case considered, we
call "old patients" the ones already present in the system (subset
{1,2,3,4}) at the fixed time zero, and "new patients" those who
will enter the system at a future time. The discrete time unit will
be the month, which we judge suitable for our purposes. Finally
let zero be the present time, that is the initial time of the prediction.

block	condition or care unit
0	External world
1 , 3	Hemodialysis Treatment
2 , 4	Post-transplantation Treatments
5	Death

Fig.2. Block-diagram representing care units in the Dialysis-
Transplantation Plan.

2.3. Number of patients at the future month t

For the old patients, if M_k is the maximum sojourn time in state k, let $\underline{N} = [N_{kr}]$, $(k = 1,2,3,4;\ r = 1,2,\ldots,M_k)$ be the bi-dimensional "information vector", where the generic entry N_{kr} represents the number of patients who presently are in state k and have been there for r months. Obviously, in this case the information vector is known, i.e., $N_{kr} = n_{kr}$ $\forall\, k,\ \forall\, r$.

Let us then define :

$c_{ki}(t \mid r)$ number of old patients who will be in state i at month t, who have been in state k for r months

$c_i(t)$ total number of old patients in state i at month t.

By definitions of $\Psi_{ki}(t \mid r)$ we have that the discrete density function of $c_{ki}(t \mid r)$, given $N_{kr} = n_{kr}$, is a binomial function of parameters n_{kr} and $\Psi_{ki}(t \mid r)$, that is

$$f(c_{ki}(t \mid r) \mid N_{kr} = n_{kr}) = B(n_{kr}, \Psi_{ki}(t \mid r))$$

where $B(n,p)$ is the binomial distribution of parameters n and p; hence we obtain :

$$E\{c_i(t) \mid N = n\} = \sum_{k=1}^{4} \sum_{r=1}^{M_k} n_{kr}\, \Psi_{ki}(t \mid r)$$

$$\text{Var}\{c_i(t) \mid N = n\} = \sum_{k=1}^{4} \sum_{r=1}^{M_k} n_{kr} \Psi_{ki}(t \mid r)[1 - \Psi_{ki}(t \mid r)]$$

For new patients, let X_m be the number of new patients to arrive at future month m, having μ_m, σ_m^2 as mean and variance according to hypothesis (D). Let then $D_i(t \mid m)$ be the number of new patients, who entered the system at month m, and will be in state i at month $t \geqslant m$ ($i = 1,\dots,4$, $m = 1,\dots,t$).

By definition of $\pi_i(t)$, we have :

$$f(D_i(t \mid m) \mid X_m = x_m) = B(x_m, \pi_i(t - m))$$

where $B(n,p)$ is the binomial density function of parameters n and p. Hence :

$$E\{D_i(t \mid m) \mid X_m = x_m\} = x_m \pi_i(t - m)$$

As in this case the conditioning event is not sure, that is : $E\{D_i(t \mid m) \mid X_m\}$ takes the values $E\{D_i(t \mid m) \mid X_m = x_m\}$ with probability $p\{X_m = x_m\}$, we obtain :

$$E\{D_i(t \mid m)\} = \mu_m \pi_i(t - m)$$

and the number $D_i(t)$ of new patients in state i at time t has the following mean and variance :

$$E\{D_i(t)\} = \sum_{m=1}^{t} \mu_m \pi_i(t - m)$$

$$\text{Var}\{D_i(t)\} = \sum_{m=1}^{t} \mu_m \pi_i(t - m)[1 - \pi_i(t - m)] + \sum_{m=1}^{t} \sigma_m^2 [\pi_i(t - m)]^2$$

Note

We assumed we knew $\pi_i(0)$, that is the probability that a new patient will enter state i.

In our case, actually, we know that $\pi_1(0) = 1$, while $\pi_i(0) = 0$ $\forall\ i \neq 1$, but this can always be obtained just adding a fictitious "arrival state" to the state space.

Finally, recalling the independence of arrivals (hypothesis (B))
and of patients movements (infinite capacity compartments), and
letting :

$R_i(t)$ = total n. of patients in state i at month t,

we have :

$$E\{R_i(t) \mid N = n\} = E\{C_i(t) \mid N = n\} + E\{D_i(t)\}$$

$$Var\{R_i(t) \mid N = n\} = Var\{C_i(t) \mid N = n\} + Var\{D_i(t)\}.$$

3. Application of the model to the nephrology ward of Padova University Children's Hospital

In this section we give the results of the application of the
model studied in section 2 to the Nephrology Ward of the P.U.C.H.
with regards to the ESRD patients.

The purpose of the application is to obtain the mean and variance of the number of patients in each considered care unit during
the next five years, in order to dimension suitably the care facilities of the ward.

3.1. Data gathering and computation of model parameters

Concerning states 1 and 3 we acted as follows :

1) probability distribution functions $H_{12}(\cdot)$, $H_{15}(\cdot)$, $H_{35}(\cdot)$
of the sojourn times, in states 1 and 3 respectively, which repre-
sent the time spent while waiting for an available transplanting
graft, are clearly dependent upon the local environment, and therefore were obtained from data regarding all patients who followed
treatment in the considered ward. We just remark that : (i) the
number of patients is presently very low, (ii) kidney transplantation
in children is a very recent medical practice. As a consequence
these functions were drawn by using the method of the "empirical

function", as no analytical fit appeared to be convenient : moreover each sojourn time came out to be in any case shorter than 100 months.

2) p_{12} and p_{34}, i.e., the probabilities that a patient will survive until a kidney transplantation is effected (and their complements, respectively p_{15} and p_{35}) were judged not to be tied to the local environment, and were drawn from the specific medical literature (Proceedings of the "European Dialysis and Transplantation Association", 1980 and 1981; Pediatrics, 1981).

Concerning states 2 and 4, related to transplanted patients, all parameters were taken from the medical literature (Proceedings of the "European Dialysis and Transplantation Association, 1980 and 1981; Pediatrics, 1981), sometimes by putting together rather heterogeneous data. Note that states 2 and 4, slightly different from the clinical point of view, were allotted the same parameters : this was actually due to the extremely low number of patients who have reached state 4 up to now.

The state transition matrix $\underline{P} = [p_{ij}]$ and the sojourn time discrete density functions $k_{ij}(\cdot)$ are given in table 1.

Initial conditions n_{kr} (see 2.3) were taken as the number of patients who, at the zero-time of the forecast (June, 1983), had been in each state k for r months. These numbers are, of course, related to the actual situation. The table with numbers n_{kr} has not been given because of its wideness.

Concerning new patients, we divided them into "local" (i.e., coming from the region where the Children's Hospital is placed, i.e., under assistance of the same health-care authority) and "external".

For "local" patients the incidence of the disease given by the number of new patients per million children population and per year was taken as constant (on physicians' advice) and equal to 3.54 (average of last 10 years); the total child population in the region for the new 5 years was taken from the official forecast (Annuario Istat, 1982).

Table 1. Transition matrix $\underline{P}$ and discrete distribution functions
$H_{ij}(t)$ of the holding times :
a) Matrix $\underline{P}$
b) Function $H_{12}(t)$
c) Function $H_{34}(t) = H_{35}(t)$
d) Function $H_{15}(t)$
e) Function $H_{23}(t) = H_{43}(t)$
f) Function $H_{25}(t) = H_{45}(t)$

a)

(i) \ (j)	1	2	3	4	5
1	.00000000	.88607716	.00000000	.00000000	.11392284
2	.00000000	.00000000	.22641510	.00000000	.77358490
3	.00000000	.00000000	.00000000	.33333333	.66666667
4	.00000000	.00000000	.22641510	.00000000	.77358490
5	.00000000	.00000000	.00000000	.00000000	1.0000000

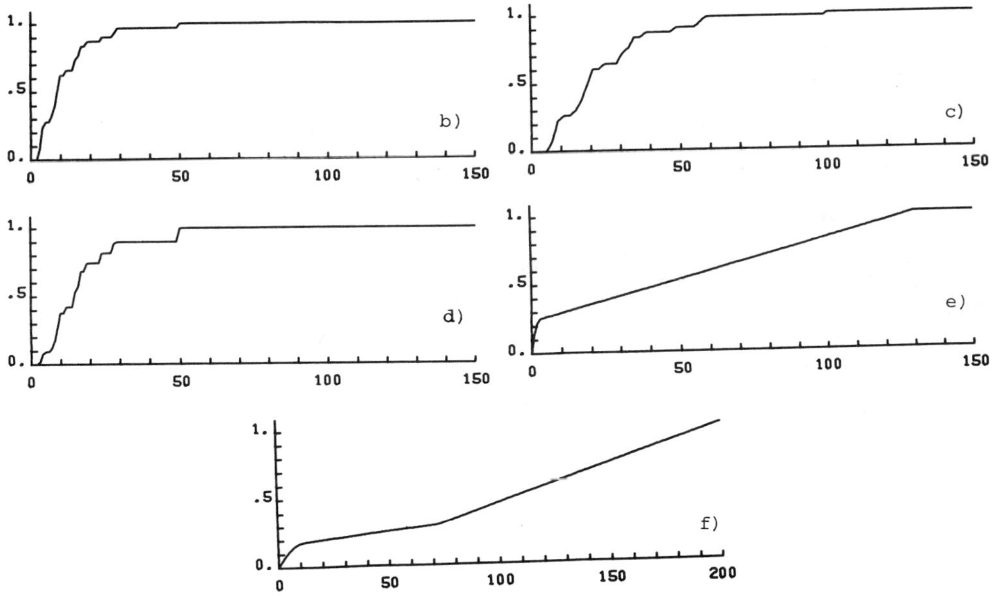

Table 2. Mean and variance of the number of patients following
          treatment at the future month t, for each of the signi-
          ficant states.

t	state 1		state 2		state 3		state 4	
	m	v	m	v	m	v	m	v
3	6.80	2.71	24.58	1.90	0.57	0.46	1.51	0.39
6	6.96	3.92	24.84	2.77	0.80	0.67	1.47	0.39
9	6.47	4.76	25.62	3.73	0.96	0.83	1.46	0.42
12	5.81	4.93	26.46	4.70	1.08	0.95	1.48	0.46
15	5.56	5.00	26.93	5.59	1.18	1.04	1.50	0.51
18	5.55	5.09	27.24	6.58	1.22	1.08	1.54	0.56
21	5.58	5.16	27.50	7.62	1.24	1.11	1.58	0.62
24	5.44	5.07	27.82	8.63	0.94	0.92	1.73	0.74
27	5.44	5.09	28.01	9.66	0.97	0.95	1.76	0.80
30	5.49	5.14	28.16	10.69	0.98	0.96	1.80	0.86
33	5.30	5.00	28.46	11.69	0.99	0.97	1.84	0.92
36	5.03	4.76	28.76	12.65	1.02	1.00	1.88	0.98
39	5.01	4.75	28.87	13.58	1.03	1.01	1.92	1.04
42	4.92	4.67	29.01	14.47	1.04	1.02	1.96	1.10
45	4.78	4.54	29.17	15.31	1.05	1.03	2.01	1.16
48	4.75	4.51	29.23	16.12	1.06	1.04	2.03	1.22
51	4.72	4.49	29.28	16.91	1.06	1.05	2.06	1.29
54	4.68	4.45	29.33	17.66	1.06	1.04	2.10	1.35
57	4.65	4.42	29.38	18.37	1.05	1.04	2.13	1.42
60	4.61	4.38	29.40	19.05	1.05	1.04	2.15	1.48

For "external" patients it was assumed that the number of new
arrivals will linearly decrease from the present average level (1.58
per year) down to zero in 25 years, according to the forecast of
the medical authorities.

3.2. Results

The model described in section 2 was implemented on a CDC 6600
computer in the language PASCAL.

The results of the implementation are presented in table 2
where the mean and variance of the number of patients for each state
are given. The sequence of numbers was shortened, so that the
time interval displayed in 3 months.

We also report, in figure 3, graphs showing, respectively, the
expected total number of patients and the expected number of patients
wearing a functioning graft.

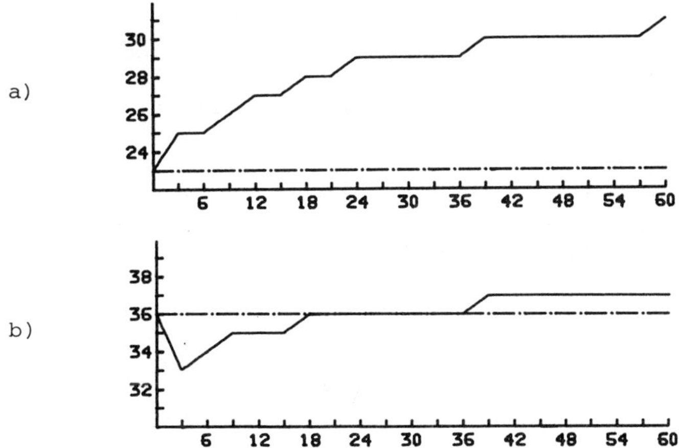

Fig.3. a) Number of patients wearing a transplanted graft in the
          forthcoming 60 months, compared with the present level.
       b) Total number of patients in the forthcoming 60 months,
          compared with the present level.

It is interesting to note that it is not expected that any
enlargement of dialysis machinery will be needed, while we foresee
a constant (though less than linear) increase of the number of
transplanted patients, to be followed with periodical checks and
possible other treatments.

4. Conclusions

A semi-Markov model was studied to describe recovery progress
of ESRD patients following the "Dialysis-Transplantation Plan" at
Padova University Children's Hospital.

The model has been implemented and used to forecast the demand
for health care (hospital facilities and medical assistance) in the
forthcoming 5 years.

The model presented can be easily extended to a bigger semi-
Markov model describing all patients with Renal Disease following
treatment in the considered Hospital ward.

Aknowledgement

We like to thank Prof. Wolfgang Runggaldier, full professor of Mathematical Statistics at Padova University for his assistance and helpful suggestions.

References

Annuario Istat (1982). Italian Institute of Statistics.

Boldy, D. (1981). Operations Research applied to health services Croom Helm, London.

Hershey, J. et al. (1981). A stochastic service network model with application to hospital facilities. Opns.Res. n°29, 1-22.

Howard, R.A. (1971). Dynamic probabilistic systems. vol.II, J.Wiley.

Kao, E.P.C. (1973). A semi-Markov population model with application to hospital planning. IEEE Trans.on System, Man and Cybernetics, vol.3, n°4, 327-336.

Kao, E.P.C. (1974). Modelling the movement of coronary patients within a hospital by semi-Markov processes. Opns.Res., n°22, 683-699.

Pediatrics (1981). 67, n°3, 413.

Polesel, R. (1083). Un modello di previsione dinamica per il programma dialisi-Trapianto in pediatria. Master Thesis in Mathematics, Univ.of Padova (Italy) (unpublished).

Proceedings of the "European Dialysis and Transplantation Association" (1980).

Proceedings of the "European Dialysis and Transplantation Association" (1981).

Weiss, E.N. et al. (1982). An iterative estimation and validation procedure for specification of semi-Markov models with application to hospital patient flow". Opns.Res., n°30, 1082-1104.

SECTION X. APPLICATIONS IN OTHER FIELDS

APPLICATIONS OF SEMI-MARKOV PROCESSES : A MISCELLANY

Valerie Isham

Department of Statistical Science, University College
Gower Street, London WC1E 6BT

1. Introduction

Many of the important fields of application of semi-Markov
processes are covered by other sessions of this Symposium. Within
this session attention is focussed on two major areas :
a) social sciences, and especially manpower planning/mobility
   studies;
b) medicine, including the analysis of survival data;
(see the papers in this volume by D. J. Bartholomew and D.R.Cox,
respectively). In this introduction a few applications in other
areas will be discussed briefly. These areas are computer science,
storage processes, social sciences (health care) and biology.

2. Computer science

Most the applications of semi-Markov processes in computer
science can be classified under one or other of the following head-
ings : queueing theory, reliability, decision theory; they therefore fall outside the scope of this introduction.
However, the papers by Leung and Wolfenden (1983) and by Leung

(1983) give an interesting application concerned with the evaluation
of the performance of a computer disc storage system.

The main features of a disc storage device are shown in figure
1. Each disc has a set of concentric tracks on each surface, with
all the tracks having the same capacity. The disc pack rotates about
a central axis. The access arm supports a set of colinear read-
write heads, with one head for each surface of each disc. A <u>cylinder</u>
consists of all the tracks accessible with the arm in a fixed position (that is, cylinder i is made up of all of the ith tracks on
each surface of each disc). Only one track of a cylinder can be
accessed at a time, but the time taken to change from one track to
another on the same cylinder can be regarded as negligible. On the
other hand, the <u>seek time</u>, which is the time taken to position the
arm correctly for the next transaction and of course depends on the

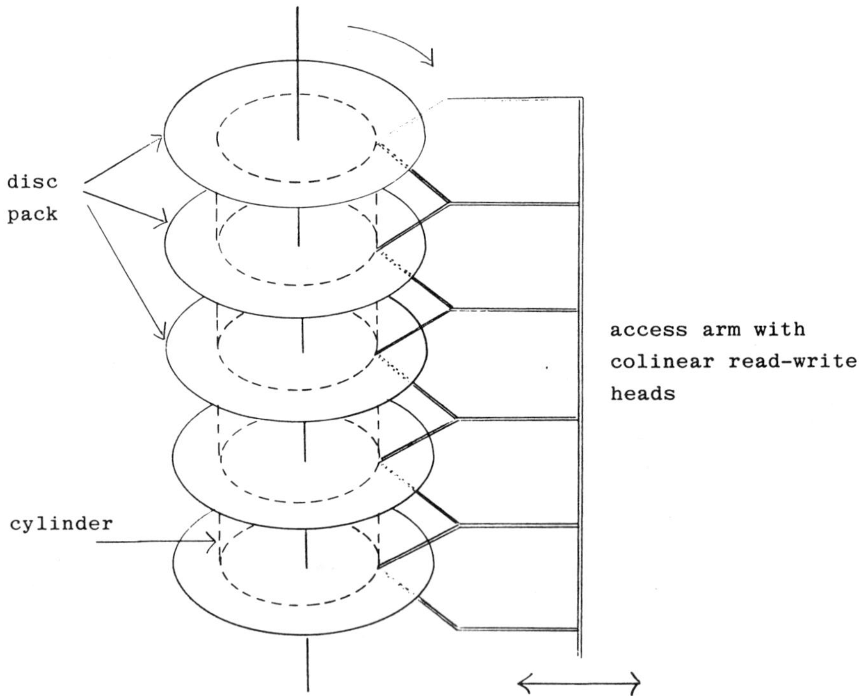

Fig.1. A computer disc storage device.

current cylinder as well as the next cylinder to be accessed, is relatively long and cannot be ignored.

Leung and Wolfenden use the following semi-Markov process to model the operation of a data base stored on a disc. Clearly, any logical structure of the data base induces a corresponding structure on the cylinders of the disc pack. The state of the process is taken to represent the cylinder currently being accessed; typically there might be 400 cylinders in the disc pack. On entry into state i, a random holding time is selected from a distribution depending upon i. This holding time represents the "useful" processing time, that is, the time taken for the particular operation on cylinder i on that visit to be carried out. This operation might be to read or write some number of blocks of information. The probability of transition to the next state depends on the current state i and the holding time in that state. A <u>penalty</u> is associated with each transition from one cylinder to the next, where the penalty is the "non-useful" processing time and consists of the seek time for the transition plus a rotational delay until the index marker reaches the read-write head.

The efficiency of a data base structure can then be measured by the total penalty incurred per unit time of useful processing, for the system in equilibrium, and provides a means of comparison of different data base structures.

In the special case of <u>locality referencing</u> (Leung, 1983) the information is arranged on the data base in such a way that transitions to non-adjacent cylinders are much less common than those to adjacent cylinders. If in the semi-Markov model above, p_{ij} denotes the transition probability that cylinder j is accessed next after cylinder i, then the aim of locality referencing is that this embedded Markov chain is such that the following conditions hold :

$$p_{i\ i-1} = p_{i\ i+1} = a > 0 , \qquad 1 < i < c ,$$

$$p_{12} = p_{1c} = p_{c1} = p_{c\ c-1} = a ,$$

$$p_{ii} = 0, \quad 1 \leqslant i \leqslant c,$$

with all other transition probabilities satisfying

$$p_{ij} = b \ll a,$$

and where c is the total number of cylinders.

As before the total penalty incurred per unit time of useful processing can be obtained, and investigated as a function of a.

3. Storage processes

Most semi-Markov models for storage processes are decision processes. However no explicit decision element is incorporated in a model considered by Puri and Woolford (1981). They base a model of the content of a semi-infinite store on a Markov renewal process $\{(X_n, T_n), n = 0, 1, \ldots\}$, where $\{T_n\}$ represents the sequence of input times, with output being regarded simply as negative input. The states $\{X_n\}$ belong to a countable set and determine the "type" of input to the store. That is, if $X_n = j$ then the nth input has distribution function F_j, where all the inputs are mutually independent. In a simple special case the X_n's might be a sequence of binary variables with, say, $X_n = 1$ indicating a (positive) input into the store and $X_n = 0$ indicating a demand.

In this paper the application is used to prompt the development of new theoretical results. In particular, the authors consider the bivariate process $\{Z(t), L(t); t \geqslant 0\}$ where $Z(t)$ is the storage level at time t and $L(t)$ is the total demand lost (that is, which cannot be met immediately due to unavailability of stock) in $(0, t]$. The process $\{Z(\cdot)\}$ is taken to have a reflecting barrier at 0 and is held on the barrier until the next positive input occurs. Any demands (negative inputs) occurring when $Z(\cdot)$ is zero are cumulated into $L(\cdot)$ as are any residual demands left unfulfilled when a demand exceeds the current stock level. The parameter ρ is defined to be the expected value of a single input quantity when the embedded Markov chain $\{X_n\}$, which is assumed to be ergodic, is in equilibrium. Puri and Woolford study the limiting distribution of $\{Z(t),L(t)\}$ as $t \rightarrow \infty$ in the three qualitatively distinct cases, when ρ is positive, negative and zero respectively. For example, in the super critical case, $\rho > 0$, $Z(t)$ and $L(t)$ are asymptotically independent and $Z(t)$ is, of course, asymptotically normally distributed, while in the subcritical, $\rho < 0$, case it is $L(t)$ that is asymptotically normal. The critical case, when $\rho = 0$, is perhaps the most interesting, and then $Z(t)$ and $L(t)$ are asymptotically dependent, each with a half-normal distribution.

4. Social sciences (health care)

There have been many applications of semi-Markov models in the field of health care. For the most part these involve a model specified in terms of a fairly small number of states, a transition matrix and a set of interval distributions. It is frequently the case that the semi-Markov model is chosen because it is the "next simplest" model once a full Markov model for the process has been rejected on the grounds that exponentially distributed intervals are inappropriate. The model tends to be used mainly as a convenient formalism within which to present and summarize the data. Simple predictions or deductions using the model may be made, for example, to aid decision-making on allocation of resources.

In many instances there is no attempt to check the various independence assumptions crucial to the semi-Markov model, which is validated only by the comparison of one or two summary measures for the observed data with those generated by the model. Weiss et al. (1982) describe in detail how the assumptions implicit in a semi-Markov model may be validated. They illustrate the procedures they discuss by applying them to data on the movements of obstetric patients between units within a large hospital. (A relevant paper playing a rather similar rôle for Markov processes is that by Tuma et al (1979)).

Some recent applications of semi-Markov models in health care
are described in Mode and Soyka (1980) and Hennessey (1980, 1983).
In each case large-scale longitudinal studies are involved. Time
is measured from the point of entry of each individual into the
study and the semi-Markov kernel is assumed to depend on the age of
the individual involved, where the complete age range is divided
into a small number of age-groups. Mode and Soyka are particularly
concerned with summarizing a vast amount of data and use non-parametric methods. In contrast, Hennessey (1980) assumes the intervals
between transitions have gamma distributions and uses maximum likelihood estimation procedures. Both these papers use comparison of
predicted and observed results as a means of judging their models.
On the other hand, in his later paper, Hennessey (1983) uses only
the first $3\frac{1}{2}$ years of a dataset covering $7\frac{1}{2}$ years, in order to fit
his model and then compares observed and predicted properties for
the next 4 years' data. In this case he fits proportional hazards
models for the instantaneous transition rates, (again using maximum
likelihood estimation), and thus incorporates covariates such as
age, primary diagnosis of the individuals and level of benefit (the
study is of the work histories of disability-insurance beneficiaries).
The aim of the study is to gain understanding of the working of the
insurance scheme, for example, with regard to work incentives.

One application in health care is described in this volume.
Polesel and Romanin Jacur fit a semi-Markov model to the flows of
child hospital patients with end-stage renal disease between states,
for example, "on haemodialysis-awaiting first transplant", "post
transplant-pre rejection" etc. Their purpose is to forecast the
demand for hospital care facilities over the next five years. It
must be noted that in any such prediction problem, in addition to
the appropriateness of the independence-type assumptions of the
semi-Markov model, it is crucial that the assumption that the process
is stationary is at least approximately true. Otherwise, of course,

changes in medical practice could render the predictions highly inaccurate.

5. Biology

Two recent papers by Becker and Kshirsagar (1981, 1982) mention applications in biology, though perhaps these are more accurately described as "potential" applications since no data analysis is attempted. In the earlier paper the authors consider the time and number of transitions before <u>termination</u> of the semi-Markov process. Various alternative definitions of termination of the process are discussed, for example, the entrance into an absorbing state or the occurrence of visits to a particular sequence of states. An application described there is that of epidemic modelling where the semi-Markov model provides an continuous time alternative to Markov chain models, and the epidemic terminates when the number of infectives falls to zero.

The later of the two papers considers the generalization of Cox and Smith's (1954) results on the superposition of renewal processes to semi-Markov processes. The applications discussed are models for human reproduction and predator-prey processes.

Hochman (1980) puts forward an interesting semi-Markov model for a process of neural discharges. In this model he assumes two independent Poisson processes, one of excitatory events and the other of inhibitory events. Each excitatory event increases a counter by one while an inhibitory event decreases the counter by an integer c, or to zero if the level of the counter does not exceed c. The value of c is taken first of all to be fixed and then, later, to be random. That is, independently for each inhibitory event, c is the observed value of an integer-valued random variable. A response (discharge) occurs when the counter reaches k and the counter returns to zero following each response. If the interval between successive events exceeds τ then the counter is again reset to zero. The density of the interval discharges is determined, albeit in terms of a matrix equation for its Laplace transform.

Finally, turning to unsolved problems, Professor Niels Keiding has drawn attention to an important biological field of potential application for semi-Markov models, that of cell kinetics. The basic idea is that during a cell cycle, each cell passes through a number of stages with cell division occurring at the end of the cycle. Transitions between the stages are, in general, not time-homogeneous and might, for example, involve periodic variation. So far, much of the modelling used in this area has been essentially deterministic and there is clearly scope for developing a suitable class of stochastic models. A discussion of cell kinetic models with an introduction to the literature on the subject is given in Keiding et al. (1984).

Aknowledgements

It is a pleasure to thank the organizers of the Symposium for all their efforts in arranging a most enjoyable and stimulating meeting.

References

Becker, M. & A.M.Kshirsagar (1981). Superposition of Markov renewal processes. S.Afr.Statist.J. 15, 13-30.

Becker, M. & A.M.Kshirsagar (1982). Terminating Markov renewal processes. Commun.Statist.-Theor.Meth. 11, 325-342.

Cox, D.R. and W.L.Smith (1954). On the superposition of renewal processes. Biometrika 41, 91-99.

Hennessey, J.C. (1980). An age dependent, absorbing semi-Markov model for work histories of the disabled. Math.Biosciences 51, 283-304.

Hennessey, J.C. (1983). Testing the predictive power of a proportional hazard semi-Markov population model. Math.Biosciences 67, 193-212.

Hochman, H.G. (1980). Models for neural discharge which lead to
     Markov renewal equations for probability density. Math.
     Biosciences 51, 125-139.

Keiding, H., Hartmann, N.R. and Moller, U. (1984). Diurnal variation
     in influx and transition intensities in the S phase of
     hamster cheek pouch epithelium cells. In Cell Cycle Clocks,
     ed. L.N.Edmunds, New York : Marcel Dekker.

Leung, C. (1983). A model for disc locality referencing. Computer
     Journal 26, 196-198.

Leung, C. & Wolfenden, K. (1983). Disc data base efficiency : a
     scheme for detailed assessment based on semi-Markov models.
     Computer Journal 26, 10-14.

Mode, C.J. & Soyka, M.G. (1980). Linking semi-Markov processes in
     time series - an approach to longitudinal data analysis. Math.
     Biosciences 51, 141-164.

Puri, P.S. & Woolford, S.W. (1981). On a generalized storage model
     with moment assumptions. J.Appl.Prob. 18, 473-481.

Tuma, N., Hannan, M.T. and Groeneveld, L.P. (1979). Dynamic analysis
     of event histories. Am.J.Sociol. 84, 820-854.

Weiss, E.N., Cohen, M.A. & Hershey, J.C. (1982). An iterative
     estimation and validation procedure for specification of semi-
     Markov models with application to hospital patient flow.
     Op.Res. 30, 1082-1104.

SOCIAL APPLICATIONS OF SEMI-MARKOV PROCESSES

D.J.Bartholomew

London School of Economics and Political Science

1. Introduction

Human populations can be divided into categories on the basis
of such things as place of residence, social class or level in a
management hierarchy. Individuals often move between categories in
an unpredictable manner so that an individual history consists of a
sequence of lengths of stay and a set of transitions between categories. There may also be losses from the system and gains to it.
These elements correspond to those of a semi-Markov process which
thus provides a possible model for such systems.

In an early paper, Vajda (1947), proposed a model for graded
manpower systems which was, in essentials, a semi-Markov model
though it was couched in the language of actuarial science.
Bartholomew (1967) also included examples of semi-Markov models
(Ch 2, Section 3 and Ch 5, Section 6) but it was not until Ginsberg
(1971) that such models were proposed in a systematic way for use
in a social context. Ginsberg's discussion was in relation to
social mobility which up to that time had often been modelled by
discrete time Markov chains. A discrete interval is more natural
for intergenerational mobility as in Prais' (1955) pioneering study
where the unit of time is the generation but it is often arbitrarily

imposed by limitations on what can be observed. For example, in Blumen, Kogan and McCarthy's (1955) work on labour mobility the data consisted of returns on occupational classification at quarterly intervals. This is typical of many situations where a continuous time process can only be observed at discrete points in time.

The starting point of many attemps to model continuous time processes has been the Markov process. But as empirical evidence has accumulated it has become clear that propensity to move from one state to another often depends strongly on length of stay in that state. An adequate model must incorporate this feature and the semi-Markov process meets the case. In practice, however, there has been a tendency to avoid the use of semi-Markov theory by redefining the state space so as to render the system approximately Markovian. For example, in manpower planning where the states are grades in a hierarchy new "grades" are formed by sub-dividing members of a grade according to their seniority. By making seniority approximately constant within each sub-grade the system can be modelled by a Markov process on the larger state-space. This approach is often adequate if one wishes to predict the aggregate behaviour of a system on certain assumptions about individual behaviour. In other applications, however, we wish to proceed in the opposite direction. We observe the aggregate behaviour of some social system and wish to deduce from it something about the behaviour of the individuals who compose it. In such cases the theory of semi-Markov processes comes into its own and provides the tool for gaining substantive knowledge. As we shall see there are severe limitations on what can be deduced from the sort of data that is usually available to us. Nevertheless broad qualitative conclusions about, for example, whether a transition rate is an increasing or a decreasing function may give valuable insights into human behaviour.

In the following sections some indications are given as to how the theory of semi-Markov processes can contribute to the understanding of social phenomena. In Sections 4 and 5 especially we introduce the idea of a "base-line" model in the sense used by sociologists (see, for example, Mayhew, 1984). Roughly speaking this is a very simple model with minimal substantive assumptions. By comparing what actually occurs with the predictions of the base-line model we hope to be able to deduce, qualitatively at least, which of the assumptions must be changed and in what direction.

2. Transition rates

A semi-Markov process may be defined by a set of transition rates $\{r_{ij}(t)\}$ $(i \neq j)$ which express the propensity of an individual with length of stay t in i to move to j. The form of these functions in any application may throw considerable light on the social or psychological processes involved. For example, McGinnis (1968) coined the term "cumulative inertia" for the situation in which propensity to leave a state declines monotonically with length of stay. If such a relationship can be identified it provides an empirical basis for theorizing about the nature of the ties which bind a person to their current category. The opposite tendency, in which propensity to move increases with sojourn time, is called cumulative stress and this might be indicative of increasing dissatisfaction with one's present location. In manpower systems transition rates sometimes rise initially to a peak before declining. This occurs when transitions are promotions. It reflects the fact that some experience is needed before promotions can be justified but after a certain point the chance decreases as it becomes clear that the individual does not have the qualities required. Within these classes of function particular interest may attach to the behaviour of the function for large t. If, with cumulative stress, the curve approaches a vertical asymptote (Vajda, 1947) there comes a time when renewal is certain whereas if it approaches a horizontal asymptote (Singer and Spilerman, 1979, see

below) behaviour reverts to that of a Markov process after a sufficiently long stay.

3. Complete data

If it is possible to observe individual behaviour in a system
continuously over a long period it is a relatively straightforward
matter to estimate the transition rates by parametric or non-parametric methods. The most extensive analyses of this kind in a
social context are those of Ginsberg (1978a & b & c, 1979) using
Scandinavian migration data. The prime object of such analyses is
to estimate the rates but something more than a simple description
is usually required. In a k-state system the number of parameters
to be estimated will often be of order k^2. What is required is
the most parsimonious description consistent with the data. If,
for example, it can be shown that propensity to move from state i
does not depend on the destination state there are considerable
gains in understanding of the basic process. In practice, when the
interval of observation is short, there may be statistical problems
arising from truncation and censoring of the kind discussed by
McLean elsewhere in this volume but these do not pose new questions
of principle.

4. Discrete observation of a continuous semi-Markov process

As we have noted above it is sometimes only possible to observe
the state of members of the system at discrete (usually equally
spaced) points of time. The question to be considered here is what
can be inferred about the model from those features which can be
observed at, perhaps, no more than two or three points. Using the
base-line approach we shall investigate how a semi-Markov process
can be distinguished from a continuous time Markov process which
here serves as the base line.

Suppose we observe the state of the system at times 0, τ, 2τ,

3τ ... and at each time we can identify the state to which each individual belongs. It will then be possible to estimate transition probability matrices relating to any interval of time which is an integer multiple of τ. If the system were a Markov process certain simple relationships would hold between those matrices. For example

$$(1) \quad \underset{\sim}{P}(r\tau) = [\underset{\sim}{P}(\tau)]^r$$

where $\underset{\sim}{P}(\cdot)$ is the matrix of transition probabilities relating to the interval indicated by its argument and where r is a positive integer. The question we now consider is how the two sides of (1) will differ if the process is semi-Markov.

In practice, starting with Blumen et al. (1955), it has often been noted that

$$(2) \quad p_{ii}(r\tau) > p_{ii}^{(r)}(\tau) \quad \text{for all} \quad i$$

where $p_{ii}^{(r)}(\tau)$ is the (i,i)th element of $[P(\tau)]^r$. Can we infer anything from the direction of the inequalities about the nature of the transition rates ? There appears to be no general answer to this question but investigation of special cases suggests the conjecture that (2) will occur if the $r_{ij}(t)$'s are monotonic decreasing functions and that the converse will hold if they are increasing functions. A simple case with two categories and decreasing rates which depend neither on the origin nor destination states was given in Bartholomew (1982, p.111). Singer and Spilerman (1979) considered an example with increasing rates given by

$$(3) \quad r_{ij}(t) = m_{ij}\lambda^2 t/(1 + \lambda t), \quad m_{ii} = 0$$

for which

$$(4) \quad \underset{\sim}{P}(\tau) = e^{-\lambda\tau}\{\cos h\lambda\tau\underset{\sim}{M}^{\frac{1}{2}} + \underset{\sim}{M}^{-\frac{1}{2}}\sin h\lambda\tau\underset{\sim}{M}^{\frac{1}{2}}\}$$

where $\underset{\sim}{M} = \{m_{ij}\}$. For sufficiently small τ it may be shown that

$$(5) \quad \underset{\sim}{P}^2(\tau) \doteqdot \underset{\sim}{P}(2\tau) + \tfrac{1}{4}\lambda\tau(1 - e^{-2\lambda\tau})(\underset{\sim}{I} - \underset{\sim}{M})$$

which implies $p_{ii}(2\tau) < p_{ii}^{(2)}(\tau)$ for all i.

It is easier to obtain somewhat weaker results of this kind by working with the trace of the transition matrices. Thus (2) implies $\text{tr } \underset{\sim}{P}(2\tau) > \text{tr } P^2(\tau)$ but not conversely. Nevertheless the trace inequality itself, if it could be established under wider conditions, would be a valuable diagnostic tool. For the example

$$(6) \quad \text{tr } P^2(\tau) = \text{tr } \underset{\sim}{P}(2\tau) + \tfrac{1}{4}e^{-2\lambda\tau} \sum_{n=1}^{\infty} \frac{(2\lambda\tau)^{2n}}{(2n)!} \sum_{i=1}^{k} \theta_i^{n-1}(1 - \theta_i)$$

where the θ's are the eigen values of M. If the θ's are real and positive, as they always appear to be in practice, and noting that $|\theta_i| \leqslant 1$ for all i then (6) establishes the trace inequality

$$(7) \quad \text{tr } \underset{\sim}{P}(2\tau) < \text{tr } \underset{\sim}{P}^2(\tau)$$

for all τ.

5. Heterogeneity

Heterogeneity is a universal feature of human behaviour. One might therefore expect that any aggregate behaviour we observe would be a mixture resulting from a variety of individual processes. The base-line approach here might be to compare the Markov process model with one which had its parameters varying between individuals. What kind of deviations from the predictions of the Markov model would such heterogeneity produce ?

We may define a Markov process as one in which changes of state are determined by a transition matrix $\underset{\sim}{P}$ at points in time which are a realisation of a Poisson process with rate parameter λ. The simplest way to introduce heterogeneity is to allow λ to vary according to some distribution $F(\lambda)$. For a given individual

$$(8) \quad \underset{\sim}{P}(\tau|\lambda) = \sum_{n=0}^{\infty} \frac{(\lambda\tau)^n}{n!} e^{-\lambda\tau} \underset{\sim}{P}^n$$

but for the aggregate population

$$(9) \quad \underset{\sim}{P}(\tau) = \int_0^\infty \underset{\sim}{P}(\tau|\lambda)dF(\lambda) = \sum_{n=0}^\infty P_n(\tau)\underset{\sim}{P}^n$$

If λ has a gamma distribution then $P_n(\tau)$ is a negative binomial distribution and the transition matrix may be formally expressed as

$$(10) \quad \underset{\sim}{P}(\tau) = (\frac{c}{c + \tau})^\nu \{\underset{\sim}{I} - \frac{\tau}{c + \tau} \underset{\sim}{P}\}^{-\nu}$$

where c and ν are the scale and shape parameters respectively of the gamma distribution. The new process is semi-Markov.

If one only has an estimate of $\underset{\sim}{P}(\tau)$ for a single value of τ there is no means of distinguishing one model from another but if data are available for intervals of different lengths discrimination becomes possible. Once again the diagonal elements of the transition matrices provide the clue. It was shown in Bartholomew (1982, p.102) that the trace inequality

$$\text{tr } \underset{\sim}{P}(2\tau) > \text{tr } \underset{\sim}{P}^2(\tau)$$

holds for any member of the family (9) provided that the eigen values of $\underset{\sim}{P}$ are real and positive.

Here we face a common dilemma in stochastic modelling in that the same inequality can be indicative of two different types of departure from the Markov model. It appears that a tendency to have more "stayers" in a category than the Markov model predicts can arise either because the longer a person stays the less likely they are to move or because the longer they stay the more likely they are to be a person with a low (but constant) propensity to leave. Heterogeneity and cumulative inertia thus lead to similar effects and cannot be distinguished if all we have to go on is the transition matrices. Although this reduces the usefulness of the trace inequality as a diagnostic tool it does focus attention on the directions which future empirical research should take.

In practice it is often possible to identify covariates which may give rise to heterogeneity. Their effect can be investigated if full data are available by expressing the transition rates as

functions of such variates. Congden (1984), following Lancaster
and Nickell (1980) assumed that

$$(11) \quad r_{ij}(t) = t^{\alpha_i} \exp \sum_h b_h x_h$$

where the x's are covariates and the b's and α's are constants.
The t^{α_i} term can cope with a monotonic increasing or decreasing
function and the size and significance of α_i shows whether any
cumulative stress or inertia remains after the heterogeneity represented by the x's has been eliminated.

Heterogeneity need not be confined to the rate parameter λ.
It is likely that $\underset{\sim}{P}$ will also vary from one individual to another.
Something is known about the effects of this for the discrete time
Markov chain (see Bartholomew, 1982, pp.33–38) where yet again the
trace inequality occurs. Theoretically the matter can be handled
by giving the elements of $\underset{\sim}{P}$ a probability distribution. This
was done for the Markov chain in Bartholomew (1975) by supposing
that the rows of $\underset{\sim}{P}$ have independent Dirichlet distributions. A
similar idea has been used in a marketing context by Goodhardt,
Ehrenberg and Chatfield (1984). Whether or not this leads to a
tractable theory remains to be investigated.

6. Future directions

The main line of approach in this paper has been to identify
ways of discriminating between a Markov process and various kinds
of semi-Markov process on the basis of estimates of transition
matrices over fixed intervals of time. In particular the pattern
of entries in the matrix $\underset{\sim}{P}(2\tau) - P^2(\tau)$ has been shown to provide
relevant information. It would be useful to know whether there are
other such matrices, possible involving $\underset{\sim}{P}(r\tau)$ for $r > 2$, which
yield further insight into the underlying process. It is also
desirable to have some sampling theory so that a judgement can be
made about the significance of any patterns observed.

We have already remarked on the need for estimation methods to deal with fragmentary data. These may be somewhat easier to handle by following the example of Thompson (1981) who used a discrete time version of the semi-Markov model for residential mobility in Ontario.

Although we have indicated the need for general theorems relating the semi-Markov model to observable properties of matrices such as $\underset{\sim}{P}(r\tau)$, the basis of our conjecture would be strengthened if further examples could be found of parametric forms of $\{r_{ij}(t)\}$ for which $\underset{\sim}{P}(\tau)$ can be explicitly evaluated. Phase-type distributions may be useful here.

It should be emphasized that the semi-Markov model is not the only generalization of the Markov model nor necessarily the most appropriate one. It is characteristic of any social environment that it is subject to change as a result of external shocks and internal evolutionary change. Non-stationarity is thus likely to be a feature of real processes that the semi-Markov model does not capture. This aspect may be detectable from transition matrices. For example, if we have a time-dependent Markov process it can be rendered stationary by a transformation of the time scale. There will then exist T_1 and T_2 with $T_2 > T_1$ but $T_2 \neq 2T_1$ such that

$$(12)\quad \underset{\sim}{P}(T_2) = \underset{\sim}{P}(T_1)\underset{\sim}{P}(T_2 - T_1)$$

If such T's can be found and the corresponding matrices estimated something can be learnt about the existence and degree of non-stationarity.

A second feature of social processes that both Markov and semi-Markov models fail to capture is the fact that individuals exert an influence over one another's behaviour. A simple way of incorporating this into a model is to suppose that individual behaviour is affected by the aggregate behaviour of others. In discrete time this gives rise to what Conlisk (1976) has called interactive Markov chains. In continuous time such models have existed for many years

in epidemic theory, for example, where they are simply referred to
as "non-linear". Non-linear processes may have several stationary
states rather than one and these may be stable or unstable. The
range of possible behaviour is then much greater as indicated, for
example, in Bartholomew (1984). There would be no difficulty in
incorporating such features into semi-Markov models.

A further use of stochastic models which has not been exploited
in the case of semi-Markov models is in suggesting suitable measures
of social phenomena. Social mobility, for example, refers to the
movement of people or family lines between social classes. Many
such measures have been proposed and the choice between them has
been much clarified by reference to the Markov chain model. This
has identified two distinct but related aspects of mobility and has
shown the eigen values of the transition matrix to be the key parameters (see Bartholomew, 1982, 2.3). In a semi-Markov model representing a mixture of Markov processes one would have to allow additionally for the varying rate at which different individuals change
state. But in any event the model serves to clarify matters to the
extent that any measure must be a function of the parameters of
the model.

References

Bartholomew, D.J. (1967). Stochastic Models for Social Processes,
    Wiley, Chichester.
Bartholomew, D.J. (1975). "Errors of prediction for Markov chain
    models". J.Roy.Stat.Soc., $\underline{37}$, 444-456.
Bartholomew, D.J. (1982). Stochastic Models for Social Processes,
    3rd ed. Wiley, Chichester.
Bartholomew, D.J. (1984). "Recent developments in non-linear sto-
    chastic modelling of social processes", Canad.J.Statistics,
    $\underline{12}$, 39-52.

Blumen, I., M.Kogan and P.J.McCarthy (1955). The Industrial Mobility
    of Labour as a Probability Process, Cornell University Press,
    Ithaca, New York.

Congden, P. (1984). Explanatory Models for Social and Occupational
    Mobility and their Application to the 1973 Irish Mobility
    Study, Ph.D.thesis, University of London.

Conlisk, J. (1976). "Interactive Markov chains", J.Math.Sociology,
    4, 157-185.

Goodhardt, G.J., A.S.C.Ehrenberg and C.Chatfield (1984). "The
    Dirichlet : A Comprehensive Model of Buying Behaviour". J.Roy.
    Statist.Soc., A 147, 621-655.

Ginsberg, R.B. (1971). "Semi-Markov processes and mobility",
    J.Math.Sociology, 1, 233-262.

Ginsberg, R.B. (1978a). "Probability models of residence histories :
    analysis of times between moves", in W.A.V.Clark and E.G.Moore
    (Eds), Population Mobility and Residential Change, Northwestern
    studies in Geography, N° 24. Evanston, Illinois.

Ginsberg, R.B. (1978b). "The relationship between timing of moves
    and choice of destination in stochastic models of migration".
    Environment and Planning, A 10, 667-679.

Ginsberg, R.B. (1978c). "Timing and duration effects in residential
    histories and other longitudinal data, II : studies of duration
    effects in Norway 1965-1971". Fels Discussion Paper n° 121.
    School of Public and Urban Policy, University of Pennsylvania.

Ginsberg, R.B. (1979). Stochastic Models of Migration : Sweden
    1961-1975, North-Holland, New York and Amsterdam.

Lancaster, A. and S.J.Nickell (1980). "The analysis of re-employment
    probabilities for the unemployed". J.Roy.Statist.Soc., A 143,
    141-165.

McGinnis, R. (1968). "A stochastic model of social mobility", Amer.
    Sociol.Rev., 33, 712-721.

Mayhew, B.H. (1984). "Base line models of sociological phenomena".
    J.Math.Sociology, 9, 259-281.

Prais, S.J. (1955). "Measuring social mobility". J.Roy.Statist.Soc., A 118, 56-66.

Singer, B. and S.Spilerman (1979). "Clustering on the main diagonal in mobility matrices". In Sociological Methodology (1979), K.Schuessler, (Ed.) Jossey-Bass, San Francisco, 172-208.

Thompson, M.E. (1981). "Estimation of the parameters of a semi-Markov process from censored records". Adv.Appl.Prob., 13, 804-825.

Vajda, S. (1947). "The stratified semi-stationary population", Biometrika, 34, 243-254.

LANGUAGE MODELLING USING A HIDDEN MARKOV CHAIN WITH APPLICATION TO

AUTOMATIC TRANSCRIPTION OF FRENCH STENOTYPY

Anne-Marie Derouault and Bernard Merialdo

IBM-France Scientific Center

36 avenue Raymond Poincaré - 75116 Paris, France

1. Introduction

Two kinds of problems arise in speech recognition : at the
acoustic level (phonetic recognition), and at the linguistic level
(lexical ambiguity).

Because of the high error rate of phonetic recognition, the
current speech recognition systems deal with a small vocabulary and
limited domain. These limitations are acceptable in some precise
applications (set of commands). However, in the future, it will be
highly desirable that speech recognition systems can handle a large
vocabulary and free syntax (for example automatic dictation).

In order to study these new difficulties, we considered the
stenotypy input because it gives good quality pseudo-phonetic data.

A possible approach is to build Markov models. They have been
used both for the acoustic and for the linguistic part of the recognition process (Jelinek, 1976). So far, language Markov models
have been built for a few thousand words English vocabulary (Bahl,
1983). Such models work at the word level. Their size increases
very rapidly with the size of the vocabulary. We have designed a

model working at the syntactic level, the size of which is independent of the number of words in the dictionary. The automatic parameter estimation allows us to build a more complete model than could be done by hand. The training on a collection of various texts gives a relative user independence to the model. The near context prediction is able to process "ungrammatical" sentences that may occur in oral speech.

Our model has been developed for French. French raises specific problems from a phonetic point of view, and for the dictionary complexity, as described in (Merialdo, 1984).

2. The stenotypy method

2.1. Description

Only one stenotypy method is used in France : the Grandjean method (see Grandjean, 1980). It allows real time coding of speech by typing on a special keyboard machine. This keyboard holds 21 keys, labelled as follows :

```
S P T * N   O A I N D
K M F R L Y E U L $ (
```

Speech in encoded pseudo-phonetically, one syllable at a time. A syllable is composed of :

(initial consonants) (vowels) (final consonants)

The 10 left keys code intial consonants. Y is used for the sound "je" and the semi-vowel "ye". Vowels are coded by the 5 keys E O A U I; E represents the sounds "é" and "è", the mute e is omitted. The 5 last keys are used for the final consonants.

The method makes the following phonetic confusions :

```
S = "se" or "ze"
K = "que" or "gue"
P = "pe" or "be"
```

```
M = "me"
T = "te" or "de"
F = "fe" or "ve"
R = "re"
N(initial) = "ne"
L(initial) = "le"
N(final) = "me", "ne", nasals
L(final) = "le" or "re"
$ = "se", "ze", "fe", "ve"
D = "de", "te", "pe", "be"
( = "que", "gue"
```

There are no separations between words, only end-of-sentence marks,
indicated by the steno symbol "*".

2.2. Transcription problem

This method is used in France mainly for recording conferences
and discussions.

After typing, the stenotypist must read the tape and transcribe
it into a written text. This process is rather slow, and the average
transcription time is 5 to 7 hours for 1 hour of speech.

The use of an automatic transcription system would reduce this
time and allow fast text input into computers.

2.3. Automatic transcription

The first tool is a steno-French dictionary. But, because of
the number of homonyms, and of the different possible boundaries for
words, the number of possible spellings for a given steno string is
very high.

For example, the steno for the French "quand nous avons" raises
the following ambiguities :

```
KAN       quand, quant ...        (homonyms)

          gant, Caen ...          (steno confusions)

NOU       nous, noue, noues ...

A         as, a ...

          avons ...               (boundary ambiguity)

FON       font, fond, vont ...
```

The average number of possible words starting at each steno syllable is 10. Therefore a given steno sentence can have several billions of possible spellings. So, for French, an automatic transcription system cannot rely only on a dictionary, and must include an algorithm to choose the correct spelling.

Our system makes this choice according to syntactic considerations. By using a Markov model of language, we can define the probability for a given sequence of words to be produced by this source. The spelling chosen by the system is the one having the highest probability.

We shall now describe every component of the system.

3. Dictionary

Our steno-French dictionary contains 140,000 entries. Each entry contains an orthographic form ("chantais" and "chantait" are two entries) together with its steno form, its possible parts of speech and frequency of occurrence.

We choose 92 parts of speech that seemed significant and useful to remove ambiguities arising in steno-French transcription. For example :

● We distinguish gender and number for nouns, adjectives, pronouns, because their corresponding forms are often homonyms.

● We put the six conjugation persons for verbs and pronouns, to write the correct endings ("je chantais, il chantait").

On the other hand, we did not take into account time or mode for verbs (except infinitive and participles), since they are not homonyms generally. Moreover, this knowledge would have no influence on the neighboring parts of speech.

4. Language model

4.1. Rigorous view and theoretical Markov model

We are interested in computing the probability of any string of words. The probability of occurrence of a word w_{n+1} should depend on the whole past string $w(1,2,\ldots,n) = w_1 w_2 w_3 \ldots w_n$. In the word strings space, $P(w(1,2,\ldots,n))$ would be the product of conditional probabilities :

$$P(w(1\ldots n)) = P(w_1) \prod_{i=2}^{n} P(w_i/w(1,2,\ldots,i-1))$$

Reliable estimates of these probabilities would require :

- a huge training corpus,
- a lot of storage.

It is natural to consider a model where the partial past string determine the state of the model. For example, we can operate a classification on the word strings. A given size corpus will be more representative for a coarser classification, but the prediction will be less precise.

4.2. A model for stenotypy transcription

We apply the equivalence :

$$w_1 w_2 \ldots w_n \rightarrow p_{n-1} p_n$$

where p_{n-1} (resp. p_n) is the part of speech (abbreviated as POS) of w_{n-1} (resp. w_n). The word production process is represented by a Markov source (Hidden Markov Chain), the states of which are the couples (p_1, p_2) of parts of speech, with initial and final state (Period,Period). The output alphabet consists of all the words in the dictionary.

From state (p_1,p_2), the probability of w_3 to be the next word is first written as :

(1) $p(w_3,p_3/p_1,p_2) = f(p_3/p_1,p_2) \times h(w_3/p_1,p_2,p_3)$

where :

• $f(p_3/p_1,p_2)$ is the relative frequency of occurrence of the part of speech p_3 following p_1p_2 in a known sample of French text :

$$f(p_3/p_1,p_2) = \frac{\text{count of } (p_1,p_2,p_3)}{\text{count of } (p_1,p_2)}$$

• $h(w_3/p_1,p_2,p_3)$ is approximated by $h(w_3/p_3)$, frequency of the word w_3, for a given part of speech p_3. We compute it by :

$$h(w_3/p_3) = \frac{\text{number of occurrences of } w_3 \text{ with p.o.s. } p_3}{\text{total number of occurrences of } p_3}$$

A French sentence now appears to be an output sequence produced from this Markov source. Its probability should be computed as the sum of the probabilities of all paths (transitions sequences) that can produce it. For computation simplicity, we estimate it by taking a maximum instead of a sum, and we carry a suboptimal search, as will be seen later.

4.3. Insufficient data problem

The corpus where the frequencies of triplets of parts of speech are collected is limited and does not cover all the possible sequences of p_1,p_2,p_3. So, any "new" sequence in a test would have a null probability if we took expression (1) for transition probability. To take care of this problem, we replace the transition probability by a linear interpolation with another distribution

which predicts the word according to the last part of speech. The
final expression of a transition probability is as follows :

$$p(w_3,p_3/p_1,p_2) = (\lambda_1 f(p_3/p_1,p_2) + \lambda_2 f(p_3/p_2)) \times h(w_3/p_3)$$

where λ_1, λ_2 are positive and of sum one. This interpretation
allows us to take advantage of both classifications : p_1, p_2 prediction is more precise and p_2 prediction is more reliable.

4.4. Parameter estimation

4.4.1. Frequencies

To get the basic frequencies we need a fairly large sample of
"labelled" text, i.e. where each word is associated with its right
part of speech in the context. To do this, we followed a semi-
automatic procedure.

We first labelled by hand a 2,000 word text, and collected
frequencies of pairs of parts of speech. Then we labelled automatically the next 18,000 words, by keeping for each sentence the
most probable sequence of POS according to these statistics (Vitterbi
alignment). With the bi-POS and tri-POS frequencies of this new
corpus (combined using fixed coefficients $\lambda_1 = \frac{1}{1.01}$, $\lambda_2 = \frac{0.01}{1.01}$, a
similar alignment allowed us to label the remaining 1.2 million
words. The error rate was estimated as less than 4%.

4.4.2. Weights

The coefficients λ_1 λ_2 can be seen as probabilities of null
transitions (i.e. producing no output) of a new Markov source :
between two states (p_1,p_2) and (p_2,p_3) let us add two fictitious
states s_1, s_2 representing the two equivalence classes (p_1,p_2)
and p_2 (see figure). The transition from s_1 (resp. s_2) to
(p_2,p_3) produces w_3, and its probability is the relative frequency

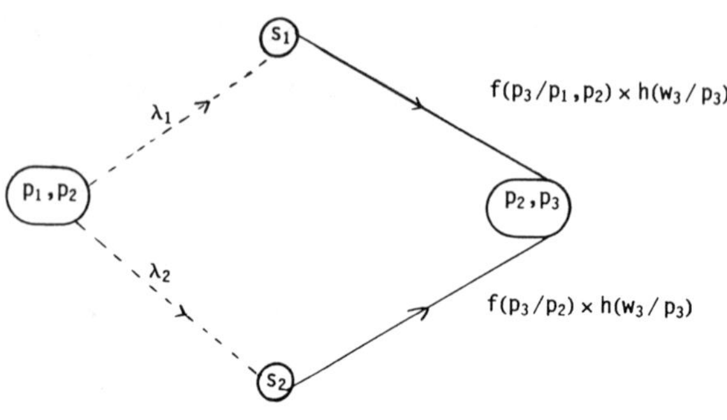

Fig. 1

$f(p_3/p_1,p_2)$ (resp. $f(p_3/p_2)$) times $h(w_3/p_3)$. Now λ_1 (resp. λ_2) is just the probability of the transition going from (p_1,p_2) to s_1 (resp. s_2). We impose that they depend only on the p_2 value (Derouault, 1984). These parameters can then be adjusted automatically from data by the Forward–Backward algorithm (Baum, 1972). They are chosen to maximize the probability of the observed output sequence (training text).

4.5. Advantages

This probabilistic method of natural language modelling has two great advantages :

- first, the model parameters are obtained automatically from real data. No a priori knowledge of the language syntax is needed.

- second, the system can accept incorrect sentences from a syntactic point of view, that are still likely to come up in spoken language.

5. Decoding

5.1. Confluent nodes

The set of all possible writings can be viewed as a graph. The nodes are points between two steno syllables, and the arcs are

words whose steno lies between two nodes. A writing is just a path
in this graph, starting from the first node and ending on the last
one.

Some nodes are used by all possible writings : they are called
confluent nodes ("CN"). They are always end of words. For example :

```
Je | passe  | par le | chemin
   | pas se | parle  | chemins
   | ...    | ...    | ...
   CN        CN        CN
```

To determine the next CN, we first search in the dictionary
for all possible words matching the beginning of the steno string.
Then we move to the right, to the first boundary of the words found,
and search again. We repeat this until we reach a point which is
the rightmost boundary among all boundaries of all words found so
far. This is a CN.

In practice, CN are very frequent (on the average every two
or three words).

5.2. Enumeration

Enumeration is carried out from left to right. At each step,
suppose we have determined two CN, namely CN_1 and CN_2, such
that :

    - the words up to node CN_1 have been fixed,

    - between CN_1 and CN_2, we have a list L_1 of sequences of
words, each sequence containing at least two words.
Then we determine CN_3 the first confluent node after CN_2 such
that every sequence of words from CN_2 to CN_3 contains at least
2 words (it is the first CN found or the second). We make a pre-
filtering of these sequences :

    - first we keep only those having the minimum number of words
or minimum +1 or minimum +2 (i.e. with "long" words).

- second we compute for each one a probability according to
bi-parts of speech, and we keep only those with "high" probability
(greater than 10^{-3} times the greatest), forming a list L_2.
We compute the probability according to the Markov model of every
concatenation of a sequence in L_1 and one in L_2. Then, the
sequence in L_1 appearing in the concatenation having the greatest
probability is chosen as the writing of the steno between CN_1 and
CN_2. We repat the process for (CN_2, CN_3) as for (CN_1, CN_2) until
the end of the text.

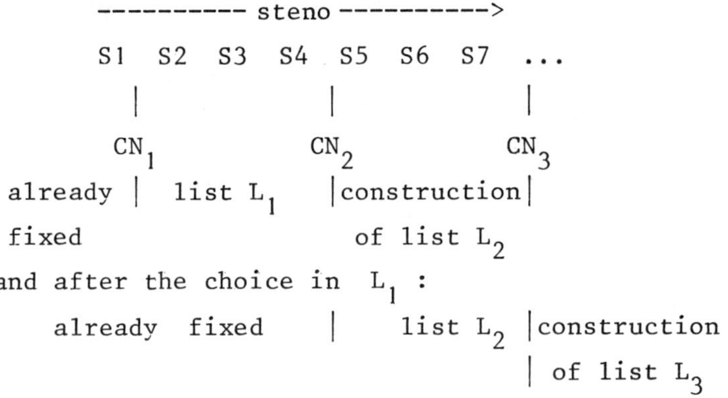

6. Results

6.1. Error rate

The system has been tested a 2,000 stenotyped words from TV
news. The error rate is 8.5% on words :
- 1.5% are due to unknown proper names.
- 1.0% are due to wrong boundaries choice.
- 2.4% are due to wrong agreement.
- 3.6% are due to bad choice of homonym.

6.2. Example

The original sentence is :

"c'est vers une heure que la fumée blanche s'est élevée au-dessus de la chapelle Sixtine après trois jours de conclave".

transcribed from its steno as :

"c'est vers une heure que la fumée blanche c'est élevé au-dessus de la chapelle six tine après trois de conclave".

In this example, we have 3 errors :
- 1 bad choice of homonym ("ce" instead of "se")
- 1 wrong agreement ("élevé" instead of "élevée")
- 1 unknown word ("six tine" instead of "Sixtine")

References

Bahl, L.R., R.L.Mercer, F.Jelinek, A.Nadas, D.Nahamoo, M.Pitcheny (1983). Recognition of isolated words sentences from a 5000 word vocabulary office correspondence task. Proceedings of the ICASSP, Boston, 1065-1067.

Baum, L.E. (1972). An inequality and associated maximization technique in statistical estimation of probabilistic functions of Markov process. Inequalities, vol.3, 1-8.

Derouault, A.-M. and B.Merialdo (1984). Language modelling at the syntactic level. Proceedings of ICPR, Montreal.

Jelinek, F. (1976). Continuous speech recognition by statistical methods. Proc IEEE 64, n°4, 532-556.

Merialdo, B. and A.-M.Derouault (1984). Recognition complexity with large vocabulary. Proceedings of ICCASP, San Diego.

SEMI-MARKOV PROCESSES IN SEISMIC RISK ANALYSIS

E. Grandori Guagenti and C. Molina

Dipartimento di Ingegneria Strutturale

Politecnico di Milano

1. Introduction

Even though the Poisson process has been widely used in the modelling of earthquake occurrences, it is well known that the temporal sequence of earthquakes in a given zone is not a sequence of independent events. In general the sequence has to be assumed to be a process with a memory of past events or, at least, non stationary.

In recent years both directions of research have been explored. The stochastic processes that have been proposed include the non-homogeneous Poisson process, the generalized Poisson process, regenerative point processes and stochastic processes with at least one step of memory, e.g. the renewal process or, more generally, semi-Markov process (Shlien and Toksoz, 1970; Vere-Jones and Ozaki, 1982; Esteva, 1982; Veneziano and Cornell, 1974; Patwardhan , Kulkarni and Tocher, 1980; Guagenti, 1979, Guagenti and Molina, 1981; Anagnos and Kiremidjian, 1984).

The choice of the model is based, in general, both on the statistical analysis of data and on the mechanisms that are supposed to govern the earthquake generation. In particular Lomnitz and

Lomnitz (1978) and Shimazaki and Nakata (1980) proposed two possible mechanisms. We can summarize them, respectively, as follows.

Case a. Slip predictable model

The energy that can be accumulated along a seismic fault has an initial lower bound. When a generic threshold of the accumulated energy is reached, the energy is suddenly released : the earthquake occurs. The correlated tectonic stress drops to a constant level after each event. The earthquakes can be of different sizes : the longer the lapsed time after an earthquake, the stronger the next one will be.

Case b. Time predictable model

The energy has an upper bound (maximum allowed energy). When this threshold is reached, the energy is suddenly released. The earthquake occurs, but the energy need not release completely : the earthquake can be of different sizes, size being measured by the coseismic slip correlated to the earthquake. In simple words, the stronger an earthquake is, the longer is the waiting time to next one.

Schematically the two behaviours are represented in figure 1, where the quantity S depends on the tectonic stress.

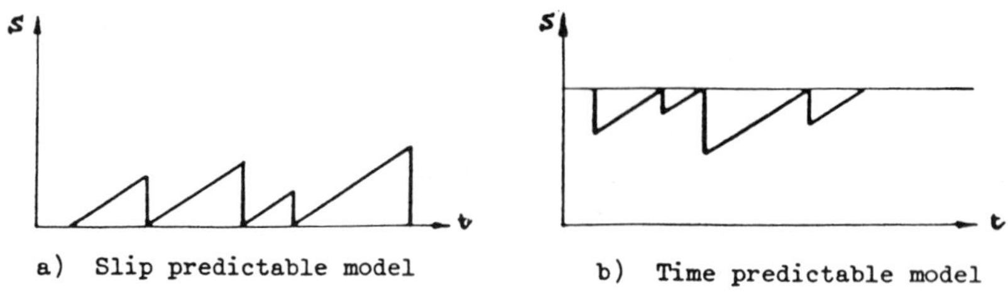

a) Slip predictable model   b) Time predictable model

Fig.1.

Both models assume a constant accumulation rate of tectonic stress.

In practice a more usual quantity, the magnitude M, is adopted as a measure of earthquake size. The relation between the interoccurrence time τ and the size M of the preceding (case b) or subsequent (case a) earthquake can be written

(1) $\tau = k \exp(aM + b)$

k, a, b being typical parameters of the fault.

The two models are different as regards the physical interpretation of the phenomenon and the prediction capacity. They formally coincide as far as the modelling of the earthquake process is concerned : if the deterministic relation (1) holds, the distribution function $F_M(m)$ of magnitude implies the distribution function $F_\tau(x)$ of interoccurrence times and vice versa.

In particular if τ is exponential (Poisson process) it can be shown (Lomnitz and Lomnitz, 1978) that

(2) $1 - F_M(m) = \exp[- k \exp(am + b)]$

A distribution of this form has been proposed independently (Guagenti, Grandori and Petrini, 1979) on the basis of statistical data; it seems a reliable size distribution, credible and stable in almost all seismogenetic regions.

The difference between the two models that are represented by (1) derives essentially from the different location of τ in respect of M (before or after). This implies that the conditional waiting times of the events of the process, given the size, do not coincide in the two models.

The present paper shows that

– when a continuous time-magnitude relation is deterministically known, the two earthquake behaviours can be modelled by a renewal process. All quantities of interest in risk analysis can be derived from it : the process of earthquakes of a given size, their hazard rate, their return period, their mean number, the cost of future

damage, and so on. It is shown that the return periods coincide in
the two cases, while the remaining quantities are different.

- when the deterministic time-magnitude relation is removed and
a finite number of sizes is assumed, the two physical behaviours
can be modelled as semi-Markov processes. However they do not coincide. The earthquake process is again a renewal process only in
the case a).

- the risk analysis is carried out in both cases.

- in all the above models the risk analysis depends on the time
instant at which the analysis is done; in particular it depends on
the elapsed time from the last earthquake. This dependence vanishes
in random starting conditions. The practical meaning of these conditions is discussed. In the time predictable model the risk analysis depends also on the size of the last earthquake.

- when the two physical behaviours cannot be treated as mutually exclusive, a more general semi-Markov model can be assumed and
again the quantities of risk analysis can be calculated.

- all the above models do not contradict the double exponential
magnitude distribution (2).

2. Slip predictable model

If the deterministic relation (1) holds (the interoccurence
times τ are assumed to be identically distributed) then (1) and
$f_\tau(x)$ define completely the earthquake process, which is a renewal
process. The magnitude distribution $F_M(m)$ is immediately derived
from (1) by expressing the generic value x of τ as a function
of m :

$$(3) \quad F_m(m) = F_\tau[x(m)].$$

In the two cases (a and b) the conditional probability density
functions (p.d.f.) $f_{M|\tau}$, $f_{\tau|M}$ are deterministic impulses for any
value of τ or M.

The return periods of events are quantities of particular interest in risk analysis. Let us now consider the return period of "strong" earthquakes, say with $M > \bar{m}$. Let $\bar{t}$ be the value of τ corresponding to $\bar{m}$ in (1) when it represents a slip predictable model. The probability p that an event is a strong earthquake is

$$(4) \quad p = 1 - F_M(\bar{m}) = \int_{\bar{t}}^{\infty} f\,(x)\,dx$$

Starting from an event, let τ_s and τ_w be the "holding times", given that the next earthquake is, respectively, a strong earthquake or a weak one $(M \leqslant \bar{m})$. Their densities h_s and h_w are :

$$(5) \quad h_s(t) = \begin{cases} \dfrac{1}{p}\,f_\tau(t) & t > \bar{t} \\[2mm] & \text{if} \\[2mm] 0 & t \leqslant \bar{t} \end{cases} \qquad h_w(t) = \begin{cases} \dfrac{1}{1-p}\,f_\tau(t) & t \leqslant \bar{t} \\[2mm] & \text{if} \\[2mm] 0 & t > \bar{t}. \end{cases}$$

The waiting time to the first strong earthquake density is

$$(6) \quad f_s(t) = f_{\tau_s}\,p + f_{\tau_w + \tau_s}(1-p)p + f_{\tau_w + \tau_w + \tau_s}(1-p)^2 p + \ldots$$

so that Laplace transform of $f_s(t)$ is $^{(*)}$

$$(7) \quad f_s^*(\gamma) = h_s^* p + h_w^* h_s^* (1-p)p + (h_w^*)^2 h_s^*(1-p)^2 p + \ldots$$

$$= \frac{p\,h_s^*(\gamma)}{1 - (1-p)h_w^*(\gamma)}$$

Equation (7) gives the density of recurrence time of strong earthquake; it holds for any value of the threshold $\bar{m}$ which is selected to define a "strong" earthquake in (4).

The r-th derivative of (7), at $\gamma = 0$, is the r-th moment of the recurrence time. In particular the return period of "strong" earthquakes is

---

(*) sometimes we left out the argument of the density functions and of their Laplace transforms.

$$(8) \quad \mu_s = - f_s^{*'}(0) = \frac{p\bar{\tau}_s + (1 - p)\bar{\tau}_w}{p} = \frac{E\{\tau\}}{p} = \frac{\mu}{p},$$

that it of great interest in seismic risk analysis. Note that $\bar{\tau}_w$ and $\bar{\tau}_s$ are the average interoccurrence times <u>before</u> a weak earthquake and <u>before</u> a strong one (μ is the average interoccurrence time).

In the analogous way we can find the density of recurrence time of "weak" earthquakes :

$$(9) \quad f_w^*(\gamma) = \frac{(1 - p)h_w^*(\gamma)}{1 - ph_s(\gamma)}$$

with return period

$$(10) \quad \mu_w = \frac{E\{\tau\}}{1 - p} = \frac{\mu}{1 - p} \ .$$

When (1) represents a time predictable model, let τ_s and τ_w be the "holding times", given that the last earthquake is, respectively, a strong earthquake or a weak one; h_s and h_w their densities. The waiting time to the first strong earthquake density is :

$$(11) \quad f_s^{(w)}(t) = f_{\tau_w} p + f_{\tau_w + \tau_w} p(1 - p) + f_{\tau_w + \tau_w + \tau_w} p(1 - p)^2 + \dots$$

or

$$(12) \quad f_s^{(s)}(t) = f_{\tau_s} p + f_{\tau_s + \tau_w} p(1 - p) + f_{\tau_s + \tau_w + \tau_w} p(1 - p)^2 + \dots$$

according to the size of the last earthquake : $f_s^{(w)}$, if the last size is weak; $f_s^{(s)}$ if the last size is strong.

The Laplace transforms are

$$(13) \quad f_s^{(w)*}(\gamma) = \frac{ph_w^*(\gamma)}{1 - (1 - p)h_w^*(\gamma)}$$

$$(14) \quad f_s^{(s)*}(\gamma) = \frac{ph_s^*(\gamma)}{1 - (1 - p)h_w^*(\gamma)}$$

Then the return period of the strong earthquakes, $- f_s^{(s)*'}(0)$, is again given by (8) in which $\bar{\tau}_w$ and $\bar{\tau}_s$ are the average inter-occurrence times <u>after</u> a weak earthquake and <u>after</u> a strong one.

Naturally, if (1) is replaced by a Bernoulli assumption, (7) (13) (14) and (9) become :

$$(15) \quad f_s^*(\gamma) = \frac{p f_\tau^*(\gamma)}{1 - (1 - p) f_\tau^*(\gamma)}$$

$$(16) \quad f_w^*(\gamma) = \frac{(1 - p) f_\tau^*(\gamma)}{1 - p f_\tau^*(\gamma)}$$

Equations (15) (16) are the formulas given by Hasofer (1974).

Note that if the earthquake process is a Poisson process, the slip (or time) predictable model selects strong (or weak) earthquakes so that they don't form a Poisson process. In fact, if τ is exponentially distributed with parameter ρ, (7) becomes

$$f_s^*(\gamma) = \frac{\rho \exp -[(\rho + \gamma)\bar{t}]}{\gamma + \rho \exp[-(\rho + \gamma)\bar{t}]}$$

that does not correspond to an exponential distribution (and analogously (13) (14); wile under the Bernoulli assumption, (15) becomes

$$f_s^*(\gamma) = \frac{p\rho}{\gamma + p\rho}$$

that leads to the well-known property that strong earthquakes randomly selected with probability p from a Poisson process form again a Poisson process with parameter $p\rho$. The return period μ_s is again given by (8).

In earthquake prediction we are interested also in the mean number of strong earthquakes.

It is well known that, for any interoccurrence time density f_τ, the mean number $H(t,t_o)$ of events in $(0,t)$ has the following Laplace transform

$$(17) \quad H^*(\gamma, t_o) = \frac{{}_f f^*(\gamma, t_o)}{\gamma \{1 - f^*_\tau(\gamma)\}}$$

Then the mean number of strong earthquakes is given by

$$(18) \quad H^*_s(\gamma, t_o) = \frac{{}_f f^*_s(\gamma, t_o)}{\gamma \{1 - f^*_s(\gamma)\}}$$

In the time predictable model, (18) leads, due to (13) (14), to two different results according to the size of the last earthquake.

For some italian seismogenetic structures a gamma distribution with shape parameter $a = 2$ (and scale parameter ρ) is suitable (Grandori, Guagenti and Petrini, 1984) [*]. In these cases, if a slip predictable model governed by (1) is assumed, the hazard rate Φ_s,

$$\Phi_s(t) = \frac{f_s(t)}{1 - f_s(t)} \ ,$$

of the strong earthquakes can be drawn from (7). In figure 2, Φ_s is plotted for the case $\rho = .2$ years^{-1} and $\bar{t} = 10$ years (this is

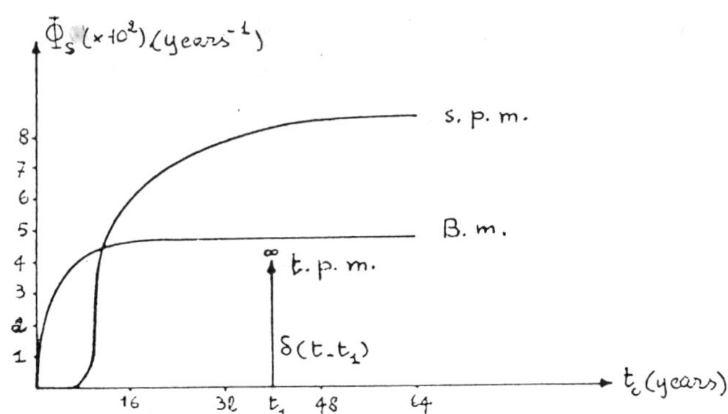

Fig.2. Hazard rate in a renewal process with slip predictable model –
time predictable model – Bernoulli model.

(*) More precisely a mixture distribution is suitable : a mixture of a lognormal distribution with a gamma distribution. The lognormal distribution represents the interoccurrence time inside the seismic crisis (foreshocks and aftershocks); the gamma distribution represents the interoccurrence time of main events.

not a strong earthquake in the physical sense) : Φ_s is much more variable than in the Bernoulli assumption (B.m.).

If a time predictable model governed by (1) is assumed, the hazard rate Φ_s, given the value m_1 of the last earthquake, should be the impulse $p\delta(t - t_1)$.

3. A more general slip predictable model

The deterministic interpretation of (1) that has been used in the preceding section seems too rigid. In order to construct a less constrained model, it is useful to express preliminarily the previous model as a semi-Markov process with a finite number of states. In other words we discretize the continuous information of (1) by defining n sizes (e.g. n = 3) of earthquakes :

$$m \leqslant m_1 \qquad \text{size 1}$$
$$m_1 < m \leqslant m_2 \qquad \text{size 2}$$
$$m_2 < m \qquad \text{size 3 (strong earthquake).}$$

Let t_1 and t_2 be the values of τ which correspond to m_1 and m_2 in (1), and Δt_j (j = 1,2,3) the time intervals $(0,t_1)$, (t_1, t_2), (t_2,∞).

The state is defined by the size of the last event.

The transition matrix elements p_{ij} depend only on the second index and they are

$$(19) \quad p_{ij} = \int_{\Delta t_j} f_\tau(x)\,dx.$$

The densities $h_{ij}(x)$ of the holding times too depend only on the second index and are normalized in Δt_j, i.e.

$$(20) \quad h_{ij}(x) = \frac{f_\tau(x)}{p_{ij}} \quad \text{in} \quad \Delta t_j.$$

All the quantities we calculated in the case of a renewal

process can be expressed in semi-Markov form limited to the defined states. In particular the interoccurrence time density $f_s(x)$ becomes the first passage time density in state 3, $f_{33}(x)$. This is given by

$$(21) \quad f_{33}^*(\gamma) = \frac{P_{i3} h_{i3}^*}{1 - P_{i1} h_{i1}^* - P_{i2} h_{i2}^*}$$

that coincides with (7), and so on.

We find again that earthquake process is a renewal process. In fact the waiting time densities do not depend on any index :

$$(22) \quad \Sigma_j \; P_{ij} h_{ij} = w_1 = w_2 = w_3 = f_\tau \; .$$

The waiting times in the states are simply the interoccurrence times.

Note that, in semi-Markov form, the conditional density $f_{M|\tau}$ is replaced by conditional transition probabilities

$$(23) \quad P_{ij}(x) = \frac{P_{ij} h_{ij}(x)}{w_i(x)}$$

Then independently of the size of the last earthquake, $p_{ij}(x)$ become unit functions in Δt_j; they take the place of previous impulses. This discrete model is a first step towards a probabilistic interpretation of (1). Moreover, it focuses the two quantities $f_\tau(x)$ and $p_{ij}(x)$ that, because of (22), define completely the semi-Markov process (together with the initial conditions). In fact the transition matrix and the holding time densities are

$$(24) \quad P_{ij} = \int_0^\infty f_\tau(x) p_{ij}(x) dx$$

$$(25) \quad h_{ij}(x) = \frac{f_\tau(x) p_{ij}(x)}{P_{ij}}$$

A further step towards a more general probabilistic approach consists in smoothing the conditional probabilites $p_{ij}(x)$. In this case the magnitude distribution (2) is still satisfied if the new transition probabilities are :

$$(26) \quad p_{i1} = F_M(m_1) \qquad p_{i2} = F_M(m_2) - F_M(m_1) \qquad p_{i3} = 1 - F_M(m_2)$$

4. A more general time predictable model

Let us introduce, as before, the same finite number of states. Now the holding times p.d.f. $h_{ij}(x)$ do not depend on the second index; p_{ij} do not depend on the first one. The waiting times w_i depend on the state and coincide with the holding times

$$(27) \quad w_i(x) = \Sigma_j \, p_{ij} h_{ij}(x) = h_{ij}(x).$$

The resulting process of all earthquakes is not a renewal process. It is an independent semi-Markov process : the conditional $p_{ij}(x)$ coincide with p_{ij}; the information on the elapsed time does not add information about the next destination. The holding time densities are derived from (2) as follows

$$(28) \quad h_{1j} = \frac{f_M[m(x)]}{F_M(m_1)} \text{ in } \Delta t_1 \; , \quad h_{2j} = \frac{f_M[m(x)]}{F_M(m_2) - F_M(m_1)} \text{ in } \Delta t_2 \; ,$$

$$h_{3j} = \frac{f_M[m(x)]}{1 - F_M(m_2)} \text{ in } \Delta t_3$$

if they are normalized in Δt_i; they are the smoothing of (28) in a more general time predictable model. The transition probability matrix is given by (26).

5. Earthquake prediction and seismic design

Prediction and prevention are two well known problems in seismic risk reduction.

The first problem mainly concerns the size and the occurrence time of the next event, while prevention is essentially achieved through a suitable seismic design of buildings.

Prediction needs a definition of risk as a function of time, based on the information available at a given instant. Prevention needs a stationary definition of risk, mainly based on the return period of strong earthquakes.

The semi-Markov process (the renewal process is a particular case) is useful in both circumstances.

The prediction about time and size of next earthquake is drawn from the core matrix elements :

$$(29) \quad p_{ij} \, {}_{f}h_{ij}(t,t_o) = p_{ij} \, \frac{h_{ij}(t + t_o)}{1 - \int_0^{t_o} w_i(x)\,dx}$$

Probabilities of an immediate event or of an event within t are also given by (29) substituting for ${}_{f}h_{ij}$ its hazard rate or its cumulative function respectively.

In general earthquake prediction incorporates in the earthquake process also the external information deriving from precursors (Guagenti and Grandori, 1983).

On the other hand a stationary risk is consistent with a random starting of the process : the life beginning of a standard, long life building may be just considered a random starting point. The seismic design can be defined on the basis of stationary risk. In fact a random starting implies that first holding time densities are

$$(30) \quad {}_{r}h_{ij}(t) = \frac{\int_t^{\infty} h_{ij}(x)\,dx}{E\{\tau_{ij}\}}$$

and consequently the expected number of j-events in [0,t] becomes simply

$$E\{N_j(t)\} = \frac{t}{\mu_j}$$

Together with mean number and return period of events, a cost-benefit analysis uses the evaluation of the cost of future damage. A discount factor γ is used to actualize a cost D that will happen at the future instant t. The actualized cost D_o is

$$D_o = D \, \exp(- \gamma t).$$

Let us calculate the expected actualized cost of all future damage in the general slip predictable and in the general time predictable model.

The semi-Markov process implies, as we saw, that the imbedded earthquakes of size j form a delayed renewal process with failure time density $f_{jj}(x)$.

Therefore, if D_j is the mean value of damage caused by an earthquake of size j, the actualized expected cost D_o of all future damage is

$$(31) \quad D_o = \sum_{1j}^{3} \sum_{1r}^{\infty} \int_0^\infty D_j \exp(-\gamma x) f_{w_r^{(j)}}(x) dx$$

where $w_r^{(j)}$ is the waiting time to r-th earthquake of size j with

$$(32) \quad f^*_{w_r^{(j)}}(\gamma, t_o) = {}_f f_{in\,j}(\gamma, t_o)[f^*_{ij}(\gamma)]^{r-1}.$$

The densities f_{jj} are the recurrence time densities in the semi-Markov process (first passage time densities). The index "in" means "initial state", i.e. the state in which the cost evaluation is done.

Therefore the expected cost D_o depends on t_o and on the size of the last earthquake. This dependence vanishes in the slip predictable model. The dependence on t_o vanishes in random starting conditions.

Taking into account (32), (31) can be written in the following more simple form

$$D_o^{(in)}(\gamma, t_o) = \sum_{1j}^{3} D_j \frac{{}^f f_{in\,j}(\gamma, t_o)}{1 - f^*_{jj}(\gamma)}$$

6. A possible general model

Presumably during the accumulation and sudden release of the energy both aspects shown in the two-models are present. It may

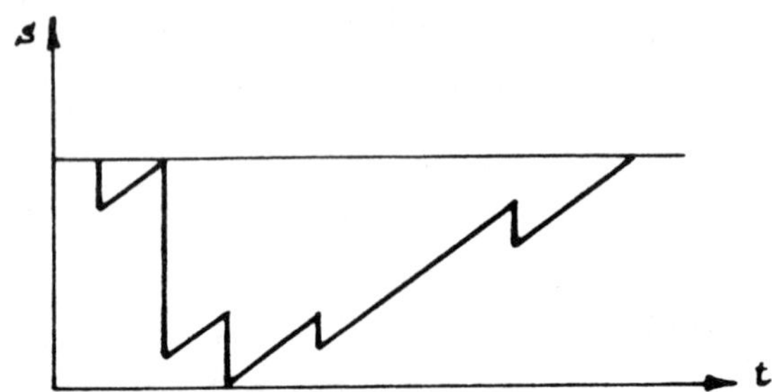

Fig.3. Holding times depending on the last earthquake and on the
       next one.

happen that some asperities in the fault do not allow a complete
liberation of energy as well as that the asperities are not sufficient to reach the complete energy accumulation. Moreover the lower
and upper bound are not deterministically known (figure 3).

It is reasonable, for modelling this double behaviour, to consider holding times τ_{ij} as dependent on both indices. Moreover
the previous definition of "state" is not suitable now because the
holding times depend also on the total amount of cumulated stress :
long holding times are not allowed near the upper bound.

A possible general model is obtained as follows.

Let the state be defined by the total amount of cumulated
stress; three states are again considered, whose total energy
release would give the previous values of M : $m \leqslant m_1$, $m_1 < m \leqslant m_2$,
$m_2 < m$. Moreover approximately $m \leqslant m_1$ means energy release inside
a state; $m_1 < m \leqslant m_2$ means energy release to the next lower state;
$m > m_2$ means energy release to the second lower state (it is
possible only from the state 3).

Assuming again the distribution (2) for M $^{(*)}$, the transition
matrix is

____

(*) truncated with an upper bound and a lower bound for M

$$
\begin{pmatrix}
F_M(m_1) & 1 - F_M(m_1) & 0 \\
F_M(m_2) - F_M(m_1) & F_M(m_1) & 1 - F_M(m_2) \\
1 - F_M(m_2) & F_M(m_2) - F_M(m_1) & F_M(m_1)
\end{pmatrix}
$$

In fact the new definition of a "state" implies that "transition" may be either an earthquake occurrence or an accumulation of stress (without any earthquake occurrence) causing a passage of the system to the next higher state. The diagonal elements p_{ii} are the probabilities of an earthquake of size 1; the elements $p_{21} = p_{32}$ are the probabilities of an earthquake of size 2; the probability p_{13} is zero because the stress accumulates gradually; p_{12} is equal to the probability that the next earthquake's size will be greater than 1; p_{23} is equal to the probability that the next earthquake's size will be greater than 2.

The holding time p.d.f. $h_{ij}(x)$ are "nearly normalized" in Δt_{ij} and $h_{3j}(x) = w_3(x)$. All quantities can be calculated in the frame of semi-Markov processes. In particular the strong earthquakes are the transitions $3 \to 1$, i.e., among the entrances in state 1, the strong earthquakes are the transitions with probability $1 - F_M(m_2) = p$. Then the strong earthquakes form a renewal process with failure time density $f_s(x)$ given by

$$
f_s^*(\mu) = \frac{p f_{11}^*(\gamma)}{1 - (1 - p) f_{11}^*(\gamma)}
$$

where f_{11} is the density of first passage time (or recurrence time) in state 1.

Aknowledgement

The authors would like to tank prof. Haresh Shah at Stanford University. This paper had its origin in a stimulating discussion with him.

Anagnos, T., A.S.Kiremidjian (1984). Stochastic slip predictable
    model, for earthquake occurrence, BSSA.

Esteva, L. (1982). Models for uncertainly and decision making for
    seismic risk conditions. National University of Mexico.

Grandori, G., E.Guagenti, V.Petrini (1984). On the use of renewal
    processes in the seismic hazard analysis. Proc. 8 WCEE,
    San Francisco.

Guagenti, G.E. (1979). Stochastic processes in the modelling of
    earthquake occurrence. In "Numerical techniques for Stochastic
    Systems", Archetti F. and M.Cugiani Eds, North Holland, 205-216.

Guagenti, G.E. (1982). Stochastic processes in seismic risk reduction.
    Rend.Sem.Mat.e Fis. di Milano.

Guagenti, G.E., G.Grandori (1983). Some probabilistic aspects of
    earthquake prediction. Meccanica, n°4.

Guagenti, G.E., G.Grandori, V.Petrini (1979). A discussion of "non-
    linear" magnitude-frequency laws". Proceedings of the 2nd U.S.
    National Conference on Earthquake Engineering, Standford
    University.

Guagenti, G.E., C.Molina (1981). Synamic semi-Markov systems in
    earthquake engineering. Rend.Sem.Mat. Torino.

Hasofer, A.M. (1974). Design for infrequent overloads. Eart.Eng.and
    Struct.Dyn., vol2, 387-389.

Howard, R.A. (1971). Dynamic probabilistic systems. J.Wiley.

Lomnitz, A.J., C.Lomnitz (1978). A new magnitude-frequency relation.
    Tectonophysics, 49, 237-245.

Patwardhan, A.S., R.B.Kulkarni, S.Tocher (1980). A semi-Markov model
    for characterizing recurrence of great earthquake. BSSA, V.70,
    n° 1, 323-347.

Shimazaki, K., T.Nakata (1980). Time predictable recurrence model
    for large earthquakes. Geoph.Res.Lett., vol.7, 279-282.

Shlien, S., M.N.Toksoz (1970). A clustering model for earthquake
    occurrence. Bull.Seismol.Soc.Am., 60, 1765-1787.

Veneziano, D., A.Cornell (1974). Earthquake model with spatial and temporal memory for engineering seismic risk. R 74-18, Structures Publ., N 385, Dept. Civil Eng., MIT.

Vere-Jones, D., Y.Ozaki (1982). Some examples of statistical estimation applied to earthquake data. Ann.Inst.Statist.Math., V, 34, Part.B, 189-207.

SECTION XI. A SECOND BIBLIOGRAPHY ON SEMI-MARKOV PROCESSES

A SECOND BIBLIOGRAPHY ON SEMI-MARKOV PROCESSES

Jozef L.Teugels

Katholieke Universiteit te Leuven, Belgium

Abe, S. (1977). The asymptotic exponential failure law for redundant
repairable systems, II. Rep.Statist.Appl.Res.Un.Japan.Sci.
Engin., 24, 15-39.

Affisoo,J. & A.Ghosal (1978). Approximation of waiting time in tandem
queues through isomorphs-1. SCIMA, J.Manag.Sci.Appl.Cybern., 7,
55-68.

Agarwal,M. & A.Kumar (1980). A review of standby redundant systems.
IEEE Trans.Reliab.R-29, 290-294.

Agarwal,M. & A.Kumar (1980). Steady-state analysis of a 2-unit
standby system with inefficient and intermittent repair. IAPQR
Trans., 5, 85-99.

Agarwal,M., R.Debabrata & A.Kumar (1978). Probability analysis of a
two-unit standby redundant system with repair efficiency and
imperfect switch-over. Intern.J.Systems Sci., 9, 731-742.

Agarwal,M., A. Kumar & S.M. Sinha (1977 and 1981). Analysis of a two-unit
priority standby system with different operative states. J.
Math.Sci., Delhi, 65-79 and 12-13.

Agnew, C.E. (1976). Dynamic modelling and control of congestion-prone
systems. Operations Res., 24, 400-419.

Agrafiotis, G.K. (1980). Modelling manpower systems. Ind.Math., 30, 103-120.

Agrafiotis, G.K. (1982). On the stochastic theory of compartments : a semi-Markov approach for the analysis of the k-compartmental systems. Bull.Math.Biol., 44, 809-817.

Agrafiotis, G.K. (1983). General distribution function for periods of downtime of a single system exceeding a limiting downtime. J.Oper.Res.Soc., 34, 1115-1118.

Agrawala, A.K. & R.L.Larsen (1983). Control of a heterogeneous two-server exponential queueing system. IEEE Trans.Software Eng., SE-9, 522-526.

Ahmedova, H.M. & T.I.Nasirova (1975). Non-stationary distribution of an inventory for a model with inventory control. Izvestija Akad.Nauk Azerbaidz.SSR, Ser.fiz.-tehn.mat.Nauk, 19-22 (Russian).

Akif, E.M., T.Chien & Y.Ho (1979). A gradient technique for general buffer storage design in a production line. Proc.1978 IEEE Conf. Decision & Control, San Diego, 625-632.

Akritas, M.G. & G.G.Roussas (1980). Asymptotic inference in continuous time semi-Markov processes. Scan.J.Stat., 7, 73-97.

Aksenov, B.E., A.M.Aleksandrov & A.N.Bakanov (1973). Application of generalized Poisson flow for the study of methods to increase the reliability. Probl.Peredaci Inform., 9, 80-86 (Russian).

Akyildiz, I.F. & G.Bolch (1982). Analyse von Rechensystemen. Analytische Methoden zur Leistungsbewertung und Leistungsvorhersage. Leitfaden der Angewandten Mathematik und Mechanik, Band 57, Teubner.

Alam, M. & R.Billiton (1978). Effect of restricted repair on system reliability indices. IEEE Trans.Reliab., R-27, 376-379.

Aleksandrov, A.M. (see Aksenov, B.E.).

Ali Khan, M.S. (1977). Infinite dams with discrete additive inputs. J.Appl.Probab., 14, 170-180.

Alj A. & A.Haurie (1979). Description and control of a continuous time population process. Proc.1978 IEEE Conf.Decision & Control, San Diego, 557-565.

Alj, A. & A.Haurie (1980). Description and control of a continuous-
    time population process. IEEE Trans.Autom.Control AC-25, 3-10.

Alj, A. & A.Haurie (1981). Hierarchical control of a population
    process with application to group preventive maintenance.
    Large scale systems, theory and applications, Proc.IFAC Symp.
    Toulouse, 1980, 333-338.

Allan, R.N., M.R.G.Al-Shakarchi & C.H.Grigg (1976). Numerical
    techniques in probabilistic load flow problems. Int.J.Numer.
    Methods Engin., 10, 853-860.

Al-Shakarchi, M.R.G. (see Allan R.N.).

Aluffi, F., S.Incerti & F.Zirilli (1980). An application of the
    diffusion approximation to queueing networks : a moving head
    disks model. Ottimizzazione non lineare e applicazioni, Atti
    Conv.int.L'Aquila, 213-222.

Anders, B. & G.Lucas (1981). Optimierungmodelle zur Regelung forst-
    lichrer Prozesse unter Beachtung von Übergangswahrscheinlich-
    keiten. Wiss.Z.Tech.Univ.Dres., 30,91-95.

Andersen, P.K. (1981). A note on filtered Markov renewal processes.
    J.Appl.Probab., 18, 752-756.

Anderson, M.Q. (1981). Monotone optimal preventive maintenance
    policies for stochastically failing equipment. Nav.Res.Logist.
    Q., 28, 347-358.

Andronov, A.M. (1980). Algorithm for finding optimal strategies in
    controlled semi-Markov birth-and-death processes. Eng.Cybern.,
    18, 67-71.

Anichkin, S.A. (1982). Coupling of renewal processes and its appli-
    cations. Stability problems of stochastic models, Proc.Semin.
    Moskva, 4-24.

Anichkin, S.A. (1983). Hyper-Erlang approximation of probability
    distributions on $(0,\infty)$ and its application. Stability problems
    of stochastic models, Proc.Semin.Moskva, 1-16.

Anisimov, V.V. (1973). On stopping a stochastic process at the time
    a certain level is reached. Soviet Math.Doklady, 14, 1500-1503.

Anisimov, V.V. (1978). Description of multidimensional limit laws
    for finite Markov chains in a series scheme : the general case.
    Theory Probab.Math.Stat., 11, 1-7.
Anisimov, V.V. (1977). Switched processes. Kibernetika, Kiev, 111-115.
Anisimov, V.V. (1979). Applications of limit theorems for switching
    processes. Cybernetics, 14, 917-929.
Anisimov, V.V. (1981). Limit theorems for processes admitting the
    asymptotic consolidation of one component. Theory Probab.Math.
    Stat., 22, 1-13.
Anisimov, V.V. (1981). Asymptotic enlargement of inhomogeneous Markov
    and semi-Markov systems with an arbitrary state space. Dokl.
    Akad.Nauk Ukr.SSR, Ser.A, 3-6 (Russian).
Anisimov, V.V. & A.I.Chernyak (1983). Limit theorems for certain
    rare functionals on Markov chains and semi-Markov processes.
    Theory Probab.Math.Stat., 26, 1-6.
Anisomov, V.V., V.I.Sereda & A.A.Vojna (1981). Choice of the optimal
    checking moments for a semi-Markov system with failures. Dokl.
    Akad.Nauk Ukr.SSR, Ser.A, 79-82 (Russian).
Anisimov,V.V. & V.N.Sityuk (1982). Asymptotic analysis of the hete-
    rogeneous Markov systems with rare semi-Markov failures. Dokl.
    Akad.Nauk Ukr.SSR, Ser.A, 66-69 (Russian).
Ansell,J., A.Bendell & S.Humble (1980). Nested renewal processes.
    Adv.Appl.Probab., 12, 880-892.
Aranda,G.J. (1982). Some relations between Markovian models of net-
    works of queues. Trab.Estad.Invest.Oper. 33, 3-29 (Spanish).
Aripov, H.M. (1974). On a generalization of the Erlang problem for
    an arbitrary distribution of the service time. Izvestija Akad.
    Nauk UzSSR, fiz.-mat.Nauk, 18, 8-12 (Russian).
Arjas, E. & V.S.Korolyuk (1980). Stationary phase extension of Markov
    renewal processes. Dopov.Akad.Nauk Ukr.SSR, Ser.A, 3-6
    (Ukrainian).
Arjas, E. & E.Nummelin (1977). Semi-Markov processes and σ-invariant
    distributions. Stochastic Processes Appl., 6, 53-64.
Arjas, E., E.Nummelin & R.L.Tweedie (1978). Uniform limit theorems

for non-singular renewal and Markov renewal processes. J.Appl. Prob., 15, 112-125.

Arjas, E., E.Nummelin & R.L.Tweedie (1980). Semi-Markov processes on a general state space : alpha-theory and quasi-stationarity. J.Aust.Math.Soc., Ser.A, 30, 187-200.

Arndt, U. & P.Franken (1979). Construction of a class of stationary processes with applications in reliability. Zastosow.Mat., 16, 379-392.

Arndt, U. & P.Franken (1979). Continuity of generalized regenerative processes. Eng.Cybern., 17, 69-72.

Arndt, U., P.Franken, D.Koenig & V.Schmidt (1982). Queues and point processes. John Wiley & Sons.

Arora, J.R. (1977). Reliability of several standby-priority-redundant systems. IEEE Trans.Reliab. R-26, 290-293.

Arsenisvili, G.L. & I.I.Ezov (1978). A certain limit theorem for semi-Markov processes of order r. Select.Translat.Math. Statist.Probab. , 14, 135-138.

Artamonov, G.T. (1976). Produktivität einer diskreten bearbeitenden Zweigerat-Taktstrase unter den Verhaltnissen von Versagern. Kibernetika, Kiev, 126-130 (Russian).

Athreya, K.B. & P.E.Ney (1978). Limit theorems for semi-Markov processes. Bull.Austral.Math.Soc., 19, 283-294.

Athreya, K.B. & K.Ramamurthy (1980). Renewal theory - A Markov process approach. J.Math.Kyoto Univ., 20, 169-177.

Athreya, K.B., D.McDonald & P.E.Ney (1978). Limit theorems for semi-Markov processes and renewal theory for Markov chains. Ann. Probability, 6, 788-797.

Avenhaus, R. & W.Haussmann (1983). Ein Markov'sches Modell für ein serielles Produktionsystem mit einen Zwischenlager. Methods Oper.Res., 46, 547-560.

Azlarov T.A. ed. (1969). Mathematical methods in material and quality control. Akad.Nauk Uzb.SSR.Inst.Mat.Romanovskogo. Tashkent.

Babadzhanyan, A.A. (1979). A certain extension of Markov processes

as basis for solution and algorithm for an optimal strategy.
Dokl.Akad.Nauk Arm.SSR, 69, 213-216 (Russian).

Baccelli, F. (1981). Analysis of a service facility with periodic
checkpointing. Acta Inf., 15, 67-81.

Bajaj, D. & A.Kuczura (1977). A method of moment for the analysis
of a switched communication network's performance. IEEE Trans.
Commun.COM-25, 185-193.

Bakanov, A.N. (see Aksenov, B.E.)

Balagopal, K. (1979). Some limit theorems for the general semi-
Markov storage model. J.Appl.Probab., 16, 607-617.

Balbo, G., S.C.Bruell & H.D.Schwetman (1977). Customer classes and
closed network models - a solution technique. Inf.Process. 77,
Proc.IFIP Congr.Toronto 1977, 559-564.

Balcer, Y. & I.Sahin (1983). A stochastic theory of pension dynamics.
Insur.Math.Econ., 2, 179-197.

Baltrunas, A. (1982). Regularity of semi-Markov processes. Lith.
Math.J., 21, 111-116.

Banach, D. & D.S.Silvestrov (1981). Uniform ergodic theorems on the
average for accumulation processes. Dopov.Akad.Nauk Ukr.SSR,
Ser.A, 34-37.

Bandura, V.N. & I.I.Ezov (1974). On the distribution of first-
passage time of a given level for a class of processes with
step-function trajectories I, II. Theor.Probab.Math.Statist.,
1, 5-18 & 19-34.

Banerjee, A.K. & G.K.Bhattacharyya (1976). Testing hypotheses in a
two-state semi-Markov process. Sankhya, Ser.A, 38, 340-356.

Baras, J.S., W.S.Levine & T.L.Lin (1979). Discrete time point
processes in urban traffic queue estimation. Proc.IEEE Conf.
Decision & Control, San Diego, 1025-1031.

Barbour, A.D. (1982). Generalized semi-Markov schemes and open
queueing networks. J.Appl.Probab., 19, 469-474.

Barbour, A.D. & R.Schassberger (1981). Insensitive average residence
times in generalized semi-Markov processes. Adv.Appl.Probab.,
13, 720-735.

Barfoot, C.B. (1983). Some mathematical methods for modelling the performance of a distributed data base system. Adequate modelling of systems, Proc.Int.Conf.Bad Honnef, 172-182.

Barlow, R.E. & F.Proschan (1976). Theory of maintained systems : Distribution of time to first system failure. Math.Oper.Res., 1, 32-42.

Basawa, I.V. (1974). Maximum likelihood estimation of parameters in renewal and Markov-renewal processes. Austral.J.Statist., 16, 33-43.

Barzily, Z. & M.Rubinovitch (1979). On platoon formation on two-lane roads. J.Appl.Probab., 16, 347-361.

Basharin, G.P., V.A.Kokotushkin & V.A.Naumov (1979). The method of equivalent substitutions for calculating fragments of communication networks for a digital computer. I. Eng.Cybern., 17, 66-79.

Bather, J.A. (1976). A control chart model and a generalized stopping problem for brownian motion. Math.Oper.Res., 1, 209-224.

Baxter, L.A., D.J.McConalogue & E.M.Scheuer (1982). On the tabulation of the renewal function. Technometrics, 24, 151-156.

Becker, M. & A.M.Kshirsagar (1981). Superposition of Markov renewal processes. S.Afr.Stat.J., 15, 13-30.

Becker,A. & A.H.Marcus (1977). Power laws in compartmental analysis. II. Numerical evaluation of semi-Markov models. Math.Biosciences, 35, 27-45.

Beichelt, F. (1976). Prophylaktische Erneuerung von Systemen. Einfuehrung in mathematische Grundlagen. Wissenschaftliche Taschenbuecher. Reihe Math.-Physik, 153, Akademie-Verlag, Berlin.

Berbee, H.C.P. (1979). Random walks with stationary increments and renewal theory. Mathematical Centre Tracts, 112.

Berg, M. (1976). A proof of optimality for age replacement policies. J.Appl.Probab., 13, 751-759.

Berg, M. (1976). Optimal replacement policies for two-unit machines with increasing running costs, I. Stochastic Processes Appl., 4, 89-106.

Berg, M. (1981). A stochastic model for the reliability of a power unit with deliberate shutdowns. Eur.J.Oper.Res., 6, 309-314.

Berman, M. (1978). Regenerative multivariate point processes. Adv. Appl.Probab., 10, 411-430.

Bertsekas, D.P. (1982). Dynamic behavior of shortest path routing algorithms for communication networks. IEEE Trans.Autom.Control AC-27, 60-74.

Bestwick, P.F. & D.Sadjadi (1979). A stagewise action elimination algorithm for the discounted semi-Markov problem. J.Oper.Res. Soc., 30, 633-637.

Beutler, F.J. & B.Melamed (1978). Decomposition and customer streams of feedback networks of queues in equilibrium. Oper.Res., 26, 1059-1072.

Beutler, F.J. & B.Melamed (1982). Multivariate Poisson flows on Markov step processes. J.Appl.Probab., 19, 289-300.

Beyer, W.A. & C.Delisi (1977). The asymptotic form of partition functions of linear chain molecules : an application of renewal theory. J.Math.Analysis Appl., 57, 416-428.

Bhalaik, H.S. & N.S.Kambo (1979). Non-homogeneous M/M/s queues in series. INFOR, Can.J.Oper.Res.Inf.Process., 17, 262-275.

Bhaskar, D. & S.K.Srinivasan (1979). Probabilistic analysis of intermittently used systems. J.Math.Phys.Sci., Madras, 13, 91-105.

Bhat, U.N. & R.E.Nance (1979). An evaluation of CPU efficiency under
Bhattacharyya, G.K. (see A.K.Benerjee).
    dynamic quantum allocation. J.Assoc.Comput.Mach., 26, 761-778.

Bieber, G. (1979). On optimal control of Markov processes with application to a class of service systems. Math.Operations-forsch.Stat., Ser.Optimization, 10, 143-162.

Billinton, R. (see M.Alam).

Birolini, A. (1977). Hardware simulation of semi-Markov and related processes. I. A versatile generator; II. Simulation of noise pulses on data channels and other applications. Math.Comput. Simulation, 19, 75-97 & 183-191.

Biswas, S., P.K.Kapur & Y.K.Mehta (1983). Generalized availability measures for repairable systems. IAPQR Trans., 8, 83-94.

Blischke, W.R. (see L.A.Baxter).

Blischke, W.R. & R.M.Scheuer (1981). Applications of renewal theory
in analysis of the free-replacement warranty. Nav.Res.Logist.Q.,
28, 193-205.

Bobbio, A. & A.Premoli (1983). Analysis of non-regenerative repair
processes through homogeneous Markov models. Int.J.Syst.Sci.,
14, 647-659.

Bobos, A.G. & E.N.Protonotarios (1978). Optimal systems for equipment
maintenance and replacement under Markovian deterioration. Eur.
J.Oper.Res., 2, 257-264.

Bodnariu, M.A. (1982). Compact Markov processes with rewards. Stud.
Cercet.Mat., 34, 211-222 (Roumanian).

Boel, R. (1981). Martingale methods for the semi-Markov analysis of
queues with blocking. Stochastics,5, 115-133.

Boel, R. (1981). Stochastic models of computer networks. Proc.NATO
ASI : Stochastic Systems, Les Arcs, 1980, 141-167.

Bogdantsev, E.N. & S.M.Brodi (1982). Limit theorem for a class of
imbedded semi-Markov processes. Cybernetics, 17, 667-693.

Bokucava, I.V. (1976). On estimates of the spectral characteristics
of sequences of intervals between the events of a stationary
semi-Markovian process. Soobscenija Akad.Nauk Gruzin.SSR 84,
17-20 (Russian).

Bolch, G. (see I.F.Akyildiz).

Boling, R. & F.S.Hillier (1977). Toward characterizing the optimal
allocation of work in production line systems with variable
operation times. Adv.Oper.Res.Proc.2nd Eur.Congres., Stockholm
1976, 649-658.

Bondarev, V.V. & V.I.Nosenko (1976). Questions of the stability and
the dissipation for stochastic differential equations with a
semi-Markov interference of the randomness. Behaviour of systems
in random media, Work Collect., Kiev, 8-15 (Russian).

Borovkov, K.A. (1980). Stability theorems and estimates of the rate

of convergence of the components of factorizations for walks
defined on Markov chains. Theory Probab.Appl., 25, 325-334.

Borozdin, O.P. & I.I.Ezov (1979). On a class of boundary functionals
for strongly regenerative random processes. Theory Probab.Math.
Statist., 18, 9-19.

Bosch, K. (1975). Erneuerungsprozesse mit Erneuerungen zu verschie-
denen Kosten. Angew.Informatik, 17, 143-150.

Bosch, K. (1981). Wartungs- und Inspektionenstrategien. Operations
Res., Proc.1980, DGOR, Essen, 229-240.

Bosch, K. & U.Jensen (1978). Deterministische Inspektionenstrategien.
Z.Operat.Res., Ser.A, 22, 151-168.

Boyse, J.W. (1974). Determining near-optimal policies for Markov
renewal decision processes. IEEE Trans.Systems Man Cybernetics
SMC-4, 215-217.

Brandwayn, A. (1976). Control schemes in queueing networks. Management Sci., 22, 810-822.

Breuer, M.A. & A.D.Friedman (1977). Diagnosis and reliable design of
digital systems. Digital System Design Series. Pitman, London.

Brodeckii, G.L. (1978). On a problem of periodic storing of results.
Kibernetica, Kiev, 70-74 (Russian).

Brodeckii, G.L. (1980). On the question of optimal organization of
recording intermediate information in the case of random failures
of a system. Eng.Cybern., 18, 61-65.

Brodi, S.M. (see E.N.Bogdantsev).

Brodi, S.M., V.S.Korolyuk & A.F.Turbin (1976). Semi-Markov processes
and their applications. J.Soviet Math., 4, 244-280.

Brown, C.B. & E.C.C.Chang (1973). Functional reliability of structures.
J.Franklin Inst., 296, 161-178.

Brown, C.C., M.H.Gail & T.J.Santner (1980). An analysis of comparative carcinogenesis experiments based on multiple times to
tumor. Biometrics, 36, 255-266.

Brown, T.C. & P.K.Pollett (1982). Some distributional approximations
in Markovian queueing networks.Adv.Appl.Probab., 14, 654-671.

Bruell, S.C. (see G.Balbo).

Buehler, G. (1976). Informationsbedarf bei Produktions-Lagerhaltungs-
     Modellen. Proc.Oper.Res., 6, DGOR, 380-389.

Buehler, G. (1979). Sicherheitsaequivalente und Informationsbedarf
     bei stochastischen dynamischen Produktions-Lagerhaltungs-
     Modellen. Haag und Herchen Verlag, Frankfurt/Main.

Burgin, T.A. & J.M.Norman (1976). A table for determining the proba-
     bility of a stock out and potential lost sales for a gamma
     distributed demand. Operat.Res.Quart., 27, 621-631.

Burman, D. (1976). Estimation of transition probabilities in a
     multiparticle semi-Markov system. J.Appl.Probab., 13, 696-706.

Burman, D. (1981). Insensitivity in queueing systems. Adv.Appl.
     Probab., 13, 846-859.

Butko, T.K. (1982). Markov renewal process of the system $M/G/1/\infty$.
     Applied problems of probability theory, Collect.Sci.Works,
     Kiev, 14-19 (Russian).

Calvert, T.W. & A.C.Sanderson (1976). Distribution coding in neural
     interaction models. Math.Biosciences, 29, 1-25.

Cao, J. & K.Cheng (1982). Analysis of M/G/1 queueing system with
     repairable service station. Acta Math.Appl.Sin., 5, 113-127
     (Chinese).

Carravetta, M. & R.de Dominicis (1981). Some applications of semi-
     Markov processes in reliability and modelling. RAIRO, Autom.
     Syst.Anal.Control, 15, 361-375.

Carravetta, M. R.de Dominicis & M.R.D'Esposito (1981). Semi-Markov
     processes in modelling and reliability : a new approach.
     Appl.Math.Modelling, 5, 269-274.

Carravetta, M. R.de Dominicis & R.Manca (1981). Semi-Markovian
     stochastic processes in social security problems. Cah.Cent.
     Etud.Rech.Oper., 23, 333-339.

Carroll, J.L., A.van de Liefvoort & L.Lipsky (1982). Solutions of
     M/G/1//N-type loops with extensions to M/G/1 and GI/M/1 queues.
     Oper.Res., 30, 490-514.

Case, J. (1979). A model to predict the quality of taxi service at
     a large municipal airport when the drivers compete. Lecture
     Notes Pure Appl.Math., 47, 187-203.

Chakravarthy, S. & M.F.Neuts (1981). A single server queue with platooned arrivals and phase type services. Eur.J.Oper.Res., 8, 379-389.

Chattergy, R. & U.W.Pooch (1978). Analysis of the availability of computer systems using computer-aided algebra. Commun.ACM 21, 586-591.

Chatterjee, A., N.D.Georganas & P.K.Verma (1980). Local congestion control in computer-communication networks with random routing. INFOR, Can.J.Oper.Res.Inf.Process., 18, 74-79.

Chauny, F., M.Chokron & A.Haurie (1984). Nursing care demand prediction based on a decomposed semi-Markov population model. Oper. Res.Lett., 2, 279-284.

Chebotarev, A.S. & S.E.Shibanov (1975). Probabilistic problem in the control of multiple allocations with common resources. Izvestija Akad.Nauk SSSR, tehn.Kibernet. 44-47 (Russian).

Cheng, K. & J.Cao (1981). Reliability analysis of a general repairable system : Markov renewal model. Acta Math.Appl.Sin., 4, 295-306.

Cherenkov, A.P. (1974). Existence theorems for a semi-Markov process with an arbitrary set of states. Math.Notes, 15, 367-373.

Cherenkov, A.P. (1977). Asymptote of additive functionals of semi-Markov processes with arbitrary set of states. Math.Notes, 21, 119-124.

Cherenkov, A.P. (1977). Attainability property and ergodicity theorems for a semi-Markov process with an arbitrary set of states. Math.Notes, 21, 167-173.

Cherenkov, A.P. (1980). Some ergodic theorems for a semi-Markov process with an arbitrary set of states. Theory Probab.Math. Statist., 19, 169-180.

Cherenkov, A.P. (1982). The law of large numbers for additive functionals of semi-Markov processes. Applied problems of probability theory. Collect.Sci.Works, Kiev, 134-141 (Russian).

Cherkesov, G.N. (1980). Semi-Markovian models of reliability of
    multichannel systems with unreplenishable reserve of time.
    Eng.Cybern., 18, 65-78.

Chernaya, M.F. (1979). A busy period analysis for a system with semi-
    Markov service. Analytical methods in probability theory,
    Collect.Sci.Works, Kiev, 98-102 (Russian).

Chernomorov, G.A. & V.G. Zhukovskij (1979). An analytic model of a
    closed service system with source by means of an enlargment of
    a semi-Markov process. Izv.Sev.-Kavk.naucn.Cent.vyss.Sk., teh.
    Nauki, 33-38 (Russian).

Chernyak, A.I. (see V.V.Anisimov).

Cherry, W.P. & R.L.Disney (1983). The superposition of two indepen-
    dent Markov renewal processes. Zastosow.Mat., 17, 567-602.

Chien, T. (see E.M.Akif).

Chikte, S.D. & S.D.Deshmukh (1976). Dynamic pricing with stochastic
    entry. Review Economic Studies, 43, 91-97.

Chokron, M. (see F.Chauny).

Choo, Q.H. & B.Conolly (1979). New results in the theory of repeated
    orders queueing systems. J.Appl.Probab., 16, 631-640.

Christer, A.H. (1979). Refined asymptotic costs for renewal reward
    processes. J.Oper.Res.Soc., 29, 577-583.

Chu, W.W. & H.Opderbeck (1976). Analysis of the PFF replacement
    algorithm via a semi-Markov model. Commun.ACM, 19, 298-304.

Chumakov, L.D. (1975). Interplay between reliability and maintenance
    costs of a technical system with periodic repair control.
    Nadezn.Dolgovecn.teh.Sist., 57-60 (Russian).

Chumakov, L.D. & V.G.Kurasov (1975). Interplay between reliability
    and maintenance costs of a technical system with exponential
    running time without failure and with periodic repair control.
    Nadezn.Dogovecn.teh.Sist., 53-57 (Russian).

Cin, M.D. (1980). Availability analysis of a fault-tolerant computer
    system. IEEE Trans.Reliab. R-29, 265-268.

Çinlar, E. (1975). Markov renewal theory. A survey. Management Sci.,
    Theory, 21, 727-752.

Çinlar, E. (1976). On the structure of semi-Markov processes compared
    with Chung processes. Ann.of Probab., 4, 402-417.

Çinlar, E. (1977). Conversion of semi-Markov processes to Chung
    processes. Ann.of Probab., 5, 180-199.

Çinlar, E. (1977). Markov additive processes and semi-regeneration.
    Proc.5th Conf.Probab.Theory, Brasov, 33-49.

Çinlar, E. (1979). On increasing continuous processes. Stochastic
    Processes Appl., 9, 147-154.

Clarotti, C.A. (1981). Limitations of minimal cut-set approach in
    evaluating reliability of systems with repairable components.
    IEEE Trans.Reliab.R-30, 335-338.

Coffman E.G. jun. & I.Mitrani (1980). A characterization of waiting
    time performance realizable by single-server queues. Oper.Res.,
    28, 810-821.

Cohen, M., J.C.Hershey & E.N.Weiss (1981). A stochastic service net-
    work model with application to hospital facilities. Oper.Res.,
    29, 1-22.

Colard, J.P. & G.Latouche (1980). Algorithmic analysis of Markovian
    model for a system with batch and interactive jobs. Opsearch,
    17, 12-32.

Collart, D. & A.Haurie (1980). On the control of care supply and
    demand in a urology department. Eur.J.Oper.Res., 4, 160-172.

Consael, R. & J.Sonnenschein (1979). Modèles pseudo-markoviens. Cah.
    Cent.Etud.Rech.Oper., 21, 315-318.

Conolly, B. (see Q.H.Choo).

Costes, A., Ch.Landrault & J.C.Laprie (1981). Parametric analysis
    of 2-unit redundant computer systems with corrective and
    preventive maintenance. IEEE Trans.Reliab., R-30, 139-144.

Courtois, P.J. (1978). Exact aggregation of queueing networks.
    Mathématiques Appl., 1er Colloq.AFCET-SMF, Palaiseau, I, 35-51.

Cox, D.R.& V.Isham (1977). A bivariate point process connected with electronic counters. Proc.Roy.Soc.London, Ser.A, 356, 149-160.

Crabill, T.B. (1974). Optimal control of a maintenance system with variable service rates. Operations Res., 22, 736-745.

Crowley, J., F.Y.Hsieh & D.C.Tormey (1983). Some test statistics for use in multistate survival analysis. Biometrika, 70, 111-119.

Cruon, R., D.Grouchko & A.Kaufmann (1975). Modèles mathématiques pour l'étude de la fiabilité des systèmes. Masson et Cie, Paris.

Cubeddu, C. (1976). Schema di una regolzione del traffico all'incrocio di due strade. Rend.Sem.Fac.Sci.Univ.Cagliari, 46, 295-300.

Cullmann, G. (1980). Les chaînes de Markov multiples. Programmation dynamique. Masson et Cie, Paris.

Cvetanov, I., J.Havel, T.Naneva & M.Vosvrda (1977). Stochastic model of supplying systems and their applications.Ekonom.-mat.Obzor, 13, 268-278 (Russian).

Daduna, H. (1982). Passage times for overtake-free paths in Gordon-Newell networks. Adv.Appl.Probab., 14, 672-686.

Dal, C.M. (1978). Self-diagnosis of interactive systems. Lect.Notes Biomath., 21, 229-244.

Dal, C.M. (1979). Fehlertolerante Systeme. Modelle der Zuverlaessig-keit, Verfuegbarkeit, Diagnose und Erneuerung. Leitfaden Ang. Math.Mech.LAMM, 50, Teubner-Stuttgart.

Danielyan, Eh.A. & A.K.Pogosyan (1980). On two-sided estimations of the probability of a rare event in a special regenerative process. Uch.Zap.Erevan.Gos.Univ., Estestv.Nauki, 1 (143), 134-136 (Russian).

Davidovich, Y.S. & V.M.Shurenkov (1974). Corrections of homogeneous Markovian processes. Teor.Verojatnost.mat.Statist.Sbornik, 10, 56-68 (Russian).

Dayan, J. & D.Rappaport (1975). On convergence of spatially in-homogeneous multiphase transport systems. SIAM J.Appl.Math., 29, 12-19.

Deb, R.K. (1976). Optimal control of batch service queues with switching costs. Adv.Appl.Probab., 8, 177-194.

Debabrata, R. (see M.Agarwal).

Degtyarev, E.K. & E.K.Kurako (1980). Investigation of the distribu-
    tion of the operative memory of a computer with the aid of semi-
    Markov processes. Izvestija Akad.Nauk SSSR, Tekn.Kibern., 3,
    73-79 (Russian).

Dekker, R. & A.Hordijk (1983). Denumerable Markov decision chains :
    sensitive optimality criteria. Proc.11th Annual Meeting Oper.
    Res., Frankfurt, 453-460.

DeLisi, C. (see W.A.Beyer).

Deshmukh, S.D. (see S.D.Chikte).

Deshmukh, S.D. & W.Winston (1977). A controlled birth and death
    process model of optimal product pricing under stochastically
    changing demand. J.Appl.Probab., 14, 328-339.

D'Esposito, M.R. (see M.Carravetta).

Deul, N. (1980). Stationarity conditions for multi-server queueing
    systems with repeated calls. Elektron.Informationsverarbeitung
    Kybernetik, 16, 607-613.

Diekmann, A. (1979). A dynamic stochastic version of the Pitcher-
    Hamblin-Miller model of collective violence. J.Math.Sociology,
    6, 277-282.

Dikarev, V.E. & N.A.Shishonok (1976). On the interval distribution
    of a class of Markov processes. Theory Probab.Math.Stat., 10,
    69-72.

Dimitrov, B.N. (1972). Sequences of renewal processes and random
    sums. Vycislit.Metody Programm., 18, 3-15 (Russian).

Dimitrov, B.N. (1972). On uniform renewal theory. Vycislit.Metody
    Programm., 18, 24-30 (Russian).

Dinse, G.E. & S.W.Lagakos (1980). The analysis of partially-censored
    data from a first-order semi-Markov model. J.Stat.Comput.
    Simulation, 11, 209-222.

Dirickx, Y.M.I. & M.R.Rao (1974). Networks with gains in discrete
    dynamic programming. Management Sci., Theory, 20, 1428-1431.

Dirickx, Y.M.I. & D.Koevoets (1977). A continuous review inventory
  model with compound Poisson demand process and stochastic lead
  time. Naval Res.Logist.Quart., 24, 577-585.

Disney, R.L. (see W.P.Cherry).

Disney, R.L., D.C.McNicke & B.Simon (1980). The M/G/1 queue with
  instantaneous Bernoulli feedback. Naval Res.Logist.Q., 27,
  635-644.

Dobrushin, R.L. & V.V.Prelov (1979). Asymptotic approach to the
  investigation of message switching networks of linear structure
  with a large number of centers. Probl.Inf.Transm., 15, 46-55.

Dobrushin, R.L. & Y.M.Sukhov (1976). Asymptotic analysis of a star-
  like network with large number of radial rays and message
  switching. Probl.Peredaci Inform., 12, 70-94 (Russian).

Doi, M. & H.Osawa (1978). A variation of the GI/M/1 queue utilizing
  idle time. Rep.Stat.Appl.Res.Union Jap.Sci.Eng., 25, 111-120.

Domanovskaya, E.F. (1976). Stay time of a semi-Markov process in
  fixed state. Control, reliability and navigation. Interuniv.
  Collect.Sci.Works, Saransk, 3, 92-94 (Russian).

Domanovskij, G.A. (1976). Total stay time of a semi-Markov process
  in a fixed set of states. Control, reliability and navigation.
  Interuniv.Collect.Sci.Works, Saransk, 3, 116-122 (Russian).

Dominicis (de), R. (see M.Carravetta).

Dominicis (de), R. (1979). Discrete-time non-homogeneous semi-
  Markov processes. Aerotec.Missili Spazio, 58, 111-113.

Dominicis (de), R. (1979). Processi stocastici semi-Markoviani non-
  omogenei conservativi. Riv.Mat.Sci.Econ.Soc., 2, 157-167.

Dominicis (de), R. (1980). Estimation of waiting average times and
  asymptotic behaviour of the semi-Markov process corresponding
  to a Markov renewal process. Statistica, 40, 467-475.

Dominicis (de), R. (1983). Spectral analysis of non linear semi-
  Markov processes. Proc.NATO ASI C, Math.Phys.Sci., 104, 407-415.

Doshi, B.T. (1979). Generalized semi-Markov decision processes.
  J.Appl.Probab., 16, 618-630.

Downs, T. (1977). An explicit form for system mean life. IEEE Trans.
     Reliab.R-26, 138-140.

Dragut, M. (1981). Non-discounted semi-Markov decision processes
     with incomplete state information. Proc.6th Conf.Probab.Theory,
     Brasov, 403-411.

Dubois, D. & J.P.Forestier (1982). Productivité et en-cours moyens
     d'un ensemble de deux machines séparées par une zone de
     stockage. RAIRO, Autom.Syst.Anal.Control, 16, 105-132.

D'Souza, C.A. & M.N.Gopalan (1975). Probabilistic analysis of a
     two-unit system with a single service facility for preventive
     maintenance and repair. Operations Res., 23, 173-177.

Duma, I. (1980). Modèles semi-Markoviens pour les canaux numériques.
     An.Univ.Bucur.,Mat., 29, 37-45.

Duyn Schouten (van der), F.A. (1983). Markov decision processes
     with continuous time parameter. Mathematical Centre Tracts,
     164, Amsterdam.

Dzirkal, E.V. & A.E.Shura-Bura (1980). Calculation of the reliability
     of a redundant group with unreliable switching and incomplete
     monitoring. Eng.Cybern., 18, 84-90.

Edwards, J.S. & R.W.Morgan (1982). Optimal control models in man-
     power planning. North-Holland Syst.Contr.Ser., 4, 143-176.

El-Affendi, M.A. & D.D.Kouvatsos (1983). A maximum entropy analysis
     of the M/G/1 and G/M/1 queueing systems at equilibrium. Acta
     Inf., 19, 339-355.

Elejko, Y.I. (1979). Ergodic theorems for semi-Markov processes with
     an arbitrary phase space. Stoch.Process.Probl.Math.Phys.,
     Collect.Sci.Works, Kiev, 73-93 (Russian).

Elejko, Y.I. (1980). Transients in the theory of multi-dimensional
     renewal with discrete time. Probab.Meth.of Infinite dimens.
     analysis, Collect.Sci.Works, Kiev, 47-60 (Russian).

Elejko, Y.I. (1981). Limit distributions for a semi-Markov process
     with arbitrary phase space. Theory Probab.Math.Stat., 23,
     55-61.

Elejko, Y.I. (1981). Limit theorems for additive functionals defined on a semi-Markov process with an arbitrary phase space. Probabilistic infinite dimensional analysis, Collect.Sci.Works, Kiev, 44-50 (Russian).

Elejko, Y.I. & V.M.Shurenkov (1980). Limit distributions of time averages for a semi-Markov process with finite number of states. Ukr.Math.J., 31, 475-479.

Ewers, T. & V. Nollau (1976, 1977). Zur Steuerung halb-Markovscher Prozesse und ihrer Anwendung bei der Ermittlung optimaler Bedienungsstrategien in einem Mehrmaschinensystem. II, III; Wiss.Z.techn. Univ.Dresden, 25, 787-790; 26, 109-113.

Ezov, I.I. (see G.L.Arsenisvili, V.N.Bandura & O.P.Borozdin).

Ezov, I.I. & V.F.Kadankov (1981). The system M//p(**r)|1| in employment conditions. Analytical Meth.Probab.Theory, Collect.Sci. Works, Kiev, 75-83 (Russian).

Ezov, I.I. & H.Sum (1977). Maximization processes with discrete time. Theory Probab.Math.Stat., 9, 87-94.

Ezov, I.I. & O.M.Zaharin (1977). On one semi-Markov system with profits. Dopovidi Akad.Nauk Ukrain.R R, Ser.A, 1120-1122 (Ukrainian).

Ezov, I.I. & O.M.Zaharin (1978). The time of stay of a complex randomized semi-Markov process in a fixed subset of states. Dopovidi Akad.Nauk Ukrain.RSR, Ser.A, 344-347 (Ukrainian).

Ezov, I.I., I.Z.Gabrovski & O.M.Zaharin (1977). On a generalization of semi-Markov processes. Theory Probab.Math.Stat., 9, 49-62.

Fajnberg, E.A. (1982). Non-randomized Markov and semi-Markov strategies in dynamic programming. Theory Probab.Appl., 27, 116-126.

Falin, G.I. (1982). State consolidation in symmetrical partially accessible circuits. Probl.Control Inf.Theory, 11, 3-12 (Russian).

Faure, R. & J.L.Lauriere (1974). Fiabilité et renouvellement des équipements. Applications élémentaires aux investissements. Coll.Programmation.Rech.Opér.Appl., 4, Gauthier-Villars, Paris.

Federgruen, A. & P.J.Schweitzer (1978). Discounted and undiscounted
    value-iteration in Markov decision problems : a survey. Proc.
    Int.Conf.Dynamic Progr.Appl., Vancouver, 23-52.

Federgruen, A. & P.J.Schweitzer (1978). The functional equations of
    undiscounted Markov renewal programming. Math.Oper.Res., 3,
    308-321.

Federgruen, A. & P.J.Schweitzer (1978). Foolproof convergence in
    multichain policy iteration. J.Math.Analysis Appl., 64, 360-368.

Federgruen, A. & P.J.Schweitzer (1977). Geometric convergence of
    value-iteration in multichain Markov renewal programming.
    Amsterdam Mathematisch Centrum, BW 80/77, 37 p.

Federgruen, A. & D.Spreen (1980). A new specification of the multi-
    chain policy iteration algorithm in undiscounted Markov renewal
    programs. Manage.Sci., 26, 1211-1217.

Federgruen, A. & H.C.Tijms (1978). The optimality equation in average
    cost denumerable state semi-Markov decision problems, recurrence
    conditions and algorithms. J.Appl.Probab., 15, 356-373.

Federgruen, A., G.de Leve & H.C.Tijms (1977). A general Markov
    decision method. I : Model and techniques; II : Applications.
    Adv.Appl.Probab., 9, 296-315 & 316-335.

Federgruen, A., A.Hordijk & H.C.Tijms (1977). A note on simultaneous
    recurrence conditions on a set of denumerable stochastic
    matrices. Amsterdam Mathematisch Centrum, BW 85/77, 7 p.

Federgruen, A., A.Hordijk & H.C.Tijms (1978). Recurrence conditions
    in denumerable state Markov decision processes. Proc.Int.Conf.
    Dynamic Progr.Appl., Vancouver, 3-22.

Federgruen, A., A.Hordijk & H.C.Tijms (1979). Denumerable state
    semi-Markov decision processes with unbounded costs, average
    cost criterion. Stochastic Processes Appl., 9, 223-235.

Federgruen, A., P.J.Schweitzer & H.C.Tijms (1977). Value-iteration
    in undiscounted Markov decision problems. II : geometric
    convergence; III : algorithms. Proc.Semin.Markov Decis.th.,
    Amsterdam, 141-151 & 153-159.

Federgruen, A., P.J.Schweitzer & H.C.Tijms (1978). Contraction
    mappings underlying undiscounted Markov decision problems.
    J.Math.Analysis Appl., 65, 711-730.

Federgruen, A., P.J.Schweitzer & H.C.Tijms (1983). Denumerable
    undiscounted semi-Markov decision processes with unbounded
    rewards. Math.Oper.Res., 8, 298-313.

Feldman, R.M. (1976). Optimal replacement with semi-Markov shock
    models. J.Appl.Probab., 13, 108-117.

Feldman, R.M. (1977). Optimal replacement with semi-Markov shock
    models using discounted costs. Math.Oper.Res., 2, 78-90.

Feldman, R.M. (1978). A continuous review (s,S) inventory system
    in a random environment. J.Appl.Probab., 15, 654-659.

Feldman, R.M. & C.Whittaker (1980). Moments for a general branching
    process in a semi-Markovian environment. J.Appl.Probab., 17,
    341-349.

Fetisov, V.N. (1979). A Markov approximation of a random sequence
    in optimal control problems. Eng.Cybern., 17, 31-37.

Fix, W. & K.Neumann (1979). Project scheduling by special GERT
    networks. Computing, 23, 299-308.

Fleischmann, K. & U.Prehn (1978). Limit theorems for spatially
    homogeneous branching processes with a finite set of types.II.
    Math.Nachr., 82, 277-296.

Forestier, J.P. (see D.Dubois).

Forestier, J.P. (1978). Optimisation de la commande des systèmes à
    représentation semi-markovienne. Math.Appl., 1er Colloq.AFCET
    SMF, Palaiseau, I, 337-346.

Fortet, R. (1977). Panels et modèles de comportements. Publ.écono-
    métr., 10, 1-39.

Foschini, G.J. (1982). Equilibria for diffusion models of pairs of
    communicating computers-symmetric case. IEEE Trans.Inf.Theory
    IT-28, 273-284.

Foster, F.G. & H.G.Perros (1979). Hierarchical queue networks with
    partially shared servicing. J.Oper.Res.Soc., 30, 157-166.

Franken, P. (see K.Arndt).

Franken, P. (1082). The point process approach to queueing theory
     and related topics. Seminarber., Humboldt-Univ.Berlin, Sekt.
     Math., 43, 72 p.

Franken, P. & A.Streller (1979). Stationary generalized regenerative
     processes. Theory Probab.Appl., 24, 79-90.

Friedman, A.D. (see M.A.Breuer).

Fries, P. (1980). Procedure for weighting failure rates in Markov
     reliability models for automatic systems. Siemens Forsch.
     Entwicklungsber., 9, 305-311.

Furukawa, N. (1980). Nearly optimal policies and stopping times in
     Markov decision processes with general rewards. Bull.Math.
     Stat., 19, 89-109.

Furukawa, N. (1982). Recurrence set-relations in stochastic multi-
     objective dynamic decision processes. Math.Operationsforsch.
     Stat., Ser.Optimization 13, 113-121.

Furukawa, N. & S.Iwamoto (1974). Correction to "Markovian decision
     processes with recursive reward functions". Bull.Math.Statist.,
     16, 127.

Gabrovsky, I.Z. (see I.I.Ezov).

Gadasin, V.A. (1975). Probabilistic properties of a system with a
     symmetric radial-ring structure. Automat.Remote Control, 35,
     1503-1509.

Gaede, K.W. (1974).Sensitivitaetsanalyse fuer einen semi-Markov
     Prozess. Z.Operat.Res., Ser.A 18, 197-204.

Gaede, K.W. (1977). Zuverlaessigkeit, mathematische Modelle. Theorie
     und Praxis des Operations Research. Carl Hanser Verlag, Munchen.

Gaede, K.W. (1977). Einiges zur Sensitivitaet stochastischer OR-
     Modelle. Quant.Wirtsch.Forsch., W.Krelle zum 60.Geb., 237-244.

Gaede, K.W. (1983). Verallgemeinerte Kontrollgrenzen bei Erzatz-
     problemen. Operations Research. Proc.11th Annual Meet.,
     Frankfurt, 433-438.

Gail, M.H. (see C.C.Brown).

Gallisch, E. (1979). On monotone optimal policies in a queueing
     model of M/G/1 type with controllable service time distribution.
     Adv.Appl.Probab., 11, 870-887.

Gamkrelidze, R.V. (ed.) (1974). Probability Theory. Mathematical
     Statistics. Theoretical Cybernetics. Band 11. "VINITI" Moskva
     (Russian).

Gardos, E. & T.Toeroek (1980). Population processes and computer
     networks. Alkalmazott Mat.Lapok, 6, 291-311 (Hungarian).

Garifullin, Z.G. & V.A.Ivnitskij (1977). On stationary distribution
     of state probabilities for a generalized scheme of birth and
     death with a symmetrical graph of transitions. Kibernetika,
     Kiev, 4, 92-96 (Russian).

Gasanenko, V.A. (1981). On thinning a semi-Markov process with a
     countable set of states. Theory Probab.Math.Stat., 22, 25-29.

Gatarek, D. (1982). On a reliability problem by stochastic control
     problems. Syst.Control Lett., 2, 248-254.

Gavish, B. & P.J.Schweitzer (1976). An optimality principle for
     Markovian decision processes. J.Math.Analysis, 54, 173-184.

Gazis, D.C. (ed). (1974). Traffic Science. Wiley Interscience
     Publication, Wiley & Sons.

Gelenbe, E. (1979). Probabilistic models of computer systems. II :
     Diffusion approximation, waiting times and batch arrivals.
     Acta Inf., 12, 283-303.

Gelenbe, E & I.Mitrani (1980). Analysis and synthesis of computer
     systems. Academic Press, London.

Gelenbe, E. & G.Pujolle (1976). The behaviour of a single queue in
     a general queueing network. Acta Informatica, 7, 123-136.

Geninson, B.A., A.A.Rubchinshij & T.M. Vinogradskaya (1984). Semi-
     Markov decision making processes with vector gains. Theory
     Probab.Appl., 28, 191-193.

Genis, Y.G. (1978). The asymptotic behaviour of a flow of rare
     events in a regenerating process. Eng.Cybern., 16, 77-84.

Genis, Y.C. (1979). Stochastic system behaviour described by semi-
     renewal process. Autom.Remote Control, 40, 668-681.

Genis, Y.G. & B.Y.Nechaev (1976). Behaviour of a complex system
     described by a semi-Markov process with a finite number of
     states. Izvestija Akad.Nauk SSSR, tehn.Kibernet., 76-84
     (Russian).

Georganas, N.D. (see A.Chatterjee).

Gergely, T. & T.L.Toeroek (1976). A stochastic process to describe
     the virtual waiting time in a discrete-time queueing system.
     Prog.Oper.Res., Eger 1974, Colloq.Math.Soc.Janos Bolyai, 12,
     417-423.

Gerrard, R. (1983). Regularity conditions for semi-Markov and Markov
     chains in continuous time. J.Appl.Probab., 20, 505-512.

Gersht, A.M. (1978). Analysis of certain low-traffic queueing
     systems. Autom.Remote Control, 39, 491-499.

Gershwin, S.B. & I.C.Schick (1979). Analytic methods for calculating
     performance measures of production lines with buffer storages.
     Proc.IEEE Conf.Decision & Control, San Diego, 618-624.

Gertsbach, I. (1976). Sufficient optimality conditions for control-
     limit policy in a semi-Markov process. J.Appl.Probab., 13,
     400-406.

Gestri, g. H.A.K.Mastebroek & W.H.Zaagman (1980). Stochastic
     constancy, variability and adaptation of spike generation :
     performance of a giant neuron in the visual system of the fly.
     Biol.Cybern., 38, 31-40.

Gheorghe, A.V. (1976). On risk-sensitive Markovian decision models
     for complex systems maintenance. Econom.Comput.econom.Cyber-
     netics Studies Res., 31-46.

Gheorghe, A.V. (1977). Partially observable Markov processes with
     a risk-sensitive decision-maker. Revue Roumaine Math.Pur.Appl.,
     22, 461-482.

Gheorghe, A.V. (1979). Systems engineering. Models and computational
     techniques. Editura Academiei Republicii Socialiste Romania.

Gheorghe, A.V. (1979). Markovian models with logical conditions for
     fault isolation with insufficient observations. Revue Roumaine
     Math.Pur.Appl., 24, 1047-1064.

Gheorghe, A.V. (1979). On maintenance policies for multifunctional
    systems using semi-Markov decision models. Revue Roum.Math.
    Pur.Appl., 24, 1337-1353.

Gheorghe, A.V. (1980). Reliability prediction of systems with semi-
    Markov structure. Revue Roum.Math.Pur.Appl., 25, 1225-1241.

Ghosal, A. (see J.Affisoo).

Gilbert, G. (1973). Semi-Markov processes and mobility; a note.
    J.Math.Sociology, 3, 139-145.

Gill, R.D. (1980). Nonparametric estimation based on censored
    observations of a Markov renewal process. Z.Wahrscheinlichkeits-
    theor.Verw.Geb., 53, 97-116.

Gillert, H. (1976). Maximum-likelihood-Schaetzungen fuer Parameter
    in semimarkovschen Prozessen. Wiss.Z.Tech.Univ.Dres., 25,
    1163-1166.

Giorgadze, A.H. & Eh.I.Kistauri (1977). Parallel decomposition of
    Markov processes. Tr.Inst.Kibern., 1, 196-206.

Gittins, J.C. (1982). Forwards induction and dynamic allocation
    indices. Proc.NATO ASI Deterministic & Stochastic Scheduling,
    125-156.

Girmes, D.H. & M.F.Ramalhoto (1977). Markov-renewal approach to
    counter theory. Proc.Eur.Meet.Stat.Grenoble, 581-590.

Glaz, J. (1981). Clustering of events in a stochastic process.
    J.Appl.Probab., 18, 268-275.

Glazebrook, K.D. (1976). Stochastic scheduling with order constraints.
    Int.J.Systems Sci., 7, 657-666.

Glazebrook, K.D. (1978). On a class of non-Markov decision processes.
    J.Appl.Probab., 15, 689-698.

Gnedenko, B.V. & D.Koenig (ed) (1983). Handbuch der Bedienungs-
    theorie I : Grundlagen und Methoden. Akademie-Verlag, Berlin.

Gnedenko, B.V. & A.D.Soloviev (1974). A general allocation problem
    with repair. Izvestija Akad.Nauk SSSR, tehn.Kibernet., 113-118
    (Russian).

Goel, A.L., T.Nakagawa & S.Osaki (1975). Stochastic behaviour of an intermittently used system. Revue Franc.Automat.Inform.Rech. Opérat., 9, 101-112.

Gondran, M. & A.Pages (1980). Fiabilité des systèmes. Editions Eyrolles, Paris.

Goodman, I.R. & S.Kotz (1981). Hazard rates based on isoprobability contours. Proc.NATO ASI Statistical Distributions in Sc.Work, 5, 289-309.

Gopalan, M.N. (see C.A.D'Souza).

Grandori Guagenti, E. (1980). Stochastic processes in the modelling of earthquake occurrence. Proc.Conf.Numer.Tech.Stoch.Systems, Gargnano, 205-216.

Grandori Guagenti, E. & C.Molina (1982). Dynamic semi-Markov systems in earthquake engineering. Rend.Semin.Mat.Torino, 119-133.

Grassmann, W. (1977). Transient solutions in Markovian queues. An algorithm for finding them and determining their waiting-time distributions. Eur.J.Oper.Res., 1, 396-402.

Grassman, W. (1981). Stochastic systems for management. North-Holland, New-York.

Graves, S.C. & J.Keilson (1981). The compensation method applied to a one-product production/inventory problem. Math.Oper.Res., 6, 246-262.

Grigg, C.H. (see R.N.Allan).

Grinold, R.C. (1976). Manpower planning with uncertain requirements. Operations Res., 24, 387-399.

Grinshpan, L.A. (19). Analysis of the stationary regimes of semi-Markov processes by the method of fictitious states.

Groenewegen, L.P.J. & K.Van Hee (1977). Markov decision processes and quasi-martingales. Proc.Eur.Meet.Stat.Grenoble, 453-459.

Grouchko, D. (see R.Cruon).

Gruver, W.A. & C.F.Klein (1981). Optimal control of Markovian queueing systems. Optim.Control Appl.Methods, 2, 23-34.

Gubenko, L.G. (1974). Optimal control of a monotone Markov process. Kibernetika, Kiev, 1, 104-106 (Russian).

Gupta, A. (1981). Markovian approach to determine optimal maintenance
　　policies for production systems. Indian J.Technol., 19, 353-356.

Gupta, S.K. & J.K.Sengupta (1975). Application of reliability pro-
　　gramming to production planning. Intern.J.Systems Sci., 6,
　　633-644.

Gupta, S. & Y.P.Gupta (1978). Chi-square goodness of fit test for
　　equilibrium Markov renewal process. Metron, 36, 187-196.

Gupta, Y.P. (see S.Gupta)

Gupta, Y.P. (1983). Counting processes within Markov renewal process-
　　es. S.Afr.Stat.J., 17, 107-119.

Gupta, Y.P. & S.C.Sharma (1976). Some results in Markov renewal
　　processes. Metron, 34, 81-92.

Gusak, D.V. & V.S.Korolyuk (1976). The asymptotic behaviour of semi-
　　Markov processes with a decomposable set of states. Theor.
　　Probab.Math.Statist., 5, 43-51.

Habel, I. (1980). Ueber einen speziellen stationaeren diskreten
　　Markovschen Entscheidungsprozess mit Fehlerbetrachtung zum
　　H-Algorithmus. Wiss.Z.Tech.Hoch-sch.Leipz., 4, 303-307.

Hannan, E. M.P.Mirabile & J.Schmee (1979). An examination of patient
　　referral and discharge policies using a multiple objective semi-
　　Markov decision process. J.Oper.Res.Soc., 30, 121-129.

Harlamov, B.P. (1976). A "correct exit" property and a limit theorem
　　for the semi-Markov processes. Zap.Naucn.Semin.Leningr.Otd.
　　Mat.Inst.Steklov, 72, 186-201 (Russian).

Harlamov, B.P. (1977). On convergence of semi-Markov walks to a
　　continuous semi-Markov process. Theory Probab.Appl., 21, 482-498.

Harlamov, B.P. (1978). The set of regeneration times of random
　　processes. J.Soviet Math., 9, 102-107.

Harlamov, B.P. (1978). Random processes with semi-Markov chains of
　　hitting times. J.Soviet Math., 9, 107-128.

Harlamov, B.P. (1980). A criterion for the Markov property for
　　continuous semi-Markov processes. Teor.Verojatn.Primen., 25,
　　535-548.

Harlamov, B.P. (1980). Additive functionals and a change of time
     which preserves the semi-Markov property of a process. Zap.
     Nauchn.Semin.Leningr.Otd.Mat.Inst.Steklov, 97, 203-216 (Russian).

Harlamov, B.P. (1981). A Markov property test for semi-Markov
     processes. Theory Probab.Appl., 25, 526-539.

Harlamov, B.P. (1982). Outleading sequences and continuous semi-
     Markov processes on the line. Zap.Nauchn.Semin.Leningr.Otd.
     Mat.Inst.Steklov, 119, 230-236 (Russian).

Harlamov, B.P. (1983). Property of "correct exit" and one limit
     theorem for semi-Markov processes. J.Soviet Math., 23,
     2352-2362.

Harrison, P.G. (1981). Transient behaviour of queueing networks.
     J.Appl.Probab., 18, 482-490.

Harrison, J.M. & A.J.Lemoine (1981). A note on networks of infinite-
     server queues. J.Appl.Probab., 18, 561-567.

Harrison, J.M. & M.I.Reiman (1981). Reflected Brownian motion on an
     orthant. Ann.Probab., 9, 302-308.

Harrison, J.M. & M.I.Reiman (1981). On the distribution of multi-
     dimensional reflected Brownian motion. SIAM J.Appl.Math., 41,
     345-361.

Harrison, J.M. & S.I.Resnick (1978). The recurrence classification
     of risk and storage processes. Math.Oper.Res., 3, 57-66.

Hastings, N.A.J. & J.M.C.Mello (1978). Decision networks. John
     Wiley & Sons.

Hatoyama, Y. (1977). Markov maintenance models with control of
     queue. J.Operations Res.Soc.Japan, 20, 164-181.

Haurie, A. (see A.Alj, F.Chauny & D.Collart).

Haurie, A. (1982). A note on "Multilayer control of large Markov
     chains". IEEE Trans.Autom.Control AC-27, 746-747.

Haussmann, W. (see R.Avenhaus).

Havel, J. (see I.Cventanov).

Hawkes, J. (1977). Intersections of Markov random sets. Z.Wahrschein-
     lichkeitstheorie Verw.Geb., 37, 243-251.

Hayne, W.J. & K.T.Marshall (1977). Two-characteristic Markov-type
manpower flow models. Navals Res.Logist.Q., 24, 235-255.

Hayre, L.S. (1983). A note on optimal maintenance policies for
deciding whether to repair or replace. Eur.J.Oper.Res., 12,
171-175.

Hee (Van), K. (see L.P.J.Groenenwegen).

Heffes, H. (1982). Moment formulae for a class of mixed multi-job-
type queueing networks. Bell.Syst.Tech.J., 61, 709-745.

Heidelberger, P. (1980). Variance reduction techniques for the
simulation of Markov processes. I. Multiple estimates. IBM J.
Res.Dev., 24, 570-581.

Heidelberger, P. & D.L.Iglehart (1979). Comparing stochastic systems
using regenerative simulation with common random numbers. Adv.
Appl.Probab., 11, 804-819.

Heidelberger, P. & M.S.Meketon (1982). A renewal theoretic approach
to bias reduction in regenerative simulations. Manage.Sci., 28,
173-181.

Helary, J.M. & R.Pedrono (1983). Recherche opérationnelle. Travaux
dirigés. Modèles stochastiques : chaînes de Markov, files
d'attente. Réseaux de files d'attente. Hermann, Paris.

Helland, I.S. (1981). Convergence to diffusions with regular
boundaries. Stochastic Processes Appl., 12, 27-58.

Hellwig, M.F. (1980). Stochastic processes of temporary equilibria.
J.Math.Econ., 7, 287-299.

Helm, W.E. & R.Schassberger (1982). Insensitive generalized semi-
Markov schemes with point process input. Math.Oper.Res., 7,
129-138.

Henderson, W. (1983). Non-standard insensitivity. J.Appl.Probab.,
20, 288-296.

Hennion, H. (1982). Transience de certaines chaînes semi-Markoviennes.
Ann.Inst.Henri Poincaré, Nouv.Sér., Sect.B, 18, 277-291.

Hershey, J.C. (see M.A.Cohen).

Hillier, F.S. (see R.W.Boling).

Hinderer, K. (1975). Neuere Resultate in der stochastischen dynamis-
    chen Optimierung. Z.Angew.Math.Mech., 55, Sonderheft, T16-T26.

Hines, W.G.S. (1976). Improving a simple monitor of a system with
    sudden parameter changes. IEEE Trans.Inform.Theory IT-22,
    496-499.

Hipp, Ch. (1983). Asymptotic expansions in the central limit theorem
    for compound and regenerative processes. Statistics, Univ.
    Cologne, Preprint, 39 p.

Hochman, H.G. (1980). Models for neural discharge which lead to
    Markov renewal equations for probability density. Math.Biosci.,
    51, 125-139.

Hoepfinger, E. (1980). Dynamic programming of stochastic activity
    networks with cycles. Lect.Notes Control Inf.Sci., 23, 309-315.

Homma, H. & K.Tanaka (1979). Non-cooperative n-person semi-Markov
    game. Sci.Rep.Niigata Univ., Ser.A, 16, 23-34.

Hofri, M. (1978). A generating-function analysis of multiprogramming
    queues. Internat.J.Computer Inform.Sci., 7, 121-155.

Ho, Y.C. (see E.M.Akif).

Hordijk, A. (see R.Dekker & A.Federgruen).

Hordijk, A. & L.C.M.Kallenberg (1981). Linear programming methods
    for solving finite Markovian decision problems. Proc.1980
    Vortr.Jahrestag DGOR, Essen, 468-482.

Hordijk, A. & R.Schassberger (1982). Weak convergence for generalized
    semi-Markov processes. Stochastic Processes Appl., 12, 271-291.

Hordijk, A., P.J.Schweitzer & H.Tijms (1975). The asymptotic beha-
    viour of the minimal total expected cost for the denumerable
    state Markov decision model. J.Appl.Probab., 12, 298-305.

Hordijk, A., O.J.Vrieze & G.L.Wanrooij (1976). Semi-Markov strate-
    gies in stochastic games. Amsterdam Mathematisch Centrum,
    BW 68/76, 9 p.

Hordijk, A., O.J.Vrieze & G.L.Wanrooij (1983). Semi-Markov strate-
    gies in stochastic games. Int.J.Game Theory, 12, 81-89.

Hsieh, F.Y. (see J.Crowley).

Huang, Z. (1982). Generalized renewal sequences, semi-p-functions
    and their F-function clusters. Acta Sci.Nat.Univ.Sunyatseni,
    2, 30-38 (Chinese).

Hughes, J.S. (1980). A note on quality control under Markovian
    deterioration. Oper.Res., 28, 421-424.

Hunter, J.J. (1982). Generalized inverses and their application to
    applied probability problems. Linear Algebra Appl., 45,
    157-198.

Hunter, J.J. (1983). Filtering of Markov renewal queues. I :
    Feedback; II : Birth-death queues. Adv.Appl.Probab., 15,
    349-375 & 375-391.

Idzik, A. (1977). On Markov policies in continuous time discounted
    dynamic programming. Trans.7th Prague Conf.Vol.A, 265-276.

Iglehart, D.L. (see Ph.Heidelberger).

Iglehart, D.L. (1976). Simulating stable stochastic systems.
    Quantile estimation. J.Assoc.Comput.Machin., 23, 347-360.

Iglehart, D.L. & G.S.Shedler (1979). Regenerative simulation of
    response times in networks of queues with multiple job types.
    Acta Informatica 12, 159-175.

Iglehart, D.L. & G.S.Shedler (1981). Regenerative simulation of
    response times in networks of queues : statistical efficiency.
    Acta Informatica 15, 347-363.

Iglehart, D.L. & G.S.Shedler (1983). Statistical efficiency of
    regenerative simulation methods for networks of queues. Adv.
    Appl.Probab., 15, 183-197.

Iglehart, D.L. & G.S.Shedler (1983). Simulation of non-Markovian
    systems. IBM J.Res.Dev., 27, 472-480.

Inaba, F.S. (1978). On stochastic entry and exit without expecta-
    tions. Review Econ.Studies, 45, 535-545.

Inagaka, T. K.Inoue, E.Sakino & I.Takami (1978). Optimal allocation
    of fault detectors. IEEE Trans.Reliab.R-27, 360-362.

Incerti, S. (see F.Aluffi).

Inoue, K. (see T.Inagaki).

Isham, V. (see D.R.Cox).

Ivchenko, G.I., V.A.Kashtanov & I.N.Kovalenko (1982). Theory of mass
    service. Moskva "Vysshaja Shkola" (Russian).

Ivnitskij, V.A. (see Z.G.Garifullin).

Ivnitskij, V.A. (1979). Connections between the stationary probabi-
    lities of the states of queueing systems at an arbitrary instant,
    the instant of placing an order, and the instant of exit of an
    order. Eng.Cybern., 17, 66-76.

Ivnitskij, V.A. (1982). On a condition of invariance of stationary
    probabilities for queueing networks. Theory Probab.Appl., 27,
    196-201.

Iwase, S., K.Tanaka & K.Wakuta (1976). On Markov games with the
    expected average reward criterion. Sci.Rep.Niigata Univ., Ser.A,
    12, 31-41.

Jagers, P. (1983).On the Maltusianness of general branching process-
    es in abstract type spaces. Probability & Mathematical Statis-
    tics, Essays in Hon.C.G.Esseen, 53-61.

Jain, G.C. & M.S.H.Khan (1980). On designing the capacity of a
    reservoir. Biom.J., 22, 359-369.

Jain, J.L. & S.G.Mohanty (1981). Busy period distributions for two
    heterogeneous queueing models involving batches. INFOR, Can.
    J.Oper.Res.Inf.Process. 19, 133-139.

Jaiswal, N.K. & J.V.Krishna (1980). Analysis of two-unit-dissimilar
    standby redundant system with administrative delay in repair.
    Int.J.Syst.Sci., 11, 495-511.

Jaiswal, N.K. & K.G.Ramamurthy (1982). Some results for pseudo
    semi-Markov process. Math.Operationsforsch.Stat. Ser.Optimi-
    zation, 13, 123-132.

Jaiswal, N.K., Karmeshu & N.S.Rangaswamy (1982). A semi-Markovian
    model for cell survival after irradiation. Biom.J., 24, 63-68.

Jannusch, E. & V.Nollau (1982). Zur Verallgemeinerung eines sukzes-
    siven Approximationsverfahrens von Van der Wal auf semi-
    Markovsche Entscheidungsprozesse. Math.Operationsforsch.Stat.,
    Ser.Optimization, 13, 299-307.

Jansen, U. (1983). A generalization of intensivity results by
    cyclically marked stationary point processes. Elektron.
    Informationsverarbeitung Kybernetik, 19, 307-320.

Jansen, U. & D.Koenig (1976). Invariante stationaere Zustandswahr-
    scheinlichkeiten fuer eine Klasse stochastischer Modelle mit
    funktionellen Abhaengigkeiten. Math.Operationsforsch.Statistik,
    7, 497-522.

Jansen, U. & D.Koenig (1980). Insensitivity and steady-state proba-
    bilities in product form for queueing networks. Elektron.
    Informationsverarbeitung Kybernetik, 16, 385-397.

Jansen, U., D.Koenig & K.Nawrotzki (1979). A criterion of insensi-
    tivity for a class of queueing systems with random marked point
    processes. Math.Operationsforsch.Statistik, Ser.Optimization,
    10, 379-403.

Jansen, U., D.Koenig, M.Kotzurek & H.Rabe (1976). Ergebnisse von
    Invarianzuntersuchungen fuer ausgewaehlte Bedienungs- und
    Zuverlaessigkeitssysteme. Math.Operationsforsch.Statistik, 7,
    523-556.

Janssen, J. (1976). Extension of a two-sided inequality of Daley and
    Moran to the completely semi-Markov queueing model. Cahiers
    Centre Etud.Rech.Opér., 18, 459-469.

Janssen, J. (1976). Some duality results in semi-Markov chain theory.
    Revue Roum.Math.Pur.Appl., 21, 429-441.

Janssen, J. (1977). Absorption problems in semi-Markov chains. Proc.
    Eur.Meet.Statist., Grenoble, 481-488.

Janssen, J. (1977). The semi-Markov model in risk theory. Proc.2nd
    Eur.Congr.Adv.Oper.Res., Stockholm, 613-621.

Janssen, J. (1978). Absorption problems in two-dimensional semi-
    Markov chains. Bull.Soc.Math.Belg., Sér.A, 30, 123-134.

Janssen, J. (1978). Some prediction problems in semi-Markov queueing
    models and related topic. Proc.2nd Symp.Oper.Res., Aachen,
    702-712.

Janssen, J. (1979). Some explicit results for semi-Markov models in
    risk theory and in queueing theory. Oper.Res.Verfahren, 33,
    217-231.

Janssen, J. (1981). Generalized risk models. Cah.Cent.Etud.Rech.
    Opér., 23, 225-244.

Janssen, J. (1982). Modèles de risque semi-Markoviens. Cah.Cent.
    Etud.Rech.Opér., 24, 261-280.

Janssen, J. (1982). Strong laws for semi-Markov chains with a general
    state space. Ric.Mat., 31, 405-414.

Janssen, J. (1982). Stationary semi-Markov models in risk and queue-
    ing theories. Scand.Actuarial J., 199-210.

Janssen, J. (1982). Détermination de la valeur de la probabilité de
    ruine avec une réserve initiale nulle pour le modèle de risque
    semi-Markovien. Mitt.Ver.Schweiz.Versicherungsmath., 275-283.

Janssen, J. (1982). Modèles de risques semi-Markoviens. Cah.Cent.
    Etud.Rech.Opér., 24, 261-280.

Janssen, J. & J.M.Reinhard (1982). Some duality results for a class
    of multivariate semi-Markov processes. J.Appl.Probab., 19,
    90-98.

Jensen, U. (see K.Bosch).

Jensen, U. (1981). Zustandsabhaengige Erneuerungsstrategien - Ein
    Modell aus der Instandhaltungstheorie. Proc.Jahrestag Oper.
    Res., DGOR, Essen, 251-252.

Jo, K.Y. (1983). Optimal service-rate control of exponential queueing
    systems. J.Oper.Res.Soc.Japan, 26, 147-165.

Jo, K.Y. & S.Stidham jun. (1983). Optimal service-rate control of
    M/G/1 queueing systems using phase methods. Adv.Appl.Probab.,
    15, 616-637.

Johansen, S.G. & S.Stidham jun.(1980). Control of arrivals to a
    stochastic input-output system. Adv.Appl.Probab., 12, 972-999.

Jolkof, S. & V.Rykov (1981). Generalized regenerative processes with
    embedded regeneration periods and their application. Math.
    Operationsforsch.Stat., 12, 575-591.

Kadankof, V.F. (see I.I.Ezov).

Kadota, Y. (1979). Countable state Markovian decision processes under the Doeblin conditions. Bull.Math.Stat., 18, 85-94.

Kadyrova, I.I. (1979). Stochastic equations for jump processes. Theory Probab.Math.Stat., 18, 67-76.

Kakumanu, P. (1977). Relation between continuous and discrete time Markovian decision problems. Naval Res.Logist.Q., 24, 431-439.

Kalashnikov, V.V. (1977). Continuity of the characteristics of regenerative processes. Izvestija Akad.Nauk SSSR, tehn.Kibernet. 87-96 (Russian).

Kalashnikov, V.V. (1979). Estimates of the stability of innovation processes. Eng.Cybern., 17, 63-67.

Kalashnikov, V.V. & V.A.Zhilin (1979). Stability estimates of regenerating processes and their application to priority service systems. Eng.Cybern., 17, 82-89,

Kallenberg, L.C.M. (see A.Hordijk).

Kallenberg, L.C.M. (1983). Linear programming and finite Markovian control problems. Mathematical Centre Tracts, 148.

Kalpakam, S. & M.J.Shahul Hameed (1983). Analysis of a modular warm standby redundant system. Cah.Cent.Etud.Rech.Opér., 25, 49-63.

Kalro, A.H. & A.Rangan (1979). Stochastic models for stock price fluctuations. Cah.Cent.Etud.Rech.Opér., 21, 331-336.

Kambo, N.S. (see H.S.Bhalaik).

Kan, V.E. & L.Ya.Peses(1975). Reliability of systems with accumulation of failures under prevention. Avtomatika vycislit.tehn., Riga, 5, 43-46.

Kander, Z. (1978). Inspection policies for deteriorating equipment characterized by N quality levels. Naval Res.Logist.Q., 25, 243-255.

Kanderhag, L. (1978). Eigenvalue approach for computing the reliability of Markov systems. IEEE Trans.Reliab.R-27, 337-340.

Kao, E.P.C. (1975). A discrete time inventory model with arbitrary interval and quantity distributions of demand. Operations Res., 23, 1131-1142.

Kapil, D.V.S. & S.M.Sinha (1981). Repair limit suspension policies for a 2-unit redundant system with 2-phase repairs. IEEE Trans. Reliab.R-30, 90.

Kaplan, M. (1980). A two-fold tandem net with deterministic links and source interference. Oper.Res., 28, 512-526.

Kaplan, E.I. & D.S.Silvestrov (1979). The theorems of the invariance principle type for semi-Markov processes with an arbitrary set of states. Analytical Methods Prob.Th., Collect.Sci.Works, Kiev, 121-125 (Russian).

Kaplan, E.I. & D.S.Silvestrov (1980). Theorems of the invariance principle type for recurrent semi-Markov processes with arbitrary phase space. Theory Probab.Appl., 24, 536-547.

Kapoor, K.R. & P.K.Kapur (1978). Stochastic behaviour of some 2-unit redundant systems. IEEE Trans.Reliab.,R-27, 382-385.

Kapoor, K.R. & P.K.Kapur (1978). Analysis of 2-unit standby redundant repairable systems. IEEE Trans.Reliab., R-27, 385-388.

Kapur, P.K. (see S.Biswas; & K.R.Kapoor).

Karapenev, H.K. (1978). Reliability estimate for a redundant system with renewal and delayed failure exposure. Izvestija Akad.Nauk SSSR, tehn.Kibernet., 107-112 (Russian).

Karlin, S. & U.Liberman (1983). Measuring interference in the chiasma renewal formation process. Adv.Appl.Probab., 15, 471-487.

Karmeshu (see N.K.Jaiswal).

Karr, A.F. (1976). Two extreme value processes arising in hydrology. J.Appl.Probab., 13, 190-194.

Karr, A.F. (1982). A partially observed Poisson process. Stochastic Processes Appl., 12, 249-269.

Kashtanov, V.A. (see G.I.Ivchenko).

Kaufman, L. (1982). Solving large sparse linear system arising in queueing problems. Lect.Notes Math., 968, 352-360.

Kaufmann, A. (see R.Cruon).

Kawai, H. & H.Mine (1976). Marginal checking of a Markovian degradation unit when checking interval is probabilistic. J.Operations Res.Soc.Japan, 19, 158-173.

Kawai, H. & H.Mine (1982). An optimal inspection and maintenance
    policy of a deteriorating system. J.Operations Res.Soc.Japan,
    1-15.

Kawashima, T. (1979). Turnaround time equations in queueing networks.
    J.Operations Res.Soc.Japan, 21, 477-485.

Kazantsev, E.N. & M.Kh.Prilutskij (1978). A class of controlled
    Markov chains. Russ.math.Surv., 33, 235-236.

Keener, R.W. (1982). Renewal theory for Markov chains on the real
    line. Ann.Probab., 10, 942-954.

Keilson, J. (see S.C.Graves).

Keilson, J. (1974). Sojourn times, exit times and jitter in multi-
    variate Markov processes. Adv.Appl.Probab., 6, 747-756.

Kekre, H.B., R.D.Kumar & H.M.Srivastava (1977). A study of varying
    efficiency multiserver queue models. Int.Symp.Oper.Theory Netw.
    Syst., Vol.II, Lubbock, 156-160.

Kelbert, M.J. (1980). The existence of the limiting distributions in
    some networks of communication of messages. Adv.Probab.Relat.
    Top., Vol.6, 397-421.

Kelbert, M.J. & Yu.M.Sukhov (1980). One class of star-shaped com-
    munication networks with packet switching. Probl.Inf.Transm.,
    15, 286-300.

Kelly, F.P. (1976). Networks of queues. Adv.Appl.Probab., 8, 416-432.

Kelly, F.P. (1979). Reversibility and stochastic networks. John
    Wiley & Sons.

Kelly, F.P. (1982). The throughput of a series of buffers. Adv.Appl.
    Probab., 14, 633-653.

Kempe, F. (1976). Extrapolation and identification of a component of
    a linear Markov process. Automat.Remote Control, 37, 501-511.

Kertz, R.P. (1978). Random evolutions with underlying semi-Markov
    processes. Publ.Res.Inst.Math.Sci., 14, 589-614.

Khan, M.S.H. (see G.C.Jain).

Kim Kwang, W. & C.C.White III (1980). Solution procedures for vector
    criterion Markov decision processes. Large Scale Syst., 1,
    129-140.

Kinugasa, M. & S.Osaki (1982). Performance-related reliability eva-
    luation of a three-unit hybrid redundant system. Int.J.Syst.
    Sci., 13, 1-19.

Kirchheim, A. (1974). Prophylaktische Reparaturtermine eines techno-
    logischen Systems. VI. Intern.Kongr.Anwend.Math.Ingenieurwiss.,
    Weimar, 1972, 178-181.

Kirchheim A. (1980). Ein spezielles Erneuerungsmodell als Semi-
    Markov-Entscheidungsprozess. Elektron.Informationsverarbeitung
    Kybernetik, 16, 569-575.

Kiryan, N.L. & Ju.A.Zak (1976). Algorithms for analysing random
    process transition regimes in the process flow diagram. Avtoma-
    tika, Kiev, 58-70 (Ukrainian).

Kistauri, Eh.I. (see A.H.Giorgadze).

Kistauri, Eh.I. (1978). Successive decomposition of semi-Markov
    processes with a discrete set of states. Soobscenija Akad.Nauk
    Gruzin.SSR 91, 577-580 (Russian).

Kistauri, Eh.I. (1980). Application of a decomposition in order to
    calculate the time in which discrete semi-Markov processes are
    in a fixed subset of states. Soobscenija Akad.Nauk Gruzin.SSR
    99, 69-72 (Russian).

Kistner, K.P. & R.Subramanian (1978). Regenerative Eigenschaften
    von Modellen der Zuverlaessigkeitstheorie. Oper.Res.Verf., 2nd
    Symp.Teil 2, Aachen 1977, 713-725.

Kistner, K.P., R.Subramanian & K.S.Venkatakrishnan (1976). Reliabi-
    lity of a repairable system with standby failure. Operations
    Res., 24, 169-176.

Kitaev, M.Yu. (1982). The existence of optimal homogeneous strate-
    gies of controllable semicontinuous semi-Markov models with
    respect to the average cost criterion. Theory Probab.Appl., 26,
    614-616.

Klega, V. (1981). The mean period of random fluctuations at a linear
    system output. Acta Tech.CSAV, 26, 1-10.

Klein, C.F. (see W.A.Gruver).

Klein Haneveld, W.K. (1980). On the behaviour of the optimal value
    operator of dynamic programming. Math.Oper.Res., 5, 308-320.

Klimov, G.P., R.Nagapetjan & S.N.Smirnov (1982). Regenerative
    processes with Markov-dependent cycles and their applications.
    Math.Operationsforsch.Stat., Ser.Optimization, 13, 287-297.

Kobrin, F.E. & R.G.Potter (1982). Effects of cyclic spouse separation
    on conception times. Math.Biosci., 59, 207-223.

Koenig, D. (see U.Arndt, V.B.Gnedenko & U.Jansen).

Koenig, D. (1974). Anwendung von stochastischen Prozessen mit
    Geschwindigkeiten zur Untersuchung von Bedienungs- und Zuver-
    laessigkeitsmodellen. Wiss.Z.Tech.n.Hoch.Otto von Guericke,
    Magdeburg, 18, 363-368.

Koenig, D. & D.Stoyan (1981). Methoden der Bedienungstheorie. Moskva :
    "Radio i Svyaz" (Russian).

Koevoets, D. (see Y.M.I.Dirickx).

Kogan, Ja.A. (1977). Analytic investigation of page replacement
    algorithms for Markov and semi-Markov program behavior models.
    Automat.Remote Control, 38, 109-111 (Russian).

Kohlas, J. (1982). Stochastic methods of operations research.
    Cambridge University Press.

Kokotushkin, V.A. (see G.P.Basharin).

Kolonko, M. (1980). Dynamische Optimierung unter Unsicherheit in
    einem Semi-Markoff-Modell mit abzaehlbarem Zustandsraum.
    Dissertation,Rhein.Fr.W.Universitaet, Bonn.

Kolonko, M. (1982). The average-optimal adaptive control of a Markov
    renewal model in presence of an unknown parameter. Math.
    Operationsforsch.Stat., Ser.Optimization, 13, 567-591.

Kolonko, M. (1982). Strongly consistent estimation in a controlled
    Markov renewal model. J.Appl.Probab., 19, 532-545.

Kolonko, M. & M.Schael (1980). Optimal control of semi-Markov chains
    under uncertainty with applications to queueing models. Proc.
    Oper.Res.Jahrestag.DGOR, Regensburg 1979, 430-435.

Kondratiev, A.S. (1976). Investigation of the asymptotic form of the
    mean failure-free operating time. Kibernetika, Kiev, 140-142
    (Russian).

Kondratiev, A.S. & G.G.Menshikov (1978). Specification of asymptotic expressions for the mean time of failure-free operation of non-recoverable redundant systems. Kibernetika, Kiev, 142-143 (Russian).

Konev, V.V. (1974). On the optimal switching of spare parts. Izvestija Akad.Nauk SSSR, tehn.Kibernet., 77-83 (Russian).

Konev, V.V. (1975). On the optimal switching for spare parts. Izvestija Akad.Nauk SSSR, tehn.Kivernet., 109-117 (Russian).

Konovalyuk, V.S. (1981). Asymptotic analysis of a two-channel system with N reserve devices. Anal.Meth.Prob.Theory, Collect.Sci. Works, Kiev, 83-96 (Russian).

Konovalyuk, V.S. (1982). Two-channel system with dependent failures. Cybernetics, 17, 808-815.

Konovalyuk, V.S. (1982). Superposition of two dependent Markov renewal processes. Ukr.Math.J., 34, 140-144.

Kopocinska, I. (1980). On system reliability under random load of elements. Zastosow.Mat., 17, 5-14.

Kopocinski, B. (see I.Kopocinska).

Korn, I. (1974). A stochastic process generated by a network with a randomly alternating switch. J.Franklin Inst., 297, 153-167.

Korolyuk, V.S. (see E.Arjas, S.M.Brodi & D.V.Gusak).

Korolyuk, V.S. (1977). Semi-Markov models of complex stochastic systems. 2nd Int.Summer Sch.Prob.Theory Math.Stat., Varna, 113-185.

Korolyuk, V.S. (1982). Superposition of Markov renewal processes. Cybernetics, 17, 556-560.

Korolyuk, V.S. & E.P.Lebedintseva (1978). Limit theorem for sojourn time of a semi-Markov process in a subset of states. Ukr.Math. J., 30, 513-516.

Korolyuk, V.S. & A.Tadzhiev (1977). Asymptotic expansion for the life time distribution of the absorbing semi-Markov process. Dopovidi Akad.Nauk.Ukrain.SSR, Ser.A, 1065-1068(Ukrainian).

Korolyuk, V.S. & A.F.Turbin (1976). Asymptotic enlarging of semi-Markov processes with an arbitrary state space. Proc.3rd Japan-

USSR Symp.Prob.Theory Taschkent 1975, Lecture Notes Math., 550, 297-315.

Korolyuk, V.S. & A.F.Turbin (1976) Semi-Markov processes and their applications. Kiev : Izdatelstvo "Naukova Dumka" (Russian).

Korolyuk, V.S. & A.F.Turbin (1978). Mathematical principles for phase extension of complex systems. Kiev : Izdatelstvo "Naukova Dumka" (Russian).

Korolyuk, V.S. & A.F.Turbin (1982). Markov renewal processes in problems of systems reliability. Kiev : Izdatelstvo "Naukova Dumka" (Russian).

Korolyuk, V.S., A.A.Tomusyak & A.F.Turbin (1979). The time of duration of a semi-Markov process in a splittable set of states. Anal.Meth.Probab.Theory, Collect.Sci.Works, Kiev, 69-79 (Russian).

Kortz, S. (see I.R.Goodman).

Kotzurek, M. (see U.Jansen).

Kouvatsos, D.D. (see M.A.El-Affendi).

Kovalenko, I.N. (see G.I.Ivchenko).

Kovalenko, I.N. (1981). Asymptotic state enlargement for random processes. Cybernetics, 16, 876-886 (Russian).

Kovalenko, I.N. (1980). Analysis of rare events in estimating the efficiency and reliability of systems. Moskva : "Sovetskoe Radio".

Kovalenko, I.N. (1982). Calculation of probabilistic characteristics of systems. Kiev : "Teknika" (Russian).

Kovalenko, I.N. & N.Yu.Kuznetsov (1981). Renewal process and rare events limit theorems for essentially multidimensional queueing processes. Math.Operationsforsch.Stat., 12, 211-224.

Kozlov, V.V. (1978). A limit theorem for a queueing system. Theory Probab.Appl., 23, 182-187.

Kozniewska, I. & M.Wlodarczyk (1978). Renewal, reliability and service models. Warszawa : Pantswowe Wydawnictwo Naukowe (Polish).

Kredentsjer, B.P. (1974). On the optimal distribution of spare parts

according to their availability. Avtomtika vycislit.Tehn.,
     Riga, 31-32.

Krishna, J.V. (see N.K.Jaiswal).

Kshirsagar, A.M. (see M.Becker).

Kshirsagar, A.M. & R.Rao Chennupati (1978). A semi-Markovian model
     for predator-prey interactions. Biometrics, 34, 611-619.

Kuczura, A. (see D.Bajaj).

Kueenle, Ch. & H.U.Kueenle (1977). Durchschnittsoptimale Strategien
     in Markovschen Entscheidungsmodellen bei unbeschraenkten
     Kosten. Math.Operationsforsch.Stat., Optimization, 8, 549-564.

Kueenle, H.U. (see Ch.Kueenle).

Kuehn, P.J. (1979). Approximate analysis of general queueing net-
     works by decomposition. IEEE Trans.Commun., COM-27, 113-126.

Kuhta, T.K. (1977). Analysis of the queueing system $GI/M/m$ by the
     method of the embedded semi-Markov processes. Kibernetika,
     Kiev, 132-134.

Kulba, V.V., A.G.Mamikonov & V.P.Pelikhov (1981). Determination of
     the probabilistic characteristics of feedback control for
     Markov and semi-Markov error processes. Autom.Remote Control,
     41, 1375-1379.

Kumar, A. (see M.Agarwal).

Kumar, A. (1977). Steady-state profit in several 1-out-of-2 : G
     systems. IEEE Trans.Reliab.R-26, 366-369.

Kumar, A. (1977). Profit in a 2-unit standby system with a general
     cost structure. Defence Sci.J., New Delhi, 27, 135-140.

Kumar, A. & R.Lal (1979). Stochastic behaviour of a two-unit standby
     system with contact failure and intermittently available repair
     facility. Internat.J.Systems Sci., 10-589-603.

Kumar, A. & R.Lal (1978). Stochastic behaviour of a standby redundant
     system. IEEE Trans.Reliab.,R-27, 169-170.

Kumar, R.D. (see H.B.Kekre).

Kumar, S. & M.F.Neuts (1982). Algorithmic solution of some queues
     with overflows. Manage.Sci., 28, 925-935.

Kunderova, P. (1979). On a mean reward of a Markov replacement process with only one isolated class of recurrent states. Acta Univ. Palacki.Olomuc., Fac.Rerum Nat., 61, Math., 18, 145-155.

Kunderova, P. (1980). On a mean reward from a common Markov replacement process. Acta Univ.Palacki.Olomuc., Fac.Rerum Nat., 65, Math., 18, 39-50.

Kunderova, P. (1981). On limit properties of the reward from a Markov replacement process. Acta Univ.Palacki.Olomuc., Fac. Rerum Nat., 69, Math., 20, 133-146.

Kupka, J. (1983). Two limit theorems for ergodically generated stochastic processes. Math.Proc.Camb.Philos.Soc., 93, 519-536.

Kurako, E.K. (see E.K.Degtyarev).

Kurasov, V.G. (see L.D.Chumatov).

Kuznetsov, N.Ya. (see I.N.Kovalenko).

Kuznetsov, N.Ya. (1979). On approximation of the probability of redundant system failure by the probability of the monotonous failure. Dopovidi Akad.Nauk Ukrain.RSR, Ser.A, 205-255 (Ukrainian).

Kuznetsov, N.Ya. (1981). Semi-Markov model for loaded duplication. Cybernetics, 16, 558-567.

Kuznetsov, V.N., A.F.Turbin & M.D.Zbyrko (1980). Semi-Markov model for analysing reliability of system with restorable protection. Autom.Remote Control, 41, 879-887.

Lacny, J. (1976). An analysis of buffering techniques in nodes of teleprocessing nets. Arch.Automat.Telemach., 21, 89-104.

Lagakos, S.W. (see G.E.Dinse).

Lagakos, S.W., C.J.Sommer & M.Zelen (1978). Semi-Markov models for partially censored data. Biometrika, 65, 311-317.

Lal, R. (see A.Kumar).

Landman, U. & M.F.Shlesinger (1980). Solutions of physical stochastic processes via mappings onto ideal and defective random walk lattices. Proc.Conf.Applied Stoch.Proc., Athens/USA, 151-246.

Landrault, Ch. (see A.Costes).

Landrault, Ch. (1978). A unified availability analysis method for systems containing design errors. IEEE Trans.Reliab., R-26, 331-334.

Langen, H.J. (1980). On weak approximation of Markov decision models. Methods Oper.Res., 37, 383-389.

Lansky, P. (1978). On a modification of Renyi's traffic models. Apl.Mat., 23, 39-51.

Laprie, J.C. (see A.Costes).

Largay, B.J., J.A.Largay III & D.Young (1982). A new computational approach to cost variance investigation problems. J.Inf.Optimization Sci., 3, 1-16.

Largay, J.A. III (see B.J.Largay).

Larsen, R.L. (see A.K.Agrawala).

Latouche, G. (see J.P.Colard).

Latouche, G. (1982). A phase type semi-Markov point process. SIAM J.Algebraic Discrete Methods, 3, 77-90, 1982.

Latyshev, Eh.eh. (1982). Reliability of systems with random streams of multistage customers. Autom.Remote Control, 43, 229-237.

Lauenstein, G. & K. Renger (1981). Haeherungsweise Ermittlung der Erneuerungsfunktion und Entwicklung von Diagrammen zur Bestimmung optimaler starr periodischer Instandhaltungszylklen bei Weibull-verteilten Laufzeiten. Wiss.Z.Tech.Hochsch.Carl Schorlemmer Leuna-Merseburg, 23, 529-545.

Lauriere, J.L. (see R.Faure).

Lave, R.E. & J.K.Satia (1973). Markovian decision processes with probabilistic observation of states. Management Sci., Theory, 20, 1-13.

Lavenberg, S.S. & M.Reiser (1980). Mean-value analysis of closed multichain queueing networks. J.Assoc.Comput.Math., 27, 313-332.

Lavenberg, S.S. & M.Reiser (1980). Stationary state probabilities at arrival instants for closed queueing networks with multiple types of customers. J.Appl.Probab., 17, 1048-1061.

Lavenberg, S.S. & M.Reiser (1981). Corrigendum to "Mean-value analysis of closed multichain queueing networks". J.Assoc.Comput. Mach., 28, 629.

Lavenberg, S.S. & C.H.Sauer (1977). Sequential stopping rules for the regenerative method of simulation. IBM J.Res.Develop., 21, 545-558.

Lavenberg, S.S., T.L.Moeller & P.D.Welch (1982). Statistical results on control variables with application to queueing network simulation. Oper.Res., 30, 182-202.

Lazar, A.A. (1983). Optimal flow control of a class of queueing networks in equilibrium. IEEE Trans.Autom.Control AC-28, 10001-1007.

Leavenworth, R.S. & R.L.Scheaffer (1979). Design of a process control scheme for defects per 100 units based on AOQL. Naval Res. Logist.,Q, 26, 463-485.

Lebedintseva, E.M. (see V.S.Korolyuk).

Lemaire, B. (1975). Etat actuel et perspectives de l'utilisation des modèles markoviens pour l'étude du comportement des consommateurs. Math.et Sci.Humaines, 50, 51-80.

Lemaire, B. (1978). Théorème de conservation des clients dans les files d'attente. RAIRO, Rech.Opérat., 12, 395-399.

Lemoine, A.J. (see J.M.Harrison).

Lemoine, A.J. (1977). Networks of queues - a survey of equilibrium analysis. Manage.Sci., 24, 464-481.

Lemoine, A.J. (1978). Networks of queues - a survey of weak convergence results. Manage.Sci., 24, 1175-1193.

Lennox, W. & B.Sahay (1974). Moments of the first passage-time for narrow-band process. J.Sound Vibration, 32, 449-458.

Le Ny, L.M. (1980). Etude analytique de réseaux de files d'attente multiclasses à routages variables. RAIRO, Rech.Opér., 14, 331-347.

Lesanovsky, A. (1982). The comparison of two-unit standby redundant system with two and three states of units. Proc.2nd Pannonian Sympt.Probability Stat.Inference, Bad Tatzmannsdorf, Austria, 239-250.

Létac, G. (1978). Chaînes colorées : trois extensions d'une formule de P.Nelson. J.Appl.Probab., 15, 321-339.

Lev, G.S. (1980) On the distribution of the absorption time for a semi-Markov multiplication process. Theory Probab.Appl., 24 876-882.

Leve (de), D. (see A.Federgruen).

Levine, W.S. (see J.S.Baras).

Levinskij, B.G. & A.F.Turbin (1981). A method for the asymptotic
    analysis of semi-Markov processes in the scheme of phase enlar-
    gement. Analyt.Meth.Probab.Theory, Collect.Sci.Works, Kiev,
    133-147.

Liberman, U. (see S.Karlin).

Lidskij, Eh.A. (1981). Reliability as the stability of the output
    characteristics of automation apparatus. Cybernetics, 16,
    450-457.

Liefvoort (van de), A. (see J.L.Carroll).

Lin, T.L. (see J.S.Baras).

Linton, D. & P.Purdue (1979). An M/G/s queue with m customer types
    subject to periodic clearing. Opsearch, 16, 80-87.

Lippman, S.A. (1976). Countable-state, continuous-time dynamic
    programming with structure. Operations Res., 24, 477-490.

Lipsky, L. (see J.L.Carroll).

Litvinov, A.L. & V.A.Popov (1978). Probabilistic model of buffer
    store for a control computer complex with group choice.
    Izvestija Akad.Nauk SSSR, tehn.Kibernet., 91-95.

Liu, M.S. & J.S.Yao (1977). The capacity of a traffic flow on a T
    form traffic intersection. Tamkang J.Math., 8, 135-143.

Lloyd, E.H. (1977). Reservoirs with seasonally varying Markovian
    inflows and their first passage times. Int.Institute Appl.
    Systems Analysis, Laxenburg, RR-77-4.

Lopatko, V.B. (1977). Communication between processes via ports with
    enclosed buffers. Mosc.Univ.Comput.Math.Cybern., 75-79.

Lucas, G. (see B.Anders).

Luss, H. (1976). Maintenance policies when deterioration can be
    observed by inspection. Operations Res., 24, 359-366.

Magazine, M.J. (1983). Optimality of intuitive checkpointing policies.
    Inf.Process.Lett., 17, 63-66.

Maisonneuve, B. (1975). Entrance-exit results for semi-regenerative processes. Z.Wahrscheinlichkeitstheorie Verw.Geb., 32, 81-94.

Malshakov, V.D., M.B.Tamarkin & A.R.Timoshenko (1980). Model of computer system on the level of intervals of resource utilization. Autom.Control.Comut.Sci., 14, 15-19.

Mamikonov, A.G. (see V.V.Kulba).

Manca, R. (see M.Carravetta).

Marcus, A.H. (see A.Becher).

Marcus, A.H. (1980). Erratum to : Power laws in compartmental analysis. Part I : A unified stochastic model. Math.Biosci., 48, 155.

Marie, R. (1978). Quelques résultats relatifs aux réseaux de files d'attente. 1er Colloq.Math.Appl., AFCET-SMF, Palaiseau 1978, I, 253-263.

Ng Stephen, F.W. & J.W.Mark (1978). Satellite packet switching with global assignments and batch Poisson arrivals. IEEE Trans. Comput., C-27, 1216-1221.

Marshall, K.T. (see W.J.Hayne).

Martynenko, O.N. & A.S.Serdakov (1982). Optimizing the checking depth in electronic equipment. Autom.Remote Control, 42, 1040-1043.

Masol, V.I. (1977). On the asymptotic behaviour of the time of arrival at a high level by a regenerative stable process with independent increments. Theory Probab.Math.Stat., 9, 139-144.

Massey, W.A. (1984). Open networks of queues : their algebraic structure and estimating their transient behaviour. Adv.Appl. Probab., 16, 176-201.

Mastebroeck, H.A.K. (see G.Gestri).

Mazo, J.E. (1982). Some extremal Markov chains. Bell Syst.Tech.J., 61, 2065-2080.

McClean, S.I. (1978). Continuous-time stochastic models of a multi-grade population. J.Appl.Probab., 15, 26-37.

McClean, S.I. (1980). A semi-Markov model for a multigraphe population with Poisson recruitment. J.Appl.Probab., 17, 846-852.

McConalogue, D.J. (see L.A.Baxter).

McDonald, D. (see K.B.Athreya).

McDonald, D. (1977). Equilibrium measures for semi-Markov processes.
     Ann.of Probab., 5, 818-822.

McDonald, D. (1978). On semi-Markov and semi-regenerative processes I.
     Z.Wahrscheinlichkeitstheorie verw.Gebiete, 42, 261-277.

McDonald, D. (1978). On semi-Markov and semi-regenerative processes II.
     Ann.of Probab., 6, 995-1014.

McKenna, J. & D.Mitra (1982). Integral representations and asymptotic
     expansions for closed Markovian queueing networks : normal usage.
     Bell Syst.Tech.J., 61, 661-683.

McKenna, J., D.Mitra & K.G.Ramakrishnan (1981). A class of closed
     Markovian queueing networks : integral representations, asymp-
     totic expansions and generalizations. Bell Syst.Techn.J., 60,
     599-641.

McNickle, D.C. (see R.L.Disney).

Meditch, J.S. (1978). A minimum principle/queueing theory approach
     to routing in message-switched networks. Comput.Electr.Eng., 5,
     305-310.

Mehlmann, A. (1979). Semi-Markovian manpower models in continuous
     time. J.Appl.Probab., 16, 416-422.

Mehta, Y.K. (see S.Biswas).

Meilijson, I. & G.Weiss (1977). Multiple feedback at a single-server
     station. Stochastic Processes Appl., 5, 195-205.

Meister, B. (1976). Models in resource allocation - trees of queues.
     Lecture Notes Econ.Math.Syst., 117, 215-223.

Meketon, M.S. (see Ph.Heidelberger).

Melamed, B. (see F.J.Beutler).

Melamed, B. (1979). Characterizations of Poisson traffic streams in
     Jackson queueing networks. Adv.Appl.Probab., 11, 422-438.

Melamed, B. (1982). On the reversibility of queueing networks.
     Stochastic Processes Appl., 13, 227.

Melamed, B. (1982). On Markov jump processes imbedded at jump epochs
     and their queueing-theoretic applications. Math.Oper.Res., 7,
     111-128.

Mello, J.M.C. (see N.A.J.Hastings).

Menselson, H. & U.Yechiali (1981). Controlling the GI/M/1 queue by
    conditional acceptance of customers. Eur.J.Oper.Res., 7, 77-85.

Menipaz, E. (1979). Cost optimization of some stochastic maintenance
    policies. IEEE Trans.Reliab., R-28, 133-136.

Mensikov, G.G. (see A.S.Kondratiev).

Mergenthaler, W. (1982). The total repair cost in a defective,
    coherent binary system. Cybern.Syst., 13, 219-243.

Michikazu, K. (1978). Some applications of reliability analysis of
    standby redundant systems with repair. J.Operations Res.Soc.
    Japan, 21, 146-169.

Mihoc, G. & S.P.Niculescu (1976). On a class of stochastic processes.
    An.Univ.Craiova, Ser.Mat.Fiz.Chim., 4, 7-10.

Mikadze, I.S. (1978). On a technique to determine operational ready-
    ness. Soobshch.Akad.Nauk Gruzin.SSR, 92, 417-420.

Mileshina, R.I. (1977). Limit theorems for a controlled random walk.
    Theory Probab.Math.Stat., 13, 125-140.

Miller, D.R. (1979). Almost sure comparisons of renewal processes
    and Poisson processes, with applications to reliability theory.
    Math.Oper.Res., 4, 406-413.

Mine, H. (see H.Kawai).

Mine, H. & S.Osaki (1977). Markovian decision processes. Moskva :
    "Nauka" (Russian).

Mirabile, M.P. (see E.Hannan).

Mironov, M.A. & V.I.Tikhonov (1977). Markov processes. Moskva :
    "Sovetskoe Radio" (Russian).

Misra, K.B. & J.Sharma (1974). Reliability optimization of a series
    system with active and standby redundancy. Internat.J.Systems
    Sci., 5, 1131-1142.

Mitra, D. (see J.Mckenna).

Mitra, D. & J.A.Morrison (1983). Asymptotic explansions of moments
    of the waiting time in closed and open processor-sharing systems
    with multiple job classes. Adv.Appl.Probab.,15, 813-839.

Mitra, D. & K.G.Ramakrishnan (1982). An overview of PANACEA, a
     software package for analysing Markovian queueing networks.
     Bell Syst.Tech.J., 61, 2849-2872.

Mitrani, I. (see E.Coffman & E.Gelenbe).

Mode, Ch.J. (1976). On the calculation of the probability of current
     family-planning status in a cohort of women. Math.Biosciences,
     31, 105-120.

Mode, Ch.I. & G.Pickens (1979). An analysis of postpartum contra-
     ceptive strategies - Policies for preventive medicine.
     Math.Biosciences, 47, 91-113.

Mode, Ch.J. & M.G.Soyka (1980). Linking semi-Markov processes in
     time series - An approach to longitudinal data analysis.
     Math.Biosciences, 51, 141-164.

Moeller, T.L. (see S.S.Lavenberg).

Mohanty, S.G. (see J.L.Jain).

Molma, C. (see G.E.Grandori).

Monahan, G.E. (1982). A survey of partially observable Markov
     decision processes : Theory, models and algorithms. Manage.
     Sci., 28, 1-16.

Morais (de), P.R. (1979). A Markovian model for a continuous-time
     inventory systems. 3rd Nat.Symp.Probab.Stat., Sao Paulo, 160-164.

Morey, R.C. (1975). Approximations for the consolidation of multi-
     echelon repair facilities. Naval Res.Logist.Q., 22, 455-460.

Morang, R.W. (see J.S.Edwards).

Mori, T.F. (1980). On the asymptotic network delay in a model of
     packet switching. Comput.Math.Appl., 7, 167-172.

Morozov, E.V. (1983). Some results for processes with continuous
     time in the system GI/G/1 with losses from the queue. I.
     Izvestija Akad.Nauk BSSR, Ser.Fiz.-Mat.Nauk, 51-55 (Russian).

Morrison, J.A. (see D.Mitra).

Morton, T.E. (1974). Decision horizons in discrete time undiscounted
     Markov renewal programming. IEEE Trans.Systems Man Cybernetics
     SMC-4, 392-394.

Motoori, M., T.Nakagawa, Y.Sawa & K.Yasui (1980). Reliability ana-
    lysis of an inspection policy and its application to a computer
    system. J.Oper.Res.Soc.Japan, 23, 273-286.

Mountford, D. (1979). An inequality in p-functions. Ann.of Probab.,
    7, 184-185.

Mueller, B. (1980). Zerlegungsorientierte, numerische Verfahren
    fuer Markovsche Rechensystemmodelle. (Dissertation). Abteilung
    Informatik der Universitaet Dortmund.

Mukhitdinova, N.M. (1978). Reservation with marginal checking and
    renewal. Dokl.Akad.Nauk UzSSR 1978, 8-11 (Russian).

Nadarajan, R. & M.F.Neuts (1982). A multiserver queue with thres-
    holds for the acceptance of customers into service. Oper.Res.,
    30, 948-960.

Nagapetjan, R. (see G.P.Klimov).

Nagapetjan, R. (1981). Use of regenerative processes with Markov
    dependent regeneration cycles. Izvestija Akad.Nauk Mold.SSR,
    Ser.Fiz.-Tehn.Mat.Nauk, 14-21 (Russian).

Nahmias, S. (1975). A sequential decision problem with partial
    information. Cah.Cent.Etud.Rech.Opér., 17, 53-64.

Naik, D.N. (1978). Steady state availability of a system with two
    types of failure. IAPQR Trans., 3, 27-33.

Naik, D.N. (1981). Estimating the parameters of a 2-out-of-3 : F
    system. IEEE Trans.Reliab. R-30, 464-465.

Nakagawa, T. (see A.L.Goel & M.Motoori).

Nagakawa, T. & S.Osaki (1976). Analysis of a repairable system which
    operates at discrete times. IEEE Trans.Reliab.R-25, 110-112.

Nakajima, S. & Y.Watanabe (1977). Optimal renewal interval of a
    part before failures - a definite part with constant renewal
    intervals. Rep.Fac.Eng., Kanagawa Univ., 15, 36-41 (Japanese).

Nakamura, M. & S.Osaki (1984). Performance/reliability evaluation
    for multi-processor systems with computation demands. Int.
    J.Syst.Sci., 15, 95-105.

Nancc, R.E. (see U.N.Bhat).

Naneva, T. (see I.Cvetanov).

Nasirova, T.I. (see H.M.Ahmedova).

Nasirova, T.I. (1978). On a model of the control of reserves. Teor. Sluchajnyck Protsessov 6, 107-119 (Russian).

Nasirova, T.I. (1980). On an ergodic theorem for certain Markov processes with delaying barrier. Theory Probab.Math.Stat., 20, 105-112.

Nasirova, T.I. (1982). Distribution of a semi-Markov walk process with delay screen. Izvestija Akad.Nauk Az.SSR, Ser.Fiz.-Tehn. Mat.Nauk., 6, 116-119.

Nasirova, T.I. & A.V.Skorohod (1981). On the asymptotic behaviour of processes in an inventory control model. Theory Probab.Math. Stat., 23, 137-142.

Nasirova, T.I. & A.V.Skorohod (1982). Limit theorems for some classes of random processes related to semi-Markov random walks. Theory Probab.Math.Stat., 25, 137-151.

Naumov, V.A. (see G.P.Basharin).

Nawrotzki, K. (see U.Jansen).

Nawrotzki, K. (1982). Ergodic theorems for random processes with embedded homogeneous flow. Theory Probab.Appl., 26, 388-392.

Nechaev, B.Ya. (see Ya.G.Genis).

Neffke, H. (1979). Renewal reward processes with linear expectation functions. Oper.Res.Verfahren, 30, 98-110.

Neffke, H. (1982). Bewertete Erneuerungsprozesse. Mathematical Systems in Economics, Hanstein.

Nesenbergs, M. (1979). A hybrid of Erlang B and C formulas and its applications. IEEE Trans.Commun.COM-27, 59-68.

Neumann, K. (see W.Fix).

Neumann, K. (1979). Recent advances in temporal analysis of GERT networks. Z.Oper.Res., Ser.A, 23, 153-177.

Neumann, K. (1980) Dynamische Optimierung und optimale Steuerung in speziellen GERT-Netsplaenen. Wiss.Z.Tech.Hochsch. Leipz., 4, 349-355.

Neumann, K. & U.Steinhardt (1979). GERT networks and the time-oriented evaluation of projects. Lecture Notes Econ.Math. Systems, 172.

Neuts, M.F. (see S.Chakravarthy, S.Kumar & R.Nadarajan).

Neuts, M.F. (1976). Moment formulas for the Markov renewal branching process. Adv.Appl.Probab., 8, 690-711.

Neuts, M.F. (1977). Some explicit formulas for the staedy-state behaviour of the queue with semi-Markovian service time. Adv. Appl.Probab., 9, 141-157.

Neuts, M.F. (1978). The second moments of the absorption times in the Markov renewal branching process. J.Appl.Probab., 15, 707-714.

Neuts, M.F. (1978). Renewal processes of phase type. Naval.Res. Logist.Q., 25, 445-454.

Ney, P.E. (see K.B.Athreya).

Ng Stephen, F.W. (see J.W.Mark).

Nicolai, W. (1979). Optimierung spezieller GERT-Netzplaene. Z.Angew. Math.Mech., 59/01, T108-T109.

Nicolai, W. (1980). On the temporal analysis of special GERT-networks using a modified Markov renewal process. Z.Oper.Res., Ser.A, 24, 263 272.

Nicolai, W. (1982). Optimization of STEOR networks via Markov renewal programming. Z.Oper.Res., Ser.A, 26, 7-19.

Nicolis, J.S. & E.N.Protonotarios (1980). Modelling communication and control between hierarchical systems. Int.J.Syst.Sci., 11, 107-127.

Niculescu, S.P. (see G.Mihoc).

Nikiforov, G.K. & Yu.I.Ryzhikov (1983). An iteration method of calculation of service of a nonhomogeneous flow in a multi-channel systel. Cybernetics, 19, 260-266.

Nishida, T. & K.Tatsuno (1982). Optimal maintenance policy with minimal repair for stochastically failing system. J.Oper.Res. Soc.Japan, 25, 238-248.

Nishida, T., F.Ohi & K.Tatsuno (1983). Extended opportunistic maintenance policy for a two-unit system. Math.Japan, 28, 569-575.

Noebels, R. (1981). A note on stochastic order of probability measures and an application to Markov processes. Z.Oper.Res., Ser.A, 25, 35-43.

Nollau, V. (see T.Ewers & E.Jannusch).

Nollau, V. (1981). Semi-Markovsche Prozesse. Deutsch-Taschenbuecher, 31, Verlag Harri Deutsch.

Norman, J.M. (see T.A.Burgin).

Nosenko, V.I. (see V.V.Bondarev).

Nummelin, E. (see E.Arjas).

Nummelin, E. (1976). Limit theorems for σ-recurrent semi-Markov processes. Adv.Appl.Probab., 8, 531-547.

Nummelin, E. (1977). On the concepts of σ-recurrence and σ-transience for Markov renewal processes. Stochastic Processes Appl., 5, 1-19.

Nummelin, E. (1978). Uniform and ratio limit theorems for Markov renewal and semi-regenerative processes on a general state space. Ann.Inst.Henri Poincaré, Sect.B, 14, 119-143.

Nummelin, E. (1979). A conservation property for general GI/G/1 queues with an application to tandem queues. Adv.Appl.Probab., 11, 660-672.

Nunen (van),J.A.E.E. (1976). Contracting Markov decision processes. Amsterdam, Mathematical Centre Tracts, 71.

Nunen (van),J.A.E.E. & M.L.Puterman (1981). On solving G/M/s queueing control systems. Proc.Jahrestag Oper.Research, DGOR, Essen, 483-490.

Nunen (van),J.A.E.E. & S.Stidham jun. (1981). Action-dependent stopping times and Markov decision process with unbounded rewards. OR Spektrum, 3, 145-152.

Nunen (van),J.A.E.E. & J.Wessels (1976). FORMASY : FOrecasting and Recruitment in MAnpower SYstems. Statistica Neerlandica, 30, 173-193.

Nunen (van),J.A.E.E. & J.Wessels (1979). On theory and algorithms for Markov decision problems with the total reward criterion. OR, Spektrum, 1, 57-67.

Nunen (van), J.A.E.E. & J.Wessels (1984). On using discrete random
    models within decision support systems. Eur.J.Oper.Res., 15,
    141-159.

Oakes, D. (1976). Random overlapping intervals - a generalization
    of Erlang's loss formula. Ann.of Probab., 4, 940-946.

Odoni, A.R. & E.Roth (1983). An empirical investigation of the
    transient behavior of stationary queueing systems. Oper.Res.,
    31, 432-455.

Ohi, F. (see T.Nishida).

Ohno, K. (1981). A unified approach to algorithms with a suboptima-
    lity test in discounted semi-Markov decision processes. J.Oper.
    Res.Soc.Japan, 24, 296-324.

Ohta, H. (1980) GERT. Analysis of the economic design of control
    chart. IAPQR Trans., 5, 57-65.

Onicescu, O., G.Oprisan & G.Popescu (1983). Renewal processes with
    complete connections. Rev.Roum.Math.Pures Appl., 28, 985-998.

Opderbeck, H. (see W.W.Chu).

Opitz, O. & M.Schader (1975). Operations Research Verfahren und
    Marketingprobleme. Math.Systems in Econ., Verlag Anton Hain.

Oprisan, G. (see O.Onicescu).

Oprisan, G. (1976). On the J-X processes. Rev.Roum.Math.Pures Appl.,
    21, 717-724.

Oprisan, G. (1977). Some limit theorems for semi-Markov processes
    with an arbitrary state space. Proc.5th Conf.Probab.Theory,
    Brasov, 1974, 247-256.

Oral, M. & S.S.Rao (1980). An objective function for inventory
    control models. Proc.Int.Conf.Stochastic Programming, Oxford,
    403-426.

Osaki, S. (see A.L.Goel, M.Kinugasa, H.Mine, T.Nakagawa & M.Nakamura).

Osaki, S. & M.Takeda (1979). A two-unit paralleled redundant system
    with bivariate exponential failure law and allowed down time.
    Cah.Cent.Etud.Rech.Opér., 21, 55-62.

Osaki, S. & F.Sugimoto (1976). Optimum repair limit replacement
policies for a system subject to intermittent blows. Cah.Cent.
Etud.Rech.Opér., 18, 485-491.

Osawa, H. (see M.Doi).

Ovchinnikov, V.N. (1976). Asymptotic behavior of the time till the
first failure of a model with inhomogeneous reservation and
rapid repair. Izvestija Akad.Nauk SSSR, Tehn.Kibernet., 82-90
(Russian).

Pages, A. (see M.Gondran).

Pandit, S.M., H.J. Steudel & S.M. Wu (1977). A multiple time series
approach to modelling the manufacturing job-shop as a network
of queues. Manage.Sci., 24, 456-463.

Pau, L.F. (1981). Failure diagnosis and performance monitoring.
Control & Systems Theory, Vol.11, Marcel Dekker, New York.

Pavlov, I.V. & I.A.Ushakov (1978). The asymptotic distribution of
the time until a semi-Markov process gats out of a kernel.
Eng.Cybern., 16, 68-72.

Pearce, C.E.M. (1982). A generalization of Rapp's formula. J.Austr.
Math.Soc., Ser.B, 23, 291-296.

Pedrono, R. (see J.M.Helary).

Pelikhov, V.P. (see V.V.Kulba).

Pellaumail, J. (1978). Probabilités stationnaires de réseaux dont
les probabilités de transfert dépendent de l'état. Proc.1er
Colloq.Math.Appl., AFCET-SMF, Palaiseau, I, 265-275.

Pellaumail, J. (1978). Formule de produit pour réseaux à probabilités de transfert dépendant de l'état. C.R.Acad.Sci.Paris, Ser.A,
287, 157-159.

Pellaumail, J. (1979). Formule du produit et décomposition de
réseaux de files d'attente. Ann.Inst.Henri Poincaré, Sect.B,
15, 261-286.

Perros, H.G. (see F.G.Foster).

Peshes, L.Ya. (see V.E.Kan).

Pezhinska, G. (1983). Estimates of expectations for the approach
    moments of renewal points for Markov renewal processes.
    Dokl.Akad.Nauk Ukr.SSR, Ser.A, 8-10 (Russian).

Phatarfod, R.M. (1982). On some applications of Wald's identity to
    dams. Stochastic Processes Appl., 13, 279-292.

Phelps, R.I. (1983). Optimal policy for minimal repair. J.Oper.Res.
    Soc., 34, 425-427.

Phillips, M.J. (1981). A characterization of the negative exponential
    distribution with application to reliability theory. J.Appl.
    Probab., 18, 652-659.

Pickens, G. (ss Ch.J.Mode).

Pinter, J. (1977). Investigations on the maximal distance of empiri-
    cal distribution series. Alkalmazott Mat.Lapok., 1, 189-195
    (Hungarian).

Pittel, B. (1979). Closed exponential networks of queues with
    saturation : The Jackson-type stationary distribution and its
    asymptotic analysis. Math.Oper.Res., 4, 357-378.

Platzman, L. (1977). Improved conditions for convergence in un-
    discounted Markov renewal programming. Operations Res., 25,
    529-533.

Pliska, S.R. (1975). A semi-group representation of the maximum
    expected reward vector in continuous parameter Markov decision
    theory. SIAM J.Control 13, 1115-1129.

Pliska, S.R. (1976). Optimization of multitype branching processes.
    Management Sci., 23, 117-124.

Plucinski, E. (1979). On some properties of discrete stochastic
    processes. Demonstr.Math., 12, 1089-1110.

Pogosyan, A.L. (see Eh.A.Danielyan).

Politsjuk, L.I. & A.F.Turbin (1977). Asymptotic expansions for some
    characteristics of semi-Markov processes. Theor.Probab.Math.
    Stat., 8, 121-126.

Pollett, P.K. (see T.C.Brown).

Pooch, U.W. (see R.Chattergy).

Popescu, G. (see O.Onicescu).

Popov, V.A. (see A.L.Litvinov).

Popov, N.N. (1977). Mass service systems directed by semi-Markov
    processes. Izvestija Akad.Nauk SSSR, tehn.Kibernet., 94-101
    (Russian).

Popovici, Al.A. (1978). A Markovian stochastic model of the profit's
    general rate formation. Econom.Comput.Econom.Cybernetics,
    Studies Res., 85-106.

Porteus, E.L. (1980). Improved iterative computation of the expected
    discounted return in Markov and semi-Markov chains. Z.Oper.Res.,
    Ser.A, 24, 155-170.

Potter, R.G. (see F.E.Kobrin).

Prehn, U. (see K.Fleischmann).

Prelov, V.V. (see R.L.Dobrushin).

Premoli, A. (see A.Bobbio).

Prilutskij, M.Kh. (see E.N.Kazantsev).

Prizva, G.I. & L.V.Tihomirova (1978). On the first passage time
    distribution for Markov processes with discrete intervention
    of chance. Theory Probab.Math.Stat., 12, 129-135.

Proschan, F. (see R.E.Barlow).

Proschan, F. & J.Sethuraman (1976). Stochastic comparison of order
    statistics from heterogeneous populations, with applications
    to reliability. J.Multivariate Analysis, 6, 608-616.

Protonotatios , E.N. (see A.G.Bobos & J.S.Nicolis).

Pruscha, H. (1983). On continuous time learning models. Lecture
    Notes Statistics, 20, 175-181.

Pujolle, G. (see E.Gelenbe).

Pujolle, G. (1979). The influence of protocols on the stability
    conditions in packet switching networks. IEEE Trans.Commun.
    COM-27, 611-619.

Purdue, P. (see D.Linton).

Purdue, P. & D.Weiner (1977). A semi-Markov approach to stochastic
    compartmental models. Commun.Stat., Theory Methods A6, 1231-1243.

Puri Prem, S. (1978). A generalization of a formula of Pollaczek
and Spitzer as applied to a storage model. Sankhya, Ser.A, 40,
237-252.

Puri Prem, S. (1979). On certain problems involving non-identifia-
bility of distributions arising in stochastic modelling. Proc.
Int.Conf.Optimizing methods in Statistics, Bombay, 1977, 403-418.

Puri Prem, S. & S.W.Wollford (1981). On a generalized storage model
with moment assumptions. J.Appl.Probab., 18, 473-481.

Puterman, M.L. (see J.van Nunen).

Pyke, R. & R.A.Schaufele (1981). Correction to : The existence and
uniqueness of stationary measures for Markov-renewal processes.
Ann.of Probab., 9, 348.

Quelle, G. (1976). Dynamic programming of expectation and variance.
J.Math.Analysis Appl., 55, 239-252.

Rabe, H. (see U.Jansen).

Racicot, R.L. (1976). Approximate confidence intervals for reliability of a series system. IEEE Trans.Reliab., R-25, 265-269.

Rajarshi, M.B. (1974). Success runs in a two-state Markov chain.
J.Appl.Probab., 11, 190-192.

Ramalhoto, M.F. (see D.H.Girmes).

Ramakrishnan, K.G. (see J.McKenna & D.Mitra).

Ramamurthy, K. (see K.B.Athreya & N.K.Jaiswal).

Ramanarayanan, R. (1981). Explicit steady state solution for a
Markovian inventory system. Stat.Neerl., 35, 215-219.

Ramanarayanan, R. & K.Usha (1981). General analysis of systems in
which a 2-unit system is a sub-system. Math.Operationsforsch.
Stat., 12, 629-637.

Ramaswami, V. (1981). Algorithms for a continuous-review (s,S)
inventory system. J.Appl.Probab., 18, 461-472.

Rangan, A. (see A.H.Kalro).

Rangaswamy, N.S. (see N.K.Jaiswal).

Rao Chennupati, R. (see A.M.Kshirsagar).

Rao, M.R. (see Y.M.I.Dirickx).

Rao, S.S. (see M.Oral).

Rappaport, D. (see J.Dayan).

Raugi, A. (1980). Quelques remarques sur le théorème de la limite
    centrale sur un groupe de Lie. C.R.Acad.Sci.Paris, Ser.A, 290,
    103-106.

Ravichandran, N. & R.Subramanian (1980). An n-unit priority redundant
    system. J.Math.Phys.Sci., 14, 85-94.

Reetz, D.A. (1976). A decision exclusion algorithm for a class of
    Markovian decision processes. Z.Operat.Res., Ser.A, 20, 125-131.

Reetz, D.A. (1980). Kontrahierende endliche Markoffsche Entscheidungs-
    prozesse. G.Marchal un H.J.Matzenbacher Wissenschaftsverlag,
    Berlin.

Reiman, M.I. (see J.M.Harrison).

Reinhard, J.M. (see J.Janssen).

Reinhard, J.M. (1980). Jeux de survie économique : un modèle semi-
    markovien. Cah.Cent.Etud.Rech.Oper., 22, 101-110.

Reinhard, J.M. (1981). A semi-Markovian game of economic survival.
    Scand.Actuarial J., 23-38.

Reinhard, J.M. (1982). Identités du type Baxter-Spitzer pour une
    classe de promenades aléatoires semi-markoviennes. Ann.Inst.
    Henri Poincaré, Sect.B, 18, 319-333.

Reiser, M. (see S.S.Lavenberg).

Reiser, M. (1977). Numerical methods in separable queueing networks.
    Algorithm.Meth.Probab., Stud.Manage.Sci., 7, 113-142.

Reiser, M. (1981). Mean value analysis and convolution method for
    queue-dependent servers in closed queueing systems. Performance
    Eval., 1, 7-18.

Renger, K. (see G.Lauenstein).

Resnick, S.I. (see J.M.Harrison).

Reznikov, S.I. (1979). A new proof for a limit theorem of the phase
    enlargement type for semi-Markov processes. Teor.Sluchajnykh
    Protsessov, 7, 85-103 (Russian).

Reznikov, S.I. (1980). A limit theorem of the type of phase enlarge-
    ment for random processes which are complemented to weakly
    inhomogeneous Markov processes. Dopov.Akad.Nauk Ukr.RSR, Ser.A,
    16-18 (Ukrainian).

Reznikov, S.I. (1980). State enlargement of random processes that are
    complemented to Markov processes. Ukr.Math.J., 31, 539-544.

Ricciardi, L.M. (1980). Stochastic equations in neurobiology and
    population biology. Lecture Notes Biomath., 39, 248-263.

Richter, K. (1982). Dynamische Aufgaben der diskreten Optimierung.
    Akademie-Verlag, Berlin.

Rieder, U. (1975). Bayesian dynamic programming. Adv.Appl.Probab.,
    7, 330-348.

Rieder, U. (1976). On the characterization of optimal policies in
    non-stationary dynamic programs. Z.Angew.Math.mech., 56,
    Sonderheft, T359-T360.

Rishel, R. (1982). A partially observed inventory problem. Lect.
    Notes Control Inf.Sci., 43, 345-353.

Ritchie, E. & P.Wilcox (1977). Renewal theory forecasting for stock
    control. Eur.J.Oper.Res., 1, 90-93.

Robinson, D. (1978). Optimization of priority queues - a semi-
    Markov decision chain approach. Manage.Sci., 24, 545-553.

Rohal-Ilkiv, B. & J.Skakala (1977). 2-unit redundant systems with
    replacement & repair. IEEE Trans.Reliab., R-26, 294-296.

Rolski, T. (1977). A relation between imbedded Markov chains in
    piecewise Markov processes. Bull.Acad.Polon.Sci., Ser.Sci.
    Math.astron.phys., 25, 181-189.

Romanov, A.K. & A.I.Terekhov (1980). Some models of the workforce
    dynamics in organizations. Optimizatsiya, 25, 124-138 (Russian).

Rosberg, Z. (1980). A positive recurrence criterion associated with
    multidimensional queueing systems. J.Appl.Probab., 17, 790-801.

Rosberg, Z. (1982). Semi-Markov decision processes with polynomial
    reward. J.Appl.Probab., 19, 301-309.

Rosemann, H. (1981). Zuverlaessigkeit und Verfuegbarkeit techniser
     Anlagen und Geraete. Mit praktischen Beispielen von Berechnung
     und Einsatz in Schwachstellenanalysen. Springer-Verlag, Berlin.

Rosenshine, M. & R.C.Rue (1981). Optimal control for entry of many
     classes of customers to an M/M/1 queue. Nav.Res.Logist.Q., 28,
     489-495.

Ross, S.M. (1976). On the time to first failure in multicomponent
     exponential reliability systems. Stochastic Processes Appl., 4,
     167-173.

Roth, E. (see A.R.Odoni).

Roussas, G.G. (see M.G.Akritas).

Rubchinskij, A.A. (see B.A.Gennison).

Rubin, I. (1979). Message delays in FDMA and TDMA communcation
     channels. IEEE Trans.Commun.COM-27, 769-777.

Rubinovitch, M. (see Z.Barzily).

Rumyantsev, N.V. (1982). Investigation of a service system with
     heating at the end of the occupation time by methods of the
     theory of semi-Markov processes. Kibernetika, 120-122 (Russian).

Rue, R.C. (see M.Rosenshine).

Ryan, T.A.jun. (1976). A multidimensional renewal theorem. Ann.of
     Probab., 4, 656-661.

Rybko, A.N. (1981). Stationary distributions of time-homogeneous
     Markov processes modelling message switching communication
     networks. Probl.Inf.Transm., 17, 49-63.

Rybko, A.N. (1982). Existence conditions for a steady-state mode
     for two types of communications networks with message switching.
     Probl.Inf.Transm., 18, 76-84.

Rykov, V. (see S.Jolkof).

Ryzhikov, Yu.I (see G.K.Nikiforov).

Ryzhikov, Yu.I. (1977). Servicing of priorityless flow of non-
     uniform requests in a multichannel Markov system. Probl.control
     Inform.Theory 6, 431-436.

Sadjaki, D. (see F.Bestwick).

Sahay, B. (see W.Lennox).

Sahin, I. (see Y.Balcer).

Sahin, I. (1978). Cumulative constrained sojourn times in semi-Markov
    processes with an application to pensionable service. J.Appl.
    Probab., 15, 531-542.

Sakino, E. (see T.Inagaki).

Sanderson, A.C. (see T.W.Calvert).

Santner, T.J. (see C.C.Brown).

Shanthikumar, J.G. (see R.G.Sargent).

Satia, J.K. (see R.E.Lave).

Satyanarayana, A. (1982). A unified formula for analysis of some
    network reliability problems. IEEE Trans.Reliab. R-31, 23-32.

Sauer, C.J. (see S.S.Lavenberg).

Sauer, W. (1980). Eine symbolische Methode zur Berechnung der
    mittleren Dauer stochastischer technologischer Prozesse. Wiss.
    Z.Tech.Univ.Dresd., 29, 261-265.

Saw, Y. (see M.Motoori).

Schader, M. (see O.Optiz).

Schael, M. (see M.Kolonko).

Schael, M. (1973). Conditions for optimality in dynamic programming
    and for the limit of n-stage optimal policies to be optimal.
    Z.Wahrscheinlichkeitstheorie verw.Gebiete, 32, 179-196.

Schael, M. (1975). On dynamic programming : Compactness of the
    space of policies. Stochastic Processes Appl., 3, 345-364.

Schael, M. (1979). On dynamic programming and statistical decision
    theory. Ann.Statist., 7, 432-445.

Schassberger, R. (see A.D.Barbour, W.E.Helm & A.Hordijk).

Schassberger, R. (1976). On the equilibrium distribution of a class
    of finite-stage generalized semi-Markov processes. Math.Oper.
    Res., 1, 395-406.

Schassberger, R. (1978). Insensitivity of steady-state distributions
    of generalized semi-Markov processes, I, II. Ann.of Probab., 5,
    87-89, 1977 & 6, 85-93.

Schassberger, R. (1978). The insensitivity of stationary probabilities
    in networks of queues. Adv.Appl.Probab., 10, 906-912.

Schassberger, R. (1978). Insensitivity of steady-state distributions
    of generalized semi-Markov processes with speeds. Adv.Appl.
    Probab., 10, 836-851.

Schassberger, R. (1979). A definition of discrete product form
    distributions. Z.Oper.Res., Ser.A 23, 189-195.

Schaufele, R.A. (see R.Pyke).

Schaufele, R.A. (1981). A class of time-reversible semi-Markov
    processes. Can.J.Stat., 9, 39-56.

Scheaffer, R.L. (see R.S.Leavenworth).

Schellhaas, H. (1974). Zur Extrapolation in Markoffschen Entschei-
    dungsmodellen mit Diskontierung. Z.Operat.Res., Ser.A 18,
    91-104.

Schellhaas, H. (1979). Ueber Semi-Markoffsche Entscheidungsprozesse
    mit endlichem Horizont. Proc.DGOR Meet.Oper.Res., Berlin,
    122-129.

Schellhaas, H. (1979). Semi-regenerative processes with unbounded
    rewards. Math.Oper.Res., 4, 70-78.

Schellhaas, H. (1980). Markov renewal decision processes with
    finite horizon. OR Spektrum 2, 33-40.

Schneeweiss, Ch. (1974). Dynamisches Programmieren. Physica Paper-
    back, Physica-Verlag, Wien.

Scheuer, E.M. (see L.A.Baxter & W.R.Blischke).

Schick, I.C. (see S.B.Gershwin).

Schmee, J. (see E.Hannan).

Schmidt, V. (see U.Arndt).

Schramm, K. (1980). Konvergenzvergleiche bei Loesungen linearer
    Differenzengleichungen mit konstanten Koefficienten. Z.Oper.
    Res., Ser.A 24, 125-136.

Schweitzer, P.J. (see A.Federgruen, B.Gavish & A.Hordijk).

Schweitzer, P.J. (1982). Solving MDP functional equation by lexico-
    graphic optimization. RAIRO, Rech.Opér. 16, 91-98.

Schweitzer, P.J. (1983). On the solvability of Bellman's functional equations for Markov renewal programming. J.Math.Analysis Appl., 96, 13-23.

Schwetman, H.D. (see G.Balbo).

Seif, F.J. (1981). Stochastic and cooperative model of excocytotic secretion of thyrotropin. Proc.6th Conf.Probability Theory, Brasov, 513-520.

Semenov, A.T. (1976). Sojourn times of a semi-Markov random walk on the semi-axis. Lith.Math.J. 15, 659-664.

Sengupta, J.K. (see S.K.Gupta).

Serdakov, A.S. (see O.N.Martynenko).

Sereda, V.I. (see V.V.Anisimov).

Serfozo, R.F. (1979). An equivalence between continuous and discrete time Markov decision processes. Oper.Res., 27, 616-620.

Sergachev, O.V. (1979). On the analysis of dynamic operating modes of deterministic automata. Autom.Control Comput.Sci., 13, 13-17.

Sethuraman, J. (see F.Proschan).

Sevcik, K.C. (1982). Queueing networks and their computer system applications : An introductory survey. Proc.NAT ASI Deterministic and Stochastic Scheduling, Durham, 211-222.

Shahul, H.M.A. (see S.Kalpakam).

Shakhbazov, A.A. (1982). Limiting distribution of the time of first entry for semi-Markov processes and its application in reliability theory. Eng.Cybern., 20, 82-90.

Shakhbazov, A.A. & A.D.Soloviev (1981). Nonhomogeneous standby with renewal. Eng.Cybern., 19, 30-38.

Shanthikumar, J.G. & R.G.Sargent (1981). An algorithmic solution for queueing model of a computer system with interactive and batch jobs. Performance Eval., 1, 344-357.

Sharma, J. (see K.B.Misra).

Sharma, S.C. (see Y.P.Gupta).

Sharma, S.C. (1976). Cumulative processes associated with Markov renewal processes in space. J.Math.Sci., 10, 63-68.

Shedler, G.S. (see D.L.Iglehart).

Shedler, G.S. & D.R.Slutz (1981). Irreducibility in closed multiclass

networks of queues with priorities : passage times of a marked
job. Performance Eval., 1, 334-343.

Shedler, G.S. & J.Southard (1982). Regenerative simulation of networks of queues with general service times : passage through
subnetworks. IBM J.Res.Dev., 26, 625-633.

Shibanov, S.E. (see A.S.Chebotarev).

Shishonok, N.A. (see V.E.Dikarev).

Shlesinger, M.F. (see U.Landman).

Shura-Bura, A.E. (see E.V.Dzirkal).

Shurenkov, V.M. (see Ju.S.Davidovic & Ya.I.Elejko).

Shurenkov, V.M. (1977). A time substitution which preserves the semi-
Markov property and regularity conditions. Theory Probab.Math.
Stat., 14, 159-163.

Sibirskaya, T.K. (1982). On a certain approach to the determination
of the characteristics of closed queueing networks in analytic
form. Izvestija Akad.Nauk Mold.SSR, Ser.Fiz.-Tekh.Mat.Nauk,
55-59 (Russian).

Siedersleben, J. (1981). Dynamically optimized replacement with a
Markovian renewal process. J.Appl.Probab., 18, 641-651.

Sigalotti, L. (1976). Particulari processi markoffiani d'arrivo che
si arrestano in intervalli finiti di tempo e che presentano la
scambiabilita degli incrementi. Rend.Ist.Mat.Univ.Trieste, 7,
173-181 (Italian).

Silvestrov, D.S. (see D.Banach & E.I.Kaplan).

Silvestrov, D.S. (1980). Mean hitting times of semi-Markov processes
and queueing networks. Elektron.Informationsverarbeitung
Kybernetik, 16, 399-415 (Russian).

Silvestrov, D.S. (1980). Exponential bounds in ergodic theorems for
regenerative processes. Elektron.Informationsverarbeitung
Kybernetik, 16, 461-463 (Russian).

Silvestrov, D.S. (1980). Synchronized regenerative processes and
explicit estimates of the convergence rate in ergodic theorems.
Dopov.Akad.nauk Ukr.RSR, Ser.A, 21-24 (Ukrainian).

Silvestrov, D.S. & G.T.Tursunov (1980). General limit theorems for sums of controlled random variables. II. Theory Probab.Math. Stat., 20, 131-141.

Simon, B. (see R.L.Disney).

Singh, C. (1980). Equivalent rate approach to semi-Markov processes. IEEE Trans.Reliab. R-29, 273-274.

Sinha, S.M. (see M.Agarwal & D.V.S.Kapil).

Sityuk, V.N. (see V.V.Anisimov).

Sityuk, V.N. (1981). On the limit behaviour of semi-Markov systems with inhomogeneous probabilities of failures. Dokl.Akad.Nauk Ukr.SSR, Ser.A, 24-27 (Russian).

Skakala, J. (see B.Rohal-Ilkiv).

Skitovitch, V.P. (1976). A system of mass maintenance with deleted and restored instruments. Control, reliability and navigation. N°3. Interuniv.Collect.Sci.Works, Saransk, 72-76 (Russian).

Sklyarevich, F.A. (1980). Allowance for structural variability of computer systems. Autom.Control Comput.Sci., 14, 52-56 (Russian).

Skorohod, A.V. (see I.I.Gikman & T.I.Nasirova).

Sladky, K. (1977). On the optimality conditions for semi-Markov decision processes. Trans.7th Prague Conf. Vol.A, 1974, 555-562.

Slutz, D.R. (see G.S.Shedler).

Smagin, V.A. (1975). Mean frequency of failures of a system with unreliable spare parts. Izvestija Akad.Nauk SSSR, Tehn.Kibernet., 118-120 (Russian).

Smirnov, S.N. (see G.P.Klimov).

Sniedovich, M. (1978). Dynamic programming and principles of optimality. J.Math.Analysis Appl., 65, 586-606.

Sobel, M.J. (1982). The variance of discounted Markov decision processes. J.Appl.Probab., 19, 794-802.

Soloviev, A.D. (see B.V.Gnedenko & A.A.Shakhbazov).

Sommer, C.J. (see S.W.Lagakos).

Sonderman, D. (1980). Comparing semi-Markov processes. Math.Oper. Res., 5, 110-119.

Sonnenschein, J. (see R.Consael).

Southard, J. (see G.S.Shedler).

Soyka, M.G. (see Ch.J.Mode).

Soyster, A.D. & D.I.Toof (1976). Some comparative and design aspects
    of fixed cycle production. Naval Res.Logist.Q., 23, 437-454.

Spaniol, O. (1979). Modelling of local computer networks. Proc.4th
    Int.Symp.Performance of comput Systems, Vienna, 503-516.

Spreen, D. (see A.Federgruen).

Spreen, D. (1981). Reducing the computational complexity of the
    multichain policy iteration algorithm in undiscounted Markov
    renewal programming. Methods Oper.Res., 44, 203-211.

Spreen, D. (1981). A further anticycling rule in multichain policy
    iteration for undiscounted Markov renewal programs. Z.Oper.Res.,
    Ser.A 25, 225-233.

Srinivasan, S.K. (see D.Bhaskar).

Srinivasan, S.K. (1974). Stochastic point processes. Ch.Griffin & Co.

Srinivasan, S.K. & R.Subramanian (1980). Probabilistic analysis of
    redundant systems. Lecture Notes in Economics & Mathematical
    Systems, 175.

Srinivasan, S.K. & N.Venugopalacharyulu (1976). Analytic solution
    of a finite dam with bounded input. J.Math.Phys.Sci., Madras,
    10, 141-164.

Srivastava, H.M. (see H.B.Kekre).

Stadje, W. (1980). Reducierte Semi-Markov Prozesse. Methods Oper.
    Res., 37, 467-476.

Staniewsky, P. & R.Weinfeld (1980). Optimization of denumerable
    semi-Markov decision processes. Syst.Sci., 6, 129-141.

Stefanski, L.A. (1982). An application of renewal theory to soft-
    ware reliability. Proc.27th Conf.Design of Experiments, Raleigh,
    101-118.

Steinhardt, U. (see K.Neumann).

Stengos, D. & L.C.Thomas (1980). The blast furnaces problem. Eur.
    J.Oper.Res., 4, 330-336.

Stepanov, S.N. (1981). Probabilistic characteristics of an incomplete
    accessible multi-phase service system with several types of
    repeated call. Probl.Control Inf.Theory, 10, 387-401 (Russian).

Steudel, H.J. (see S.M.Pandit).

Stidham, S.jun (see Y.Jo Kyung, S.G.Johansen & J.van Nunen).

Stiffler, J.J. (1978). The reliability of a fault-tolerant configuration having variable coverage. IEEE Trans.Comput.C-27, 1195-1197.

Stokalski, A. (1976). Optimization of the control schedule for Markov models of an operation system. Podstawy Sterowania, 6, 57-70 (Polish).

Stoyan, D. (see D.Koenig).

Stoyan, D. (1977). Qualitative Eigenschaften und Abschaetzungen stochastischer Modelle. R.Oldenbourg Verlag, Muenchen.

Streller, A. (see P.Franken).

Subramanian, R. (see K.P.Kistner, N.Ravichandran & S.K.Srinivasan).

Sugimoto, F. (see S.Osaki).

Sukhov, Yu.M. (see R.L.Dobrushin & M.Ya.Kelbert).

Sum, H. (see I.I.Ezov).

Szasz, D. (1974). A limit theorem for semi-Markov processes. J.Appl. Probab., 11, 521-528.

Tadzhiev, A. (see V.S.Korolyuk).

Tadzhiev, A. (1978). Asymptomatic expansion for the distribution of absorption time of a semi-Markov process. Ukr.Math.J., 30, 331-335.

Tadzhiev, A. (1978). Asymptomatic expansion for the distribution of absorption time of a semi-Markov process. The case of several ergodic subsets. Izvestija Akad.Nauk Turkm.SSR, Ser.Fiz.-teh. him.geol.Nauk., 12-18 (Russian).

Takacs, L. (1976). On seome recurrence equations in a Banach algebra Acta Sci.Math., 38, 399-416.

Takacs, L. (1977). On the ordered partial sums of real random variables. J.Appl.Probab., 14, 75-88.

Takacs, L. (1977). Combinatorial methods in the theory of stochastic processes, 2nd edition. R.E.Krieger Publ.Co., New York.

Takacs, L. (1978). On fluctuations of sums of random variables.
     Studies in Probab.Ergodic Theory, Adv.Math., Suppl.Stud.Vol.,
     2, 45-93.

Takahashi, H. & M.Woodroofe (1981). Asymptotic expansions in non-
     linear renewal theory. Commun.Stat., Theory Methods A 10,
     2113-2135.

Takami, I. (see T.Inagaki).

Takeda, M. (see S.Osaki).

Tamaki, M. (1979). Recognizing both the maximum and the second
     maximum of a sequence. J.Appl.Probab., 16, 803-812.

Tamaki, M. (1982). An optimal parking problem. J.Appl.Probab., 19,
     803-814.

Tamarkin, M.B. (see V.D.Malshakov).

Tambouratzis, D.G. (1979). A generalization of the M/G/1 queue.
     Bull.Greek Math.Soc., 20, 106-120.

Tammela, Y. (1980). Optimal control of production processes with
     complex structure. A model with complete information.
     Izvestija Akad.Nauk Ehst.SSR, Fiz.,Math., 29, 201-207 (Russian).

Tan, H.H. (1975). An algorithm to compute the resolvent of a sto-
     chastic matrix with application to Markov decision processes.
     IEEE Trans.Systems Ma; Cybernetics, SMC-5, 617-620.

Tanaka, K. (see H.Homma & S.Iwase).

Tanaka, K. & K.Wakuta (1976). On Markov games with the expected
     average reward criterion, II. Sci.Rep.Niigata Univ., Ser.A 13,
     49-54.

Tanaka, K. & K.Wakuta (1976). On semi-Markov games. Sci.Rep.Niigata
     Univ., Ser.A 13, 55-64.

Tanaka, K. & K.Wakuta (1978). On continuous time Markov games with
     countable state space. J.Operations Res.Soc.Japan, 21, 17-28.

Taylor, H.M. (1975). Optimal replacement under additive damage and
     other failure models. Naval Res.Logist.Quart., 22, 1-18.

Tatsuno, K. (see T.Nishida).

Teghem, J. jun. (1979). Use of discrete transforms for the study of a GI/M/s queue with impatient customer phenomena. Z.Operat.Res., Ser.A 23, 95-106.

Ten Hoopen, M. (1977). Incidental distortions in a point processes. Biom.J., 19, 3-7.

Teplickij, M.G. (1977). Problem of the optimal control of a Markov chain with incomplete information. Izvestija Akad.Nauk.SSSR, tehn.Kibernet., 89-94 (Russian).

Tepleckij, M.G. (1978). On some problems concerning control of a finite Markov chain. Izvestija Akad.Nauk SSSR, tehn.Kibernet., 96-100 (Russian).

Terekhov, A.I. (see A.K.Romanov).

Teugels, J.L. (1974). On Markov success chains. Proc.Belgo-Israeli Coll.Oper.Res., Louvain, 1972, 251-264.

Teugels, J.L. (1976). A bibliography on semi-Markov processes. J.Comput.Appl.Math., 2, 125-144.

Thomas, L.C. (see D.Stengos).

Thomas, L.C. (1981). Second order bounds for Markov decision processes. J.Math.Anal.Appl., 80, 294-297.

Thomas, L.C. (1982). The Wijngaard-Stidham bisection method and replacement models. IEEE Trans.Reliab. R-31, 482-484.

Thomas, L.C. & L.Yeh (1983). Adaptive control of M/M/1 queues - continuous-time Markov decision processes approach. J.Appl. Probab., 20, 368-379.

Thompson, M.E. (1981). Estimation of the parameters of a semi-Markov process from censored records. Adv.Appl.Probab., 13, 804-825.

Thorin, O. (1982). Probabilities of ruin. Scand.Actuarial J., 65-102.

Tien, T.M. (1977). Generalized piece-wise linear Markov processes for queueing and reliability theory. Math.Operationsforsch. Stat., Ser.Optimization, 8, 419-431.

Tihomirova, L.V. (see G.I.Prizva).

Tijms, H.C. (see A.Federgruen & A.Hordijk).

Tijs, S.H. (1980). Stochastic games with one big action space in
    each state. Methods Oper.Res., 38, 161-173.

Tikhonov, V.I. (see M.A. Mironov).

Timoshenko, A.R. (see V.D.Malshakov).

Toeroek, T. (see E.Gardos & T.Gergely).

Toeroek, T. (1979). Interleaved memory systems with Markovian.
    requests. Proc.4th Int.Symp.Performance Comp.Systems, Vienna,
    421-428.

Tomko, J. (1981). A limit theorem for Markov renewal processes.
    Proc.1st Pannonian Symp.Math.Statistics, Bad Tatzmannsdorff.
    Lecture NotesStatist., 8, 277-283.

Tomko, J. (1981). Semi-Markov analysis of an (E//K)/G/1 queue with
    finite waiting room. Colloq.Math.Soc.Janos Bolyai, 24, 381-389.

Tomusÿak, A.A. (see V.S.Korolyuk).

Toof, D.I. (see A.D.Soyster).

Tormey, D.C. (see J.Crowley).

Towsley, D. (1980). Queueing network models with state-dependent
    routing. J.Assoc.Comput.Mach., 27, 323-337.

Treier, V.N. (1974). Kinetic foundations for the theory of life--
    length and reliability of machines equipment. Doklady Akad.
    Nauk BSSR 18, 333-336 (Russian).

Trueman, R.E. (1977). An introduction to quantitative methods for
    decision making, 2nd edition. Holt, Rinehart and Winston,
    New York.

Tsaregradskij, I.P. (1982). On a queueing system with an infinite
    number of locally interacting devices. Teor.Veroyatn.Primen.,
    27, 583-587.

Tsur, S. (1978). Analysis of queueing networks in which processes
    exhibit locality-transition behaviour. Inform.Processing
    Letters, 7, 20-23.

Turbin, A.F. (see S.M.Brodi, V.S.Korolyuk, V.N.Kuznetsov,
    B.G.Levinskij & L.I.Politsjuk).

Tursunov, G.T. (see D.S.Silvestrov).

Tweedie, R.L. (see E.Arjas).

Unkelbach, H.D. (1979). Convergence to equilibrium in a traffic
    model with restricted passing. J.Appl.Probab., 16, 881-889.

Usha, K. (see R.Ramanarayanan).

Usha, K. (1979). Reliability and availability of 2-unit warm-standby
    system with random transition rates. IEEE Trans.Reliab. R-28,
    333.

Ushakov, I.A. (see I.V.Pavlov).

Ushakov, I.A. (1980). An approximate method of calculating complex
    systems with renewal. Eng.Cybern., 18, 76-83.

Varaiya, P. & J.Walrand (1980). Interconnections of Markov chains
    and quasi-reversible queueing networks. Stochastic Processes
    Appl., 10, 209-219.

Varaiya, P. & J.Walrand (1080). Sojourn times and the overtaking
    condition in Jacksonian networks. Adv.Appl.Probability, 12,
    1000-1018.

Venkatakrishnan, K.S. (see K.P.Kistner).

Venugopalacharyulu, N. (see S.K.Srinivasan).

Verma, P.K. (see A.Chatterjee).

Vermes, D. (1981). Optimal stochastic control under reliability
    constraints. Lect.Notes Contr.Inf.Sci., 36, 227-234.

Vickson, R.G. (1980). Generalized value bounds and column reduction
    in finite Markov decision problems. Oper.Res., 28, 387-394.

Vinogradskaya, T.M. (see B.A.Geninson).

Vjazemskii, V.O. (1977). On the transformation of recurrent random
    flows of signals in the path of an information-measuring system.
    Izvestija Akad.Nauk SSSR, tehn.Kibernet., 142-152 (Russian).

Vjugin, O.V. (1977). Asymptotic moments for branching processes
    close to critical ones. Theory Probab.Appl., 21, 825-833.

Vojna, A.A. (see V.V.Anisimov).

Vojna, A.A. (1980). Occupation time of a fixed set of states and
    consolidation of stochastic processes with arbitrary phase
    space. Theory Probab.Math.Stat., 20, 35-44.

Vojna, A.A. (1983). Asymptotic analysis of systems with a continuous
    component. Cybernetics, 18, 516-524.

Vosvrda, M. (see C.Cvetanov).

Vranceanu, G. (1978). Models for non-euclidean hyperbolic geometry.
    Studii Cerc.Mat., 30, 477-478 (Roumanian).

Vrieze, O.J. (see A.Hordijk).

Wagner, H.M. (1973). Principles of operations research. With appli-
    cations to managerial decisions. Moskva Verlag Mir (Russian).

Wakuta, K. (see S.Iwase & K.Tanaka).

Wakuta, K. (1981). Semi-Markov decision processes with incomplete
    state observation. J.Oper.Res.Soc.Japan, 24, 95-109.

Wakuta, K. (1982). Semi-Markov decision processes with incomplete
    state observation. Discounted cost criterion. J.Oper.Res.Soc.
    Japan, 25, 351-362.

Waldmann, K.-H. (1982). On two-state quality control under Markovian
    deterioration. Metrika, 29, 249-260.

Walrand, J. (see P.Varaiya).

Walrand, J. (1982). On the equivalence of flows in networks of
    queues. J.Appl.Probab., 19, 195-203.

Walrand, J. (1982). Poisson flows in single class open networks of
    quasireversible queues. Stochastic Processes Appl., 13, 293-303.

Walrand, J. (1983). A discrete-time queueing network. J.Appl.Probab.,
    20, 903-909.

Wanrooij, G.L. (see A.Hordijk).

Watanabe, Y. (see S.Nakajima).

Weeda, P.J. (1974). On the relationship between the cutting operation
    of generalized Markov programming and optimal stopping.
    Amsterdam Mathematisch Centrum, BW 39, 15 p.

Weeda, P.J. (1974). Some computational experiments with a special
    generalized Markov programming model. Amsterdam Mathematisch
    Centrum, BW 37, 25 p.

Weeda, P.J. (1976). Sensitive time and discount optimality in Markov
    renewal decision problems with instantaneous actions.
    Amsterdam Mathematisch Centrum, BW 59, 20 p.

Weeda, P.J. (1979) . Finite generalized Markov programming.
Amsterdam Mathematical Centre Tracts, 92.

Weiner, D. (see P.Purdue).

Weinfeld, R. (see P.Staniewski).

Weiss, E.N. (see M.A.Cohen).

Weiss, G. (see I.Meilijson).

Welch, P.D. (see S.S.Lavenberg).

Wessels, J. (see J.van Nunen).

White, C.C. III (see K.W.Kim).

White, C.C. III (1976). Procedures for the solution of a finite-
horizon, partially observed, semi-Markov optimization problem.
Operations Res., 24, 348-358.

White, C.C. III (1975). Cost equality and inequality results for a
partially observed stochastic optimization problem. IEEE Trans.
Systems Man Cybernetics SMC-5, 576-582.

White, D.J. (1978). Finite dynamic programming. An approach to finite
Markov decision processes. John Wiley & Sons.

White, D.J. (1981). Isotone optimal policies for structured Markov
decision processes. Eur.J.Oper.Res., 7, 396-402.

Whitt, W. (1980). Continuity of generalized semi-Markov processes.
Math.Oper.Res., 5, 494-501.

Whitt, W. (1981). Comparing counting processes and queues. Adv.Appl.
Probab. 13, 207-220.

Whitt, W. (1982). Approximating a point process by a renewal
process. I : Two basic methods. Oper.Res., 30, 125-147.

Whittaker, C. (see R.M.Feldman).

Wickwire, K.H. (1979). The Poisson disorder problem for large intensities. Int.J.Syst.Sci., 10, 1353-1358.

Wijngaard, J. (1982). On the calculation of the total expected cost
in skip-free Markov chains : the matrix case. Operations
Research, Proc.10th Annual Meet., Goettingen, 449-453.

Wilcox, P. (see E.Ritchie).

Wilkins, C.W. (1980). First passage times of semi-Markov processes. Southeast Asian Bull.Math., 4, 94-100.

Winston, W. (see S.D.Deshmukh).

Witsenhausen, H.S. (1975). On policy independence of conditional expectations. Inform. and Control, 28, 65-75.

Wlodarczyk, M. (see I.Kozniewska).

Wolff, M.-R. (1980). Dynamische Instandhaltungsplanung fuer einfache Produktionssysteme. Proc.Jahrestag Oper.Res., DGOR, Regensburg, 412-418.

Wolfson, D.B. (1976). A Lindeberg-type theorem. Stochastic Processes Appl., 4, 203-213.

Wolfson, D.B. (1977). Limit theorems for sums of a sequence of random variables defined on a Markov chain. J.Appl.Probab., 14, 614-640.

Woodroofe, M. (see H.Takahashi).

Woolford, S.W. (see P.S.Puri).

Wu, R. (1981). Distribution of a class of remainder times. Acta Math. Sin., 24, 494-503 (Chinese).

Wu, S.C. (1982). A semi-Markov model for survival data with covariates. Math.Biosci., 60, 197-206.

Wu, S.M. (see S.M.Pandit).

Yao, J.S. (see S.Liu Mu).

Yasuda, M. (1978). Semi-Markov decision processes with countable state space and compact action space. Bull.Math.Statist., 18, 35-54.

Yasui, K. (see M.Motoori).

Yechiali, U. (see H.Mendelson).

Yeh, M. (see L.C.Thomas).

Young, D. (see B.J.Largay).

Yurenkov, K.E. (1976). A stochastic exchange model for a two-level-memory. Verojatn.Metody Kibern., 91-102 (Russian).

Yushkevich, A.A. (1973). On a class of strategies in general Markov decision models. Theory Probab.Appl., 18, 777-779.

Yushkevich, A.A. (1981). On controlled semi-Markov processes with
average reward criterion. Lect.Notes Control Inf.Sci., 36,
235-238.

Yushkevich, A.A. (1982). On the semi-Markov controlled models with
the average reward criterion. Theory Probab.Appl., 26, 796-803.

Zaagman, W.H. (see G.Gestri).

Zaharin, O.M. (see I.I.Ezov).

Zaharin, O.M. (1979). An ergodic theorem for singly complex randomized semi-Markov processes. Ukr.Math.J., 30, 610-611.

Zaharin, O.M. (1979). Complex randomized semi-Markov processes.
Cybernetics, 14, 903-907.

Zaharin, O.M. (1980). Some Markov models of systems with partial
aftereffect. Cybernetics, 16, 175-188.

Zak, Yu.A (see N.L.Kiryan).

Zakusilo, O.K. (1977). On the convergence of sums of a random number
of random variables defined on an ergodic process. Theory
Probab.Math.Stat., 9, 95-103.

Zaslavskij, A.E. (1979). A multidimensional Markov renewal theorem
and an infinite system of particles of countably many types.
Theory Probab.Math.Stat., 18, 47-58.

Zasushin, V.S. (1978). Adaptive control of semi-Markov processes.
Izvestija Akad.Nauk SSSR, tehn.Kibernet., 207-213 (Russian).

Zbyrko, M.D. (see V.N.Kuznetsov).

Zelen, M. (see S.W.Lagakos).

Zhutovskij, V.G. (see G.A.Chernomorov).

Zhilin, V.A. (see V.V.Kalashnikov).

Zirilli, F. (see F.Aluffi).

Zlatorunskii, N.K. (1974). Interaction of stochastic systems with
finite state space. Kibernetika, Kiev, 35-39 (Russian).

Zieu, B.K. (1978). A scheme of improvement of a strategy by parts
for semi-Markovian decision process. Vestnik Leningrad.Univ.,
19, Mat.Men.Astron., 137-138 (Russian).

Zijlstra, M. (1981). Renewal replacement policies for one unit systems.
Eur.J.Oper.Res., 8, 289-293.

Zoelzer, G.A. (1976). Anwendung linearer Programmierung bei Markov-
Modellen fuer Personalstruktur-Probleme. Proc.Annual Meet.Oper.
Res., DGOR, 70-76.

Zuckerman, D. (1978). Optimal stopping in a semi-Markov shock model.
J.Appl.Probab., 15, 629-634.

Zuckerman, D. (1978). Optimal replacement policy for the case where
the damage process is a one-sided Levy process. Stochastic
Processes Appl., 7, 141-151.

Zuckerman, D. (1980). Optimal stopping in terminating one-sided
processes with unbounded reward functions. Z.Oper.Res., Ser.A,
24, 145-153.

Concluding remark : this bibliography complements a first part which
appeared in 1976 in J.Comp.Appl.Math., 2, 125-144. The number of
papers covered here is close to a thousand from more than 800
authors. There is no claim whatever on completeness. The reader's
attention is drawn on translations made of journals published in
Russian.